SEED ECOLOGY

Published Proceedings of Previous Easter Schools in Agricultural Science

SOIL ZOOLOGY
Edited by D. K. McE. Kevan (Butterworths, London, 1955)

THE GROWTH OF LEAVES
Edited by F. L. Milthorpe (Butterworths, London, 1956)

CONTROL OF THE PLANT ENVIRONMENT
Edited by J. P. Hudson (Butterworths, London, 1957)

NUTRITION OF THE LEGUMES
Edited by E. G. Hallsworth (Butterworths, London, 1958)

THE MEASUREMENT OF GRASSLAND PRODUCTIVITY
Edited by J. D. Ivins (Butterworths, London, 1959)

DIGESTIVE PHYSIOLOGY AND NUTRITION OF
THE RUMINANT
Edited by D. Lewis (Butterworths, London, 1960)

NUTRITION OF PIGS AND POULTRY
Edited by J. T. Morgan and D. Lewis (Butterworths, London, 1961)

ANTIBIOTICS IN AGRICULTURE
Edited by M. Woodbine (Butterworths, London, 1962)

THE GROWTH OF THE POTATO
Edited by J. D. Ivins and F. L. Milthorpe (Butterworths, London, 1963)

EXPERIMENTAL PEDOLOGY
Edited by E. G. Hallsworth and D. V. Crawford (Butterworths, London, 1964)

THE GROWTH OF CEREALS AND GRASSES
Edited by F. L. Milthorpe and J. D. Ivins (Butterworths, London, 1965)

REPRODUCTION IN THE FEMALE MAMMAL
Edited by G. E. Lamming and E. C. Amoroso (Butterworths, London, 1967)

GROWTH AND DEVELOPMENT OF MAMMALS
Edited by G. A. Lodge and G. E. Lamming (Butterworths, London, 1968)

ROOT GROWTH
Edited by W. J. Whittington (Butterworths, London, 1969)

PROTEINS AS HUMAN FOOD
Edited by R. A. Lawrie (Butterworths, London, 1970)

LACTATION
Edited by I. R. Falconer (Butterworths, London, 1971)

PIG PRODUCTION
Edited by D. J. A. Cole (Butterworths, London, 1972)

SEED ECOLOGY

PROCEEDINGS OF THE NINETEENTH EASTER SCHOOL IN
AGRICULTURAL SCIENCE, UNIVERSITY OF NOTTINGHAM, 1972

Edited by

W. HEYDECKER, M.SC., PH.D., N.D.H.

Senior Lecturer in Horticulture,
Department of Agriculture and Horticulture,
University of Nottingham School of Agriculture,
Sutton Bonington, Loughborough, Leicestershire

THE PENNSYLVANIA STATE UNIVERSITY PRESS

UNIVERSITY PARK AND LONDON

First published in 1973 by
Butterworth & Co (Publishers) Ltd, London

Published in the United States of America by
The Pennsylvania State University Press

© The several contributors named in list of contents, 1973

ISBN 0-271-01158-0

Library of Congress Catalogue Card Number 73-1459

Library of Congress Cataloging in Publication Data

Easter School in Agricultural Science, 19th,
University of Nottingham, 1972.
Seed ecology; proceedings.

Includes bibliographies.
1. Seeds—Congresses. 2. Botany—Ecology—Congresses. I. Heydecker, W., ed.
II. Title.

QK661.E25 1972   582'.03'32   73-1459

ISBN 0-271-01158-0

Filmset and Printed in England by Page Bros (Norwich) Ltd., Norwich, England.

PREFACE

Care with the seed, joy with the harvest—Psalm 126

The University of Nottingham Easter School in Agricultural Science is traditionally a kind of international interdisciplinary market place, which has the object of attracting persons interested in as many facets as possible of a single, and at first sight, narrow theme. Success depends on a spirit of adventure, critical openmindedness and co-operative give and take: a barter of knowledge and expertise, an eagerness to do justice to allied aspects of one's subject of which one may not have been sufficiently aware, and the willingness to challenge and help one another to take one more vital step forward.

In the event, the Easter School on *Seed Ecology* was attended by representatives of the whole spectrum of the seed-centred world—geneticists, plant biochemists and physiologists, seed technologists, soil scientists, environmental physicists, plant pathologists, agronomists, forestry specialists, horticultural scientists, seedsmen, growers, brewers, conservation experts and plant ecologists— -over 200 in all from 23 countries.

It had originally been the intention to confine the subject of this Easter School to crop seed ecology, omitting such- -agronomically speaking—undesirables as weeds and dormancy. But seed dormancy is an important ruse of nature, and weed seeds, being naturally robust systems, continue to be with us. So we widened the scope and made it *Seed Ecology*, contributions dealing with seeds of both cultivated and non-cultivated plants, thus giving a much wider insight into the modern aspects of seed research and development. Through the wide range of topics presented, this volume aims at heightening the all-round awareness of the many and complex relationships of seeds with their environment, and provides a stimulus for the cross fertilisation of ideas between disciplines which ought to be, but are not often enough, aware of their mutual significance.

W. Heydecker

ACKNOWLEDGEMENTS

The success of this Easter School is due to all participants who created a most friendly atmosphere conducive to free and relaxed discussion; but to a number of persons in particular: those whose suggestions helped to shape the programme, Mr. R. B. Austin, Mr. T. M. Clucas, Dr. R. K. Scott and in particular Professor M. Evenari, Hebrew University of Jerusalem; all the speakers who distilled the outcome of years of work into informative and thought-provoking papers, and to all those who contributed to the discussions (which can only be reported in a very curtailed form); to Professor J. D. Ivins, Deputy Vice-Chancellor, University of Nottingham who opened the Proceedings and presided over the first session; to all the other sessional chairmen: Mr. R. B. Austin, Plant Breeding Institute, Cambridge; Professor E. W. Simon, Queen's University, Belfast; Professor H. E. Street, University of Leicester; Professor E. C. D. Cocking, Professor J. L. Monteith, F.R.S., Professor H. Smith, all of the University of Nottingham; and Professor J. P. Hudson, Director, University of Bristol Research Station, Long Ashton, who also gave us an inspiring address at the conference dinner. Our thanks are due to those who took so much trouble to contribute and stage the many exhibits which enriched the conference and to whose contents we can unfortunately do only scant justice by merely summarising them in these Proceedings; and to Dr. M. Spurný, Czechoslovak Academy of Sciences, for the loan of three fascinating photomicrographic time lapse films.

A special word of thanks should go to all the efficient workers behind the scenes, and particularly to Miss Grace E. Fox for all the clerical organisation.

The University of Nottingham is grateful to the following organisations for their generosity which enabled us to invite speakers from overseas:

ACKNOWLEDGEMENTS

Amalgamated Seed Merchants (ASMER Seeds) Ltd.,
Anglo-Maribo Seed Company Ltd.,
BASF United Kingdom Ltd.,
Bush Johnsons Ltd.,
Dobies of Chester,
Erin Peat Products Ltd.,
Marks and Spencer Ltd.,
Mommersteeg Seed Company Ltd.,
Rothwell Plant Breeders Ltd.,
Charles Sharpe and Company Ltd.,
British Association of Seed Analysts.

Finally, the Editor would like to express his cordial appreciation
to all the many friends whose help lightened his burden in the
preparation of these Proceedings. Special mention should be made of
Professor J. F. Harrington and Professor J. L. Monteith who made
suggestions for the Glossary, Dr. Margaret E. Marston and Dr.
Winifred M. Dullforce for help in many ways, of Miss Gillian Manison
for indefatigably typing manuscripts, and notably of Mr. R. L.
Gulliver for his many helpful suggestions; and of Betty Heydecker
for her saintly patience.

CONTENTS

CONTENTS

CONTENTS

1

SEED ECOLOGY

W. HEYDECKER

Department of Agriculture and Horticulture,
University of Nottingham School of Agriculture

A seed is an end and a beginning; it is the bearer of the essentials of inheritance; it symbolises multiplication and dispersal, continuation and innovation, survival, renewal and birth.

Ecology has been a fashionable branch of science only since it has become clear that biological systems can be endangered when their environmental setting is disturbed. But since nothing can live *in vacuo*, ecology has always been a vital corollary to biology: ecology is fundamentally the science of where living things are at home; but a seed has no one home, and indeed, should it remain in its original one, it would be likely to fail in its 'purpose'. Seed ecology is therefore largely concerned with the hazards and the opportunities of migration, and of settling down anew.

We now know a good deal about many aspects of seed physiology— about many of the processes which take place inside seeds. But few systematic attempts have been made to relate these processes to the time structure of the environment on the one hand and, on the other, to the often abrupt change-over of conditions when seeds are moved from one environment to another.

In nature, seeds pass from the plant and the vicinity where they have been produced, through the very temporary environment of the distributing agency, to a place on or in the ground where they will probably lie dormant for a while and where they will, under the right conditions, ultimately germinate and grow into new plants.

In cultivation *(Figure 1.1)* seeds are deliberately produced, harvested, tidied up, stored, sometimes processed further and finally sown in a seed bed carefully prepared in order to encourage germination and rapid subsequent growth. Each of these steps is as important as it is highly unnatural.

We should be under no illusion about obstacles to an understanding of seed ecology. First of all, there are many large gaps in our knowledge because some ecological questions have *seemed* too naive to be worth asking at all; secondly, most environmental factors can have more than one effect on the performance of a seed; thirdly, seeds are

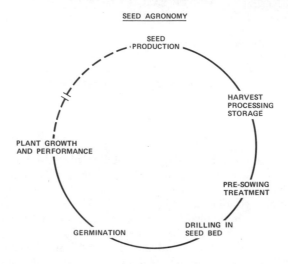

Figure 1.1

influenced by 'memories' of their place of birth; and fourthly, their ingenious mechanisms often outwit our teleological imagination.

It is never easy (and it is largely irrelevant) to define the borderline between ecology—the relation of organisms to their environment—and physiology. The concept of an 'internal environment' ceases to be a paradox if one considers an organism as essentially a genetic code within surroundings that determine which parts of that code shall be switched on and which suppressed.

The schematic diagram (*Figure 1.2*) of the factors involved in seed ecology, though vastly oversimplified, shows at least that interactions and feedback mechanisms are likely to abound. We should at the outset humbly realise that any purely biochemical, purely physical, or indeed any other 'pure' explanation of the events from seed genesis to seed germination would miss the point: that, in order to succeed in bridging the gap between two generations of growing plants, seeds have to be equipped to manoeuvre themselves out of fallaciously enticing situations, and make the best of desperately hazardous ones. To be successful, a seed has to form, and has to acquire dormancy,

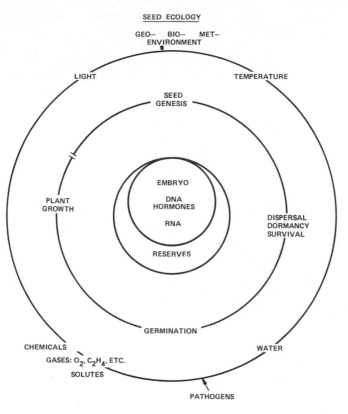

Figure 1.2

under conditions that favour immediate growth; remain outwardly inactive until the time and place are right for a new start; and then negotiate the transition from the sheltered existence of 'a plant in a nutshell' to that of a self-supporting entity in a minimum of time so as to avoid failure during the intervening, highly vulnerable heterotrophic phase. In order to understand how seeds cope successfully, the seed ecologist has to subordinate the physiological bits and pieces to the holistic approach which takes account of the constant interplay of the organism with all the factors of the environment.

This volume, therefore, shows some of the mechanisms by which seed physiology is shaped by, and tuned to, seed environment, some of the requirements which are imposed on seeds of cultivated plants, and some of the steps that can be taken to help seeds to come up to expectation.

2

GENETIC REGULATION OF GERMINATION

W. J. WHITTINGTON

Department of Physiology and Environmental Studies,
University of Nottingham School of Agriculture

INTRODUCTION

This paper is concerned with genetically controlled quantitative variation in germination behaviour. Initially, germination will be equated to field emergence but later the definition of the germinated state will be based on time to emergence of the radicle, or in cereals the coleorhiza.

It must be recognised at once that there are difficulties in attempting to study the genetic control of germination. Environmental factors during seed development and seed storage have a considerable effect on subsequent germination, while from the genetic point of view the testa and other parts which may enclose the seed are of maternal genotype, the endosperm is usually two-thirds maternal and one-third paternal in origin and the embryo is equally divided between the paternal and maternal genotypes. Genetic control of variability in the time to germination may, however, be recognised in three ways, namely from the recognition of consistent differences between species or varieties, from the results of controlled crosses and from the effects of deliberate selection experiments designed to modify germination behaviour. Some previously published results and recent experimental evidence will be considered in relation to these headings.

GENETIC EFFECTS ON GERMINATION

It is useful to assess the extent of the genetic component in the

5

variation associated with establishing plant populations. An examination of a National Institute of Agricultural Botany (N.I.A.B.) trial (Willey, 1970) on field emergence of four monogerm sugar beet varieties, at three centres over four years, showed that the varietal variation was significant but of less importance than the effects of years or sites. The variation due to genetic differences was less than that due to the environment but there was nevertheless a highly significant genetic difference in the material available to the farmer (*Table 2.1*). Of even greater interest, however, is that the year ×

Table 2.1 ANALYSIS OF VARIANCE FOR PERCENTAGE FIELD EMERGENCE (TRANSFORMED TO LOG_{10}) IN SUGAR BEET (DATA FROM WILLEY (1970) FOR THE CAMBRIDGE, BURY ST. EDMUNDS AND TERRINGTON SITES)

Component of variation	Variance ratio
Years	129·4***
Sites	39·7***
Varieties	10·8***
$Y \times S$	40·4***
$Y \times V$	2·7*
$S \times V$	2·0

*   P = 0·05
*** P = 0·001

variety interaction term was significant. This latter result was further examined by submitting the published, and later, results (transformed to logarithms) to analysis by the regression technique of Finlay and Wilkinson (1963). In this method the performance of each variety is plotted against the mean performance of all varieties studied at each site. The log site means for germination, in ascending order, form the independent variables and the log variety germination values are the dependent variables. Linear regression lines indicate the stability in performance of the individual varieties over the sites. Regression coefficients below 1·0 represent above-average stability and values above 1·0 represent below-average stability. The results for two only of the varieties showed that one, Bush Mono, was more stable in field emergence than the other, Sharpe's Klein Monobeet (*Figure 2.1*). The differences between the varieties were enhanced in years or at sites when the percentage emergence was low. The correct choice of genotype becomes of greater importance to the farmer under adverse conditions. The difference in the behaviour of the two varieties was not related to a different ploidy since both the varieties mentioned were diploids. Triploid varieties in the trial had intermediate stabilities. Perdok (1970), however, reported that

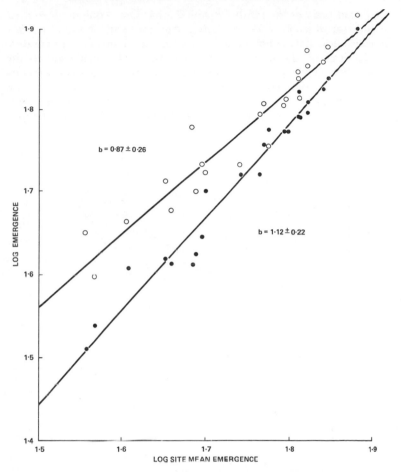

Figure 2.1   Regression of log emergence on log site mean emergence for two monogerm sugar beet varieties analysed by the method of Finlay and Wilkinson (1963). Data re-analysed from Willey (1970) and unpublished data for 1971 at the Cambridge, Bury St. Edmunds and Terrington trial sites. The varieties were Bush Mono (○) and Sharpe's Klein Monobeet (●)

under adverse conditions of storage diploids survived better than triploids and tetraploids.

Again, the data of Bunting and Willey (1957) with two varieties of maize, namely Kuma and Rhodesian, may be expressed in the same way and show the same relationship and close fit of the points to the regression line (*Figure 2.2*). In fact, they are so close as to enable one to predict performance at levels of field emergence lower than the main body of the results. *Figure 2.2* has been composed from 47 results (of which only 34 are shown) but one result showed such low germination that the points could not, for reasons of scale, be included. Predictions on the basis of the linear regression coefficient of the points in question showed that the expected values for the two varieties, respectively, should have been 2·5 and 15·5 per cent germination, whereas those observed were 2·0 and 20·0 per cent. It should be noted that Bunting and Willey (1957) made reference

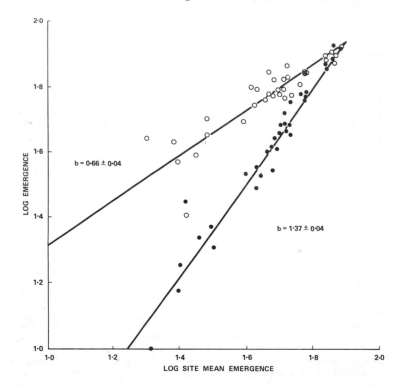

Figure 2.2   Regression of log emergence on log site mean emergence for two maize varieties Kuma (○) and Rhodesian (●). Data re-analysed from Bunting and Willey (1957)

to the fact that the difference in emergence between the varieties was greatest under the worst conditions.

The genetic differences between the varieties in these two examples probably involve many loci but specific genes can affect establishment. Hull (1937) wrote: 'It is well known to gardeners that under conditions unfavourable to germination round-seeded varieties of pea give a better stand of plants than those varieties which have wrinkled seeds.' Re-analysis of Hull's data on the emergence, in the absence of seed dressing, of the varieties Pilot (round) and Gradus (wrinkled) from different sowing dates emphasises that this conclusion is correct (*Figure 2.3*). It appears that the percentage emergence of

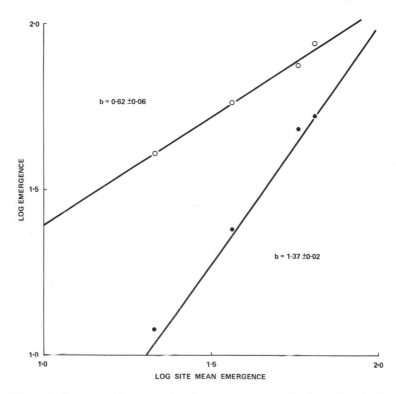

Figure 2.3   Regression of log emergence on log site mean emergence for the round pea variety Pilot (○) and the wrinkled variety Gradus (●). Data re-analysed from Hull (1937)

genotypes may be related to the chemical composition of the seeds. Wrinkled seeds are high in stachyose rather than starch (Haldane, 1952) and have a higher sucrose content (Kooistra, 1962).

Maize endosperm is triploid in constitution and a single gene difference can determine the survival of seedlings during germination. Haskell (1952) gave results which showed that the germination capacity of maize decreased as the proportion of the sugary allele (su) in the endosperm decreased relative to the (Su) starchy allele

(*Table 2.2*). Mather and Haskell (1949) also noted that emergence under cold conditions was higher for starchy rather than sugary segregates.

There is little doubt that the failure of certain genotypes to emerge, particularly when unprotected by seed dressing, is related to the

Table 2.2   THE EFFECT OF VARYING PROPORTIONS OF THE SU (STARCHY ENDOSPERM ALLELE) RELATIVE TO SU (SUGARY ENDOSPERM) (AFTER HASKELL, 1952)

Endosperm genotype			Percentage germination
Su	Su	Su	93
Su	Su	su	87
Su	su	su	67
su	su	su	46

interaction with cold wet conditions and pathogenic organisms (Harper, Landragin and Ludwig, 1955). It is likely that many single gene differences, particularly those with severe effects or concerned with carbohydrate and amino acid composition of the endosperm, might affect the longevity of the seed in storage, the time to radicle emergence or seedling establishment. Some published results support this view. Weiss and Wentz (1937) found that of six *luteus* genes in maize, two (2 and 4) were associated with decreased viability and decreased vigour on emergence. Crane (1964) reported that the allele (ae) lowered percentage germination in maize, while Angell (1950) reported a greater reduction of seedling emergence from green than from yellow peas. There seems uncertainty in this last report as to whether the colour difference was genetic in origin.

It is generally agreed that the grain of white-grained wheat is non-dormant, while red-grained wheats, although variable between themselves, are dormant (Everson and Hart, 1961). Red grain colour was shown by Nilsson-Ehle (1909) to be determined by any one of three genetic loci, and variation in the number of loci in the dominant form may account for the different behaviour of red-grained types. According to Everson and Hart (1961) and Belderok (1961), the difference between red- and white-grained wheats in their tendency to sprout in the ear was also reported by Nilsson-Ehle (1914), and later references to variability in sprouting were made by Harrington (1932), Jonard (1933) and Åkerman (1936). One might wonder why white-grained varieties, if they are prone to sprout, were common in North-West Europe before and after 1900. I am grateful to Professor Sir F. L. Engledow, F.R.S., and Dr. G. D. H. Bell,

F.R.S., for their views on this matter. It is clear from their comments that other considerations were more important in determining the success of a variety and, since there was a place for white wheats, where they were good in other respects, their liability to sprout was of secondary importance. Certainly, the association of 'sprouting resistance' with grain colour was known to breeders, many farmers and millers by 1928. Tests on sprouting were made by the breeders and, at their request, by the seed-testing department of the N.I.A.B. about 1935. The tests showed amongst red and white wheats different degrees of liability to sprout, but they emphatically confirmed the greater liability of the white-grained varieties. Resistance to sprouting then began to influence breeding policy but characterisation of varieties on the basis of resistance to sprouting began much later (about 1950) in N.I.A.B. tables of merit. References continued to occur to variability in sprouting (Harrington and Knowles, 1940; Greer and Hutchinson, 1945; Ellerton, 1946; Gfeller and Svejda, 1960) but it is still uncertain whether the greater tendency of white wheats to sprout is without exception and whether the association is pleiotropic or due to linked genes. It should be noted that Nilsson-Ehle (1914) emphasised that red grains differ structurally from white grains. This formed the basis of the analysis by Wellington (1956) of the difference in sprouting characteristics between red- and white-grained wheats. Miyamoto, Tolbert and Everson (1961), however, suggested an explanation in terms of inhibitory effects.

Other reports on variability in germination characters which do not implicate single genes are given by Jones (1926) and Lindstrom (1942). Cumming (1959) discussed differences between species of Chenopodiaceae, and Justice (1941), Timson (1965) and Hammerton 1967a; 1967b) found variation between strains and species of *Polygonum*. Chromosomal variation rather than gene variation can also influence germination. Seeds of *Nicandra physaloides*, containing a single isochromosome rather than paired isochromosomes, exhibited delayed germination (Darlington and Janaki-Ammal, 1945). On the other hand, Vosa (1966) reported that germination was faster in leek seeds with a low number of B chromosomes, i.e. chromosomes in addition to the normal number, while those with a higher number germinated later. Moss (1964) showed that in rye the possession of four B chromosomes resulted in larger grain and a slower germination. Williams (1970) showed no significant differences in rye grass between the different B chromosome number classes but the results showed small non-significant differences, with odd and even numbers of chromosomes showing related effects. The mean germination was faster in seeds giving plants with one or three B chromosomes than in seeds forming plants with none.

It is possible that differences might occur in the rapidity of germination of male and female seeds of dioecious species. Rosa (1925) found that if seeds of spinach were thickly sown and the small plants were removed the survivors showed an excess of female plants. The effect here may not, of course, be due to an earlier germination of the female seeds but be related to a difference in embryo size or growth rate, and Sneep (1958) states that no sex effects on germination occur in spinach. A similar example where the validity of the evidence on differences in germination behaviour is in doubt occurs with single as opposed to double flowered stocks.

An approach which seemed to be promising in relating variation in time-to-germination with specific chromosomes was the use of wheat chromosome substitution lines. The lines are formed by substituting each homologous pair of chromosomes from one variety into another. Since wheat has 21 pairs of chromosomes, 21 lines can be formed numbered 1–7, respectively, for genomes A, B and D, and used to detect the chromosomes on which the genetic differences which separate the parents are located. This material has been successfully used by Law (1967). An analysis was made of the effects on germination of the substitution lines of the variety Hope into Chinese Spring. The results (Whittington and Fierlinger, unpublished data) showed that there were no marked differences between the varieties or lines for the duration of post-harvest dormancy which lasted five weeks. The results of several experiments in which seeds were germinated in petri dishes showed that although all germination was virtually completed in 16 h there were significant differences between the parents and substitution lines in the mean time to germination. There was a direct relationship between seed size and time-to-germination, but Hope grains germinated before those of Chinese Spring, despite being heavier. The fact that the parental varieties differed encouraged the belief that the early germination of Hope might also occur in one of the substitution lines and thus indicate that a specific chromosome is involved. In fact, no line resembled Hope in combining rapid germination and large grain weight, but analyses of grains selected for uniform weight suggested that chromosome 2 A might shorten, and 4 D might increase, the period to germination relative to Chinese Spring. The fact that no particular chromosome was clearly related to time to germination could, of course, be explained if more than one chromosome was involved. The substitution lines effectively screen only one chromosome pair at a time and therefore effects due to two unlinked genes cannot be located using the initial lines. There is no doubt, however, that the lines differ in their mean grain weights and, since a significant correlation exists between grain weight and time to germination, an association could easily be established between a particular chromosome and mean time to germination.

The association between grain size and rapidity of germination may explain an unexpected result from these experiments. This was that a correlation existed between the number of hours to germination and the number of days from sowing to ear emergence. Such an association was surprising in that the variability for germination extended over a few hours, while the variability for ear emergence extended over a six-week period. It appears likely that the association was due to the fact that, with the exception of Hope, the mean grain size, at least in the sample used, of the early-flowering types was lower than that for the late-flowering lines.

There are, however, other reports in the literature concerned with an association between germination and flowering. Dore (1955) and Roberts (1961) referred to the fact that in rice there is much variation between varieties for the duration of the post-harvest dormant phase. They quote examples of previous evidence that the length of the dormant phase was positively correlated with the time to flowering and carried out further experiments to substantiate the relationship, but they were unable to do so.

EVIDENCE ON GENETIC EFFECTS FROM CROSSES

Having ended the last section on a suitably definite note there is, in fact, further evidence about related germination and flowering characteristics from a controlled cross. Shifriss and George (1965) crossed a non-dormant day neutral North American variety of cucumber (Marketer) with a short-day Indian strain named, by them, Baroda which had marked seed dormancy under their conditions but not when humidity was high. The pattern of germination behaviour in the crosses between the varieties showed reciprocal differences in the F_1 generation, intermediacy of the F_2 and a backcross generation whose mean was intermediate between the F_2 and the recurrent parent (Baroda). Under long days Marketer flowered much earlier than Baroda and the F_2 results showed that the delayed flowering was due to a single recessive gene. The F_2 and backcross results showed that early-germinating and late-germinating progeny were not distributed equally with respect to the flowering gene. There was a significant excess of early-flowering segregates amongst early germinators and late-flowering plants were in excess of expectation amongst late germinators. In fact, the late-flowering types tended to occur at more or less the same frequency at each period of observation throughout the period of germination, so that they occurred too infrequently initially and too frequently at the end of the germination period. The results did not support the view that a single gene influenced both time to flowering and time to germination, but the authors suggested that the results might indicate either a three-locus control of germination or linkage between a gene for time to flowering and a gene for rapidity of germination.

There are relatively few examples in which germination characteristics of the segregating generations from controlled crosses have been followed. The results of Whittington *et al.* (1965) were expressed in terms of means and variances and were not considered to be related to single gene effects. Elkins, Hoveland and Donnelly (1966) studied *Vicia sativa*, *Vicia angustifolia* and F_4 and F_5 lines derived from them. *Vicia angustifolia* showed a high proportion of hard seeds, while *V. sativa* showed few. Scarified seeds of *V. sativa* germinated the faster at a high temperature and *V. angustifolia* faster at a low temperature. The derived lines showed wide variation for both the frequency of hard seeds and the temperature effects, but the number of lines studied was too small to allow any analysis of segregation ratios. The analysis of late generations has advantages over analysis of F_2 data in self-fertilising species because simple ratios, e.g. the 3:1, will only occur in the F_2 if maternal effects are unimportant. In later generations, after self-fertilisation, the different tissues have identical genetic constitutions. Whittington *et al.* (1970) analysed time to germination in 48 late-generation families from crosses between *Avena fatua* and *A. ludoviciana*. These species are widely different in their reaction to light and temperature. It was concluded tentatively that the poor germination of *A. ludoviciana* at 18°C may have been due to a single recessive gene, whereas the poor germination of *A. fatua* at 5°C may have been due to three further recessive genes.

Earlier, Garber and Quisenberry (1923) found that delayed germination was recessive in crosses between *A. fatua* and *A. sativa*, and Johnson (1935) suggested that three loci were involved with the triple recessive showing the greatest dormancy.

A very convenient way to study the quantitative aspects of genetic control of germination is to study the results from diallel crosses between inbred lines. In these, a number of varieties are crossed in every combination and conclusions can be reached about the extent of maternal or paternal effects, additivity, dominance and the extent of gene interaction. Whittington and Fierlinger (1972) studied a 6 × 6 diallel in tomatoes. The varieties were five commercial types of *Lycopersicon esculentum* and the decorative variety Red Currant (*L. pimpinellifolium*) which has very small fruit and seeds. The time to germination of F_1 seeds from crossing Red Currant with any of the others always resembled the maternal parent. *Table 2.3* shows a representative extract from the results of crossing, reciprocally, the

Table 2:3

	Female parent	
	Red Currant	Moneymaker
Male parent		
Red Currant	43·8	69·1
	0·98	3·18
Moneymaker	51·4	76·5
	1·23	3·49

L. pimpinellifolium tomato variety (Red Currant) with varieties (in this instance, Moneymaker) from *L. esculentum.* The upper value for each cross is time (h) to 80 per cent germination and the lower value is mean seed weight (mg). The F_1's with Red Currant as female parent were slightly slower to germinate than the parent itself, whereas the times to germination of the hybrids between *L. esculentum* varieties were more erratically distributed around the values for their maternal parent. This variation was probably related to differences in seed size between samples, since this can be affected by fruit size and the number of seeds set per fruit (Luckwill, 1939).

The F_1 results showed highly significant maternal effects when analysed by Topham's (1966) method. She suggested the relationship

$$\frac{\text{Variance of column (varieties as female parents) totals}}{\text{Variance of row totals (varieties as male parents)} + \text{Variance of column totals}} \times 100$$

to assess the effect: values over 50 per cent indicate maternal control. The results for the tomato, including or excluding the Red Currant

Table 2.4   VARIANCE RATIOS FOR THE SIGNIFICANCE OF MATERNAL EFFECTS AND PERCENTAGE OF VARIANCE DUE TO MATERNAL EFFECTS

	Variance ratio	Percentage of variance
Tomato F_1	10·6***	83
Tomato (omitting the variety Red Currant)	0·49	53
Swede	4·5*	60
Sugar beet*	6·45***	92
Crucifers	212·51	96

*   P = 0·05
*** P = 0·001

crosses, sugar beet (Battle and Whittington, 1971), swedes and cruciferous species (Witcombe and Whittington, unpublished data) (*Table 2.4*) show the extent of the maternal effects at values over 50 per cent. They are most obvious in the tomatoes, where the inclusion of Red Currant had a marked influence on seed size; in sugar beet, where the seed is enclosed in a large quantity of maternal tissue; and in the cruciferous species, where wild turnip as a maternal parent caused delayed germination of its hybrids. There were, however, still some maternal differences in the swede crosses. However, in commercial practice involving outbreeding species or the production of F_1 hybrids, the maternal effects may have little importance. The varieties used in crosses are not likely to show extreme size variation, or be partially dormant, whilst in sugar beet, where clusters are rubbed, graded and pelleted, the maternal influence is reduced. Furthermore, environmental effects on seed size are likely to obscure the influence of the different parents.

The 6 × 6 tomato diallel referred to earlier was carried further and the 15 F_1 hybrids were intercrossed with each other. The seeds were germinated and the results were so arranged that a comparison could be made between the germination of hybrids receiving further genes from the Red Currant or from the commercial varieties through the pollen.

The seeds receiving the Red Currant genes germinated significantly earlier than those receiving *L. esculentum* genes (*Table 2.5*). The table shows the mean time to 80 per cent germination (h) in intercrosses between F_1 hybrids arranged to compare the effect of male parents with and without genes of *L. pimpinellifolium* Pi. The

Table 2.5

F_1 *male parents*	*Female parent* *Moneymaker* × *Plumpton King* F_1
Pi × M	72·9
Pi × O	79·4
Pi × Po	81·3
Pi × Pk	73·4
Pi × A	75·0
Mean of:	
M × Po, O, A	82·4
O × Po, M, A	85·0
Po × M, O, Pk, A	78·0
Pk × M, O, Po, A	82·2
A × M, O, Po, Pk	80·0

L. esculentum varieties Moneymaker, Outdoor Girl, Potentate, Plumpton King and Amateur are denoted by M, O, Po, Pk and A, respectively. In both F_1's and the F_1 intercrosses the time to germination was related to seed weight. The smaller seeds germinated earlier than the larger seeds and the reciprocal differences in seed size accounted for the marked maternal effects in the F_1 generation. The paternal effect could not be accounted for by seed weight differences. In tomato, germination occurs after the radicle has emerged through the endosperm, and it was suggested earlier (Whittington *et al.*, 1965) that this process actually involves a digestion of the tissue. Whereas most of the variation in time to germination in tomato seeds resides in the maternal parent, through its effect on seed size or quantity of endosperm overlying the embryo, the paternal effects shown by

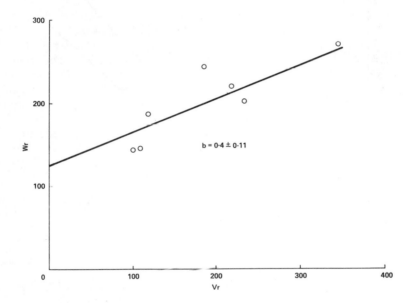

Figure 2.4   Diallel analysis by the Wr/Vr regression technique (Whitehouse et al., 1958) of the percentage germination (transformed to angles) from an F_2 diallel in subterranean clover (Morley, 1958)

Red Currant are less likely to be related to quantity of endosperm and might be related to a more rapid enzymatic degradation of the endosperm.

A simple way of analysing diallel crosses is by regression of the covariance (Wr) of parental and progeny means against the variance

(Vr) of parents and their progeny (Jinks, 1954; Whitehouse, Thompson and Do Valle Ribeiro, 1958). A regression coefficient of 1·0 indicates absence of gene interaction. Evidence of interaction increases as the value falls and an increasing degree of dominance is shown as the intercept of the regression line moves from high positive to low values on the Wr axis.

Figure 2.4 shows the Wr/Vr regression graph for the data of Morley

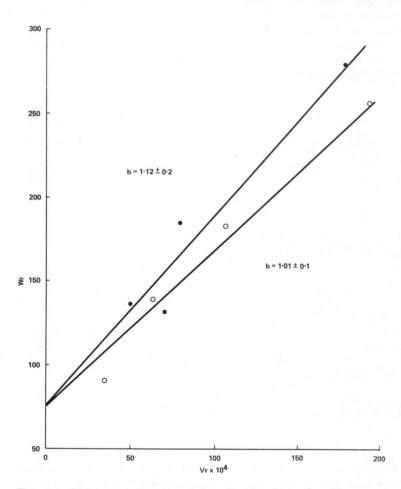

Figure 2.5   Diallel analysis for two separate 4 × 4 F₂ diallel analyses (● and ○, respectively) for seed size in Phaseolus vulgaris. *(Data from Boast, unpublished.) The varieties were* ○ = *Masterpiece, Mexico 120, Higuerillo, Panameno;* ● = *Chimbolo, Panameno, Mexican Black, San Jorgé*

(1958), who carried out a 7 × 7 F_2 diallel with subterranean clover. The slope of the line is much less than 1, indicating gene interaction, and Morley in his analysis noted the strong evidence for specific combining ability in the data. Interaction effects in the germination of seeds from diallels were also noted by Simmonds (1964), using potatoes, and by Battle and Whittington (1971) with sugar beet. The effects can be traced to the action of particular varieties by inspection of the data and recalculation after omission of the suspect parent and its crosses. Morley (1958) found that additive effects predominated after removal of the variety Bacchus March from his results. Unfortunately, little attention has been given to either establishing the consistency of these effects with particular crosses or understanding the physiological mechanisms by which they originate. Perhaps the most remarkable result is that reported by Harper and McNaughton (1960) and McNaughton and Harper (1960), who studied the germination characteristics between poppy species. In particular, they found that dormancy was marked in *Papaver lecoquii* and *P. dubium*, that it was maintained in the cross where *P. dubium* was the maternal parent but disappeared where *P. lecoquii* was used as the female parent.

The explanation in the above example may be related to the balance of substances concerned in stimulating or inhibiting germination, but elsewhere structural effects may be important. In two 4 × 4 F_2 diallels of varieties of *Phaseolus vulgaris*, the inheritance of seed weight showed additivity without interaction (*Figure 2.5*),

Table 2.6

	Masterpiece	Mexico 120	Higuerillo	Panameno
Masterpiece	<u>49</u>	94	65	40
Mexico 120		<u>22</u>	56	58
Higuerillo			<u>28</u>	40
Panameno				<u>40</u>

but the values recorded for time to 50 per cent germination were most erratic, as shown in *Table 2.6* which gives mean time to 50 per cent germination (h) in an F_2 diallel cross of four varieties of *Phaseolus vulgaris*. Underlined values on the diagonal are for the parents (Whittington and Boast, unpublished data).

In germination experiments in which the seeds were placed on moist cotton wool, many of the hybrids behaved like 'hard' seeds and did not, or were slow to, take up water (*Table 2.7*). The table gives the percentage of unimbibed seeds after 216 h from supplying

Table 2.7

	Masterpiece	Mexico 120	Higuerillo	Panameno
Masterpiece	<u>0</u>	92	58	0
Mexico 120		<u>10</u>	66	25
Higuerillo			<u>0</u>	0
Panameno				<u>0</u>

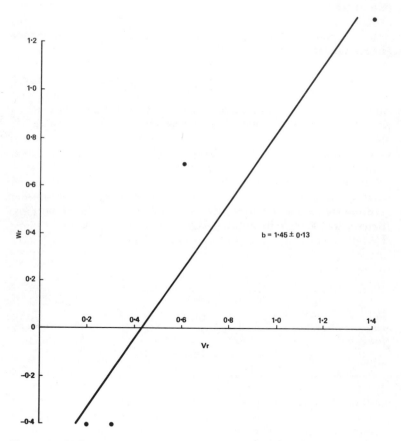

Figure 2.6   *Diallel analysis for time to 50 per cent germination in the* F_2 *diallel of* Phaseolus vulgaris, *after removing part of the testa*

water in an F_2 diallel cross of four varieties of *Phaseolus vulgaris*. Underlined values on the diagonal are for the parents. Hardness may be much affected by the environmental conditions during the ripening period (Hyde, 1954; Quinlivan, 1968). All the hybrid seeds were raised at one time but the majority of the parental seeds were sown and harvested rather later. It is possible that the greater evidence of hardness in the hybrids than the parents was therefore due to environmental factors but the hybrids did differ amongst themselves. The 'hard' seeds seemed particularly to occur in the F_2 generation from crosses between large- and small-seeded parents. The 'hard' seed effect in certain of the hybrids naturally raised the length of time before they germinated so that hybrid values were sometimes larger and not always intermediate between their parents. The points on the diallel were thus very erratic in their distribution. The reduction of the 'hard' seed effect by removing part of the testa had the effect of improving the distribution of the points, although the regression line was still not unity (*Figure 2.6*).

THE EFFECTS OF SELECTION

The results from the experiments on quantitative variation show that additive genetic effects predominate and thus selection to modify germination behaviour should be relatively easy. There is a belief that in the evolution of crop plants from wild ancestors the extent of the dormancy has been reduced. That this might occur has been demonstrated by Witcombe (Witcombe and Whittington, in press). He selected for an increased capacity to germinate in charlock, a species with marked dormancy. In this species a few seeds always germinate when supplied with water, but their capacity to do so is largely environmentally determined. A response to selection depends upon taking those seeds that have a genetically determined low degree of dormancy. Witcombe achieved his response by selecting over several generations from those seeds which germinated in a decreasing concentration of gibberellic acid. The experiment is similar to that on genetic assimilation in *Drosophila melanogaster* (Waddington, 1961). The germination of seeds of the last generation when supplied with water was significantly higher than in a control population (*Figure 2.7*).

Simmonds (1964) similarly showed that selection was effective in increasing or decreasing the dormancy of both diploid and tetraploid potato seed (*Figure 2.8*). He also showed that there was a correlation between the extent of seed and tuber dormancy. Mather and Haskell (1949) had more practical considerations in mind when they successfully attempted to make sweet corn more suitable for English conditions by improving germination under cold conditions (*Figure 2.9*). The stable performance of the sugar beet variety Bush Mono,

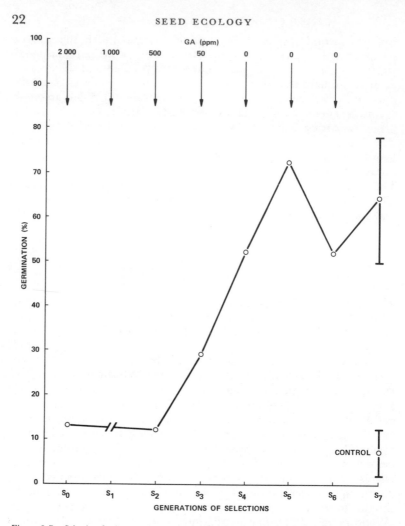

Figure 2.7   Selection for increased percentage germination of charlock (Sinapis arvensis *L.*) *in water over seven generations. The concentration of gibberellic acid used to break the dormancy of the subsequent generation is shown. (Data from Witcombe and Whittington, in press)*

previously referred to, owes at least part of its performance to selection. Dr. S. Ellerton, in a personal note, wrote, 'when monogerm varieties became the order of the day . . . and these were designed to save labour, it was clear that the best way to do this was to drill to a stand. More selection than ever before was applied in the direction of good germination. This was achieved in two ways; (1) by increasing considerably the proportion of potential selections rejected on the score of poor germination at optimum temperature, and (2) by retesting all the potentially accepted material in $1\frac{1}{2}$ in of compost at

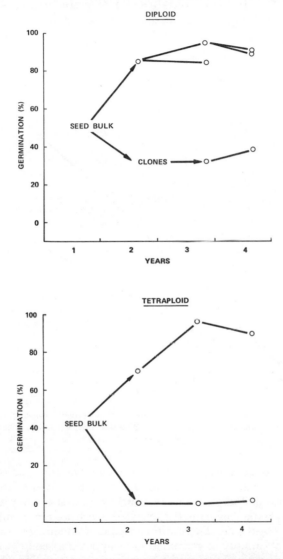

Figure 2.8   Effect of selecting for increased or decreased percentage germination in diploid and tetraploid potatoes. Figure re-drawn from Simmonds, 1964

6°C. We found considerable differences between lots in germination at this low temperature, even between lots which germinated equally well at optimum temperature.'

It is, however, neither agriculturally nor biologically useful for seeds to be too ready to germinate. The maintenance of some dormancy prevents seeds from germinating before they can be

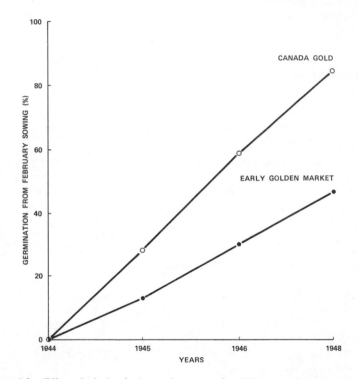

Figure 2.9   *Effects of selection for improved emergence from February sowings of sweet corn (figure re-drawn from Mather and Haskell, 1949). Note that selection lines were continued from March sown plants in 1944*

harvested or dispersed from the fruit, although such germination may occur under certain conditions in, for example, maize crucifers, legumes and tomatoes. Selection for early germination may also be expected to produce unwanted correlated effects as, for example, in reducing the seed size. There are contending advantages and disadvantages of rapidity of germination as opposed to embryo size at emergence. Certainly the earlier germination of the Red Currant tomato does not compensate for its small embryo: the later germinating but large seeds of commercial varieties have the larger seed-lings at any stage. In natural populations one would expect that the germination characteristics and the initial embryo size are related to the competitive conditions between seedlings at emergence. It is also likely that the ability to germinate rapidly may be correlated with a more rapid decline in germination capacity either under storage or under conditions of imposed dormancy in the soil.

Selection is sometimes practised for decreased rates of germination. Burrows (1964) reported on the successful attempt to transfer the dormancy of wild oats (*Avena fatua*) to cultivated oats. Essentially, the objective was to sow the so-called 'dorm oats' in autumn because cultivated oats will not survive the winter as green plants and it is often difficult to get on to the land sufficiently early to sow spring oats. Natural germination of the winter-dormant selections would then occur in the spring without the necessity for early cultivations. Similarly, Bennett (1959) has carried out selections for an increased proportion of hard seeds in crimson clover. In this species, the proportion of hard seeds is small and in North America germination often occurs in the autumn in the absence of sufficient water to sustain subsequent seedling growth. Mass selection raised the percentage germination successively from 1 to 4, 12, 16, 20, 23, 48, 49 to 63 in nine years.

CONCLUSIONS

The conclusions to be drawn from the three types of evidence considered are that germination characteristics are at least partially under genetic control and the particular pattern of behaviour in a cultivated or wild species is likely to have been the result of selection. It may be argued that variability in behaviour is of adaptive importance but it should be remembered that variability cannot fail to occur once any form of dormancy mechanism has been adopted because seeds may so easily be affected by ecological and develop-mental variables. Koller (1964) has given a note of warning against concluding too readily that variability is necessarily advantageous. He stated: 'The nature of many germination-regulating mechanisms in seeds lends itself to an interpretation in terms of survival value for the species. This, however, is no more than an indication, and direct experimental evidence is needed regarding the role played by the entire germination-regulatory equipment of the seeds in relation to the spectrum of combinations of variables which make up the environmental complex.'

Broadly, one may classify genetic effects on germination into those due to major genes, possibly controlling specific inhibitors, and polygenic effects, perhaps related to seed size. There is no difficulty, however, in seeing why so much of the variability in germination characteristics is likely to appear to be due to polygenic effects. The causative agents may be:

1. Quantitative effects during the destruction of inhibiting substances or of tissues imposing dormancy.
2. Quantitative effects on seed size and the time required to acquire a critical water level or breakdown of enveloping tissues.
3. Quantitative effects when laying down inhibitors through variation in the quantity of tissue involved or in seed size.

Furthermore, environmentally determined variation has the effect of blurring the sharp discontinuities between genetic classes so that even single gene differences may be mistaken for polygenic distributions. There has not yet been an attempt to adopt suggestions such as Thoday's (1961) to discover the location of the polygenes involved.

The evidence discussed earlier has been concerned with quantitative differences amongst essentially normal plants. One might speculate whether or not single-gene mutants might occur with a more dramatic effect on the controlling mechanisms of germination. First, do mutant alleles occur which entirely prevent seeds from germinating? Assuming these to be recessive their frequency in populations will be very low because of the intense selection pressure against homozygotes. Whether or not seeds with such defective physiology would develop after fertilisation, sustained by the maternal parent, only to fail after dispersal, or whether they would abort in the embryo stage, is a matter for debate. We know nothing about such matters for we do not analyse the reasons why the occasional seed, unaffected as yet by old age, dies or remains comatose. Second, single-gene mutants might also occur at those loci concerned in the control of germination through the production or breakdown of inhibitory substances. If the mutant condition was imposed by the female parent then all the seeds produced by the plant would behave as mutants irrespective of their own genotype. Alternatively, if the mutant condition was not maternally affected then the behaviour of the seeds should vary amongst themselves. The results might well be lethal if the effects were to cause seeds to germinate before dispersal or at the wrong moment of the year, and then the allele would have a low frequency. If the effects were less severe, then the gene would rise in frequency. This might have happened in varieties of plants which have lost an initial dormancy or differ markedly from one another.

What is certain is that physiological investigations with genetically variable material can be very rewarding. Single-gene effects have been proposed to account for differences in reaction to temperature but there has been surprisingly little attention paid to genetic analyses of light requirements. If one assumes that for certain species it is advantageous only to germinate in light then, accepting the concept of the evolution of dominance, one would expect that requirement for light would be dominant. This, in fact, was found by Honing (1930) for *Nicotiana* and by Howard and Lyon (1945) for watercress seeds. The stimulus of light in both instances was received by the embryo. The lack of available genetic evidence here is in contrast to that available from physiological studies and it is quite clear that we require far more analyses combining genetic and physiological techniques to understand fully any one of the contrasting reasons for variability in germination. We have established the differences but we do not always understand them.

ACKNOWLEDGEMENTS

I am grateful to Emeritus Professor Sir F. L. Engledow, F.R.S., Dr. G. D. H. Bell, C.B.E., F.R.S., Dr. F. Ellerton and Dr. K. W. Dent for their helpful comments on points raised in this paper. My thanks are also due to Mr. L. A. Willey for providing unpublished results for analysis and Dr. Alice Evans for providing seeds of *Phaseolus vulgaris*.

REFERENCES

ÅKERMAN, Å. (1936). 'Über die Keimungsverhältnisse und Auswuchsneigung rot-und weisskörniger Weizensorten', *Züchter*, **8**, 25–29

ANGELL, H. R. (1950). 'Seedling blight. 1. Seed colour and soil in relation to pre-emergence blight of William Massey peas', *Aust. J. agric. Res.*, **1**, 33–42

BATTLE, J. P. and WHITTINGTON, W. J. (1971). 'Genetic variability in time to germination of sugar-beet clusters', *J. agric. Sci. Camb.*, **76**, 27–32

BELDEROK, B. (1961). 'Studies on dormancy in wheat', *Proc. int. Seed Test. Ass.*, **26**, 697–760

BENNETT, H. W. (1959). 'The effectiveness of selection for the hard seeded character in crimson clover', *Agron. J.*, **51**, 15–16

BUNTING, E. S. and WILLEY, L. A. (1957). 'The emergence of maize from field sowings in Great Britain. 1. The effect of date of sowing on the extent and speed of emergence of different varieties', *J. agric. Sci. Camb.*, **48**, 447–456

BURROWS, V. D. (1964). 'Seed dormancy. A possible key to high yield of cereals', *Agric. Inst. Rev.*, **19**, 33–35

CRANE, P. L. (1964). 'Effect of the gene *ae* on seed quality in maize', *Crop Sci.*, **4**, 359–360

CUMMING, B. G. (1959). 'Extreme sensitivity of germination and photoperiodic reaction in the genus *Chenopodium* (Tourn) L', *Nature, Lond.* **184**, 1044–1045

DARLINGTON, C. D. and JANAKI-AMMAL. E. K. (1945). 'Adaptive isochromosomes in *Nicandra*', *Ann. Bot., N.S.*, **9**, 267–281

DORE, J. (1955). 'Dormancy and viability of padi seed', *Malay agric. J.*, **38**, 163–173

ELKINS, D. M., HOVELAND, C. S., and DONNELLY, E. D. (1966). 'Germination of *Vicia* species and interspecific lines as affected by temperature cycles', *Crop Sci.*, **6**, 45–48

ELLERTON, S. (1946). 'Harvests in peril', *Fmrs' Wkly.*, **25**, 32–33

EVERSON, E. H. and HART, S. H. (1961). 'Varietal variation for dormancy in mature wheat', *Mich. State Univ. Quart. Bull.*, **43**, 820–829

FINLAY, K. W. and WILKINSON, G. N. (1963). 'The analysis of adaptation in a plant breeding programme', *Aust. J. agric. Res.*, **14**, 742–752

GARBER, R. J. and QUISENBERRY, K. S. (1923). 'Delayed germination and the origin of false wild oats, *J. Hered.*, **14**, 267–274

GFELLER, F. and SVEJDA, F. (1960). 'Inheritance of post-harvest seed dormancy and kernel colour in Spring wheat lines', *Can. J. Plant Sci.*, **40**, 1–6

GREER, E. N. and HUTCHINSON, J. B. (1945). 'Dormancy in British grown wheat', *Nature, Lond.*, **155**, 381

HALDANE, J. B. S. (1952). *The Biochemistry of Genetics*. Allen and Unwin Ltd., London

HAMMERTON, J. L. (1967a). 'Studies on weed species of the genus *Polygonum* L. IV. Variations in seed weight and germination behaviour in *P. lapathifolium*', *Weed Res.*, **7**, 1–21

HAMMERTON, J. L. (1967b). 'Studies on weed species of the genus *Polygonum* L. V. Variations in seed weight, germination behaviour and seed polymorphism in *P. persicaria* L.', *Weed Res.*, **7**, 331–348

HARPER, J. L., LANDRAGIN, P. A. and LUDWIG, J. W. (1955). 'The influence of environment on seed and seedling mortality. II. The pathogenic potential of the soil', *New Phytol.*, **54**, 119–131

HARPER, J. L. and MCNAUGHTON, I. H. (1960). 'The inheritance of dormancy in inter- and intra-specific hybrids of *Papaver*', *Heredity, Lond.*, **15**, 315–320

HARRINGTON, J. B. (1932). 'Comparative resistance of wheat varieties to sprouting in the stook and windrow', *Scient. Agric.*, **12**, 635–645

HARRINGTON, J. B. and KNOWLES, P. F. (1940). 'The breeding significance of after harvest sprouting in wheat', *Scient. Agric.*, **20**, 402–413

HASKELL, G. (1952). 'Genetics of cold tolerance in maize and sweet corn seed', *Heredity, Lond.*, **6**, 377–385

HONING, J. A. (1930). 'Nucleus and plasma in the heredity of the need for light for germination in *Nicotiana* seeds', *Genetica*, **12**, 441–468

HOWARD, H. W. and LYON, A. G. (1945). 'Effect of light on the germination of watercress seeds', *Nature, Lond.*, **168**, 253–254

HULL, R. (1937). 'Effect of environmental conditions and more particularly of soil moisture upon the emergence of peas', *Ann. appl. Biol.*, **24**, 681–689

HYDE, E. O. C. (1954). 'The function of the hilum in some *Papilionaceae* in relation to ripening of the seed and the permeability of the testa', *Ann. Bot., N.S.*, **18**, 241–256

JINKS, J. L. (1954). 'The analysis of continuous variation in a diallel cross of *Nicotania rustica* varieties', *Genetics, Princeton*, **39**, 767–788

JOHNSON, L. P. V. (1935). 'The inheritance of delayed germination of hybrids of *Avena fatua* and *Avena sativa*', *Can. J. Res.*, **13**, 367–387

JONARD, P. (1933). 'Relation entre la couleur du grain des diverses variétés de blé et leur aptitude à germer en moyettes', *Sélectionneur*, **2**, 41–48

JONES, J. W. (1926). 'Germination of rice seeds as affected by temperature, fungicides and age', *J. Amer. Soc. Agron.*, **18**, 576–592

JUSTICE, O. L. (1941). 'A study of dormancy in seeds of *Polygonum*', *Mem. Cornell Univ. Agric. Exp. Stn.*, **235**

KOLLER, D. (1964). 'The survival value of germination regulating mechanisms in the field', *Herb. Abstr.*, **34**, 1–7

KOOISTRA, E. (1962). 'On the differences between smooth and three types of wrinkled peas', *Euphytica*, **11**, 357–373

LAW, C. N. (1967). 'The location of genetic factors controlling a number of quantitative characters in wheat', *Genetics, Princeton*, **56**, 445–461

LINDSTROM, E. W. (1942). 'Inheritance of seed longevity in maize inbreds and hybrids', *Genetics, Princeton.*, **27**, 154

LUCKWILL, L. C. (1939). 'Observations on heterosis in *Lycopersicum*', *J. Genet.*, **37**, 421–440

MCNAUGHTON, I. H. and HARPER, J. L. (1960). 'The comparative biology of closely related species living in the same area. III. The nature of barriers isolating sympatric populations of *Papaver dubium* and *P. lecoquii*', *New Phytol.*, **59**, 129–137

MATHER, K. and HASKELL, G. (1949). 'Breeding cold hardy sweet corn in Britain', *J. agric. Sci., Camb.*, **39**, 56–63

MIYAMOTO, T., TOLBERT, N. E. and EVERSON, E. H. (1961). 'Germination inhibitors related to dormancy in wheat seeds', *Pl. Physiol., Lancaster*, **36**, 739–746

MORLEY, F. H. W. (1958). 'The inheritance and ecological significance of seed dormancy in subterranean clover (*Trifolium subterraneum* L.)', *Aust. J. biol. Sci.*, **11**, 261–74

MOSS, J. P. (1964). 'The adaptive significance of B-chromosomes in rye, in *Chromosomes Today*, (ed. Darlington, C. D. and Lewis, K. R.) Oliver and Boyd, London and Edinburgh

NILSSON-EHLE, H. (1909). *Kreuzungsuntersuchungen an Hafer und Weizen. Lund.*

NILSSON-EHLE, H. (1914). 'Zur Kenntnis der mit der Keimungsphysiologie des Weizens in Zusammenhang stehenden inneren Faktoren', *Z. Pflzücht.*, **2**, 153–187

PERDOK, E. A. (1970). 'The influence of gradual deterioration of polyploid beet seed on the ploidy pattern', *Proc. int. Seed Test Ass.*, **35**, 813–814

QUINLIVAN, B. J. (1968). 'The softening of hard seeds of sand plain lupin (*Lupinus varius* L.)', *Aust. J. agric. Res.*, **19**, 507–515

ROBERTS, E. H. (1961). 'Dormancy of rice seed. 1. The distribution of dormancy periods', *J. exp. Bot.*, **12**, 319–329

ROSA, J. J. (1925). 'Sex expression in Spinach', *Hilgardia*, **1**, 259–274

SHIFRISS, O. and GEORGE, W. L. (1965). 'Delayed germination and flowering in cucumber', *Nature, Lond.*, **206**, 424–425

SIMMONDS, N. W. (1964). 'The genetics of seed and tuber dormancy in the cultivated potatoes', *Heredity, Lond.*, **19**, 489–504

SNEEP, J. (1958). 'The present position of spinach breeding', *Euphytica*, **7**, 1–8

THODAY, J. M. (1961). 'Location of polygenes', *Nature, Lond.*, **191**, 368–370

TIMSON, J. (1965). 'Germination in *Polygonum*', *New Phytol.*, **64**, 179–186

TOPHAM, P. B. (1966). 'Diallel analysis involving maternal and maternal interaction effects', *Heredity, Lond.*, **21**, 665–74

VOSA, C. G. (1966). 'Seed germination and B chromosomes in the leek (*Allium porrum*), in *Chromosomes Today*, (ed. Darlington, C. D. and Lewis, K. R.) Oliver and Boyd, London and Edinburgh

WADDINGTON, C. H. (1961). 'Genetic assimilation', *Adv. Genet.*, **10**, 257–290

WEISS, M. G. and WENTZ, J. B. (1937). 'Effects of luteus genes on longevity of seed in maize', *J. Am. Soc. Agron.*, **29**, 63–75

WELLINGTON, P. S. (1956). 'Studies on the germination of cereals. 2. Factors determining the germination behaviour of wheat grains during maturation', *Ann. Bot., N.S.*, **20**, 481–500

WHITEHOUSE, R. N. H., THOMPSON, J. B. and DO VALLE RIBEIRO, M. A. M. (1958). 'Studies in the breeding of self pollinating cereals. 2. The use of a diallel cross analysis in yield prediction', *Euphytica*, **7**, 147–69

WHITTINGTON, W. J., CHILDS, J. D., HARTRIDGE, J. M. and HOW, J. (1965). 'Analysis of variation in the rates of germination and early seedling growth in tomato', *Ann. Bot., N.S.*, **29**, 59–71

WHITTINGTON, W. J. and FIERLINGER, P. (1972). 'The genetic control of time to germination in tomato', *Ann. Bot.* (in press)

SEED ECOLOGY

WHITTINGTON, W. J., HILLMAN, J., GATENBY, S. M., HOOPER, B. E. and WHITE, J. C. (1970). 'Light and temperature effects on the germination of wild oats', *Heredity, Lond.*, **25**, 641–650

WILLEY, L. A. (1970). 'Trials of commercial varieties of sugar beet', *Brit. Sugar Beet Rev.*, 165–167

WILLIAMS, P. (1970). 'Genetical effects of B-chromosome in *Lolium*', *Ph.D. Thesis*, University College of Wales

WITCOMBE, J. R. and WHITTINGTON, W. J. (1971). 'The effects of selection for reduced dormancy in Charlock (*Sinapis arvensis* L.)', *Heredity, Lond.* **29**, 37–49

DISCUSSIONS

Harrington: It is desirable to select for longevity of seeds but care is needed in the ageing technique used for selection.

Gordon: Our data suggest some overlap in the degree of dormancy of red and white grained wheats.

Whittington: Red grained types of wheat vary in their post-harvest dormancy, but evidence is lacking for any dormancy in white types. However, exceptions can usually be found when looked for.

E. H. Roberts: The correlation reported between time from sowing to germination (often measured in hours) and from germination to flowering (measured in weeks) in rice, wheat, cucumbers, *Arabidopsis* (see Ratcliffe and Evans, Appendix 2), and between dormancy of seeds and tubers of potato may conceivably be due to growth substance (e.g. GA or phytochrome) mechanisms that affect both processes.

3

GEOGRAPHICAL ADAPTATION OF SEEDS

P. A. THOMPSON

Jodrell Laboratory, Royal Botanic Gardens, Kew

INTRODUCTION

Innumerable observations on germination responses have established that control of the process is exercised through physical factors in the environment such as light and temperature. Experiments have also established that those parameters of these factors which favour germination vary from species to species, so that it may be possible to describe the responses of particular taxa qualitatively in ways related to their germination requirements; for example, as light-requiring or as high-temperature dependent. General reviews of the range of germination responses by Vegis (1963), Stokes (1965) and Lang (1965), amongst many others, testify to the range and variety of responses involved.

It is reasonable to assume that these responses are adaptive, and that success or failure of a population of a species in a particular locality depends on the way in which its germination responses fit a specification for survival determined by the ecological constitution of the habitat. (It is not necessary at this stage to consider facets of the plant's physiological responses other than germination, nor to define what is implied by the concept of an ecological fit.) Several workers have reported variations in the germination requirements of different populations of particular species which involve taxa as widely separated as *Tsuga canadensis* (Stearns and Olson, 1958), *Typha* spp. (McNaughton, 1966) and *Amaranthus retroflexus* (McWilliams, Landers and Mahlstede, 1968). Other workers have reported distinctive characters of the germination response which they have associated with the particular nature of distinct habitats, such as desert regions with intermittent, unpredictable levels of

31

precipitation (Went, 1949; Koller, 1955; Evenari, 1949) or with regions with well-marked regular annual seasons, such as the Mediterranean (Thompson, 1968). Comparisons between a group of species within the sub-family Silenoideae of the Caryophyllaceae have suggested that, within Europe at least, it may be possible to correlate germination responses to temperature with the ranges of particular species defined in geobotanical terms (Thompson, 1970a), and that species occurring within a locally restricted region may possess germination responses which are dissimilar (Thompson, 1970b), but which can be correlated with the overall range of each species rather than with the local characteristics of the area in which the collections were made.

An understanding of the biological significance of germination responses depends on the way in which comparisons between different taxa may be made to establish germination behaviour as a character, which can be examined in relation to the effects of modifying factors. The latter may include the locality of collection or previous history of the parent plants, the duration of storage and storage conditions prior to sowing, and physical factors of the environment after sowing, such as light quality, intensity or duration, moisture tension or the nature of the germination substrate.

This paper is concerned with experiments in which these modifying factors are simplified as far as possible to provide a model designed to demonstrate, in a very narrowly defined way, germination responses to temperature, which may be correlated with particular geographical ranges of wild plants, weeds and strains of cultivated vegetables. In terms of this model the germination character is defined here as the response to temperature over the range 0°C to 45°C of seeds stored after harvest for a minimum of six months at room temperature and provided with a daily exposure to low intensity light. This definition avoids consideration of varying degrees of post-harvest dormancy, of induced dormancy, or of the effects of chilling treatments or fluctuating temperatures, amongst a variety of other complicating factors. The results as presented are correlated with natural distributions expressed in gross geobotanical terms, but they are not considered in relation to the ecological requirements of different species, nor to seasonal growth patterns since these depend on data derived from an interacting milieu of factors which have been deliberately excluded from this study.

The method adopted is to consider successively the germination responses, first, of a group of wild species distributed naturally in Europe, then of a group of species which occur either as weeds or in association with man's activities, and finally of a group of vegetables widely grown in temperate regions. Representatives of the wild and weed species are chosen entirely from members of the sub-family Silenoideae of the Caryophyllaceae. The comparison between the responses of this relatively close-knit group of species with vegetables which are taxonomically distant one from another and from the Silenoideae is deliberate, and is designed to emphasise the level at which generalisations applicable to comparisons between germination responses and geographic distribution may be made.

MATERIALS AND METHODS

Seeds of species used in these experiments had been collected from plants growing wild in various parts of Europe, and were stored at laboratory temperatures for a minimum of six months, and in most cases for at least one year, before being tested. A complete list of the species used and their provenances is provided in *Table 3.1*. Seeds of

Table 3.1 COLLECTIONS OF SEED FROM WILD PLANTS, DETAILING COUNTRIES OF ORIGIN AND COLLECTION DATA

Species	Accession No. R.B.G., Kew	Country of origin	Collection data
Silene vulgaris	000-69-12786*	Switzerland	Collected from wild plants growing at about 1500 m.s.m.† in Valais
Silene vulgaris	000-69-14322	Germany	Collected near Rudolstadt, East Germany
Silene v. subsp. *commutata*	000-69-13721	Mallorca	Collected near Pollensa
Silene v. subsp. *maritima*	000-69-12785*	Norway	Wild plants at sea level near Asker, Akershus
Silene italica	000-69-12783*	Turkey	Collected from wild plants in western Turkey
Silene italica	000-69-14308*	France	Wild plants near Bras, Var, in the Alpes Maritimes
Silene italica	384-69-03252*	Bulgaria	Collected at about 1000 m.s.m. in Pirin Planina, near Simitli
Gypsophila perfoliata	174-70-01560	Romania	From Botanic Garden, Bucharest
Gypsophila scorzonerifolia	174-70-01561	Romania	From Botanic Garden, Bucharest
Gypsophila paniculata	152-70-01388	Hungary	From Botanic Garden, Vacratot
Gypsophila acutifolia	174-70-01557	Romania	From Botanic Garden, Bucharest

Table 3.1 *continued*

Species	Accession No. R.B.G., Kew	Country of origin	Collection data
Lychnis flos-cuculi	—	U.S.S.R.	Collected from wild plants near Dolina in the Carpathian mts
Lychnis flos-cuculi	—	Czechoslovakia	Collected near Brno
Lychnis flos-cuculi	000-69-12487*	Sweden	Collected in Øg, near Stockholm
Lychnis flos-cuculi	317-69-02622	Portugal	Wild plants near Antanhol, Coimbra
Petrorhagia prolifera	—	Hungary	Received from Hungarian University, Budapest '*in loco natali lecta*'
Petrorhagia prolifera	000-69-12718*	Germany	Collected near Jocketa, Vogtland
Petrorhagia prolifera	000-69-12730	France	From wild plants near Art-sur-Meurthe, M. et M., at about 210 m.s.m.
Silene conica	—	Germany	Collected near Greifswald, East Germany
Silene conica	000-69-12706*	Hungary	*In loco natali lecta*
Silene conica	000-69-12766*	Hungary	Collected near Godollo
Silene conica	000-69-12731*	Portugal	Collected from wild plants near Bonafles
Silene gallica	—	Canary Islands	Gran Canaria, near Beo de Los Tilos, about 600 m.s.m
Silene gallica	317-69-02642	Portugal	Collected from wild Plants near Amtuzede, Coimbra
Silene gallica	—	South Africa	Wild plants near Pretoria
Agrostemma githago	000-69-11773*	Poland	Collected from wild plants at Strykowo near Lodz
Agrostemma githago	347-71-03193*	Greece	Collected from about 30 plants on N.W. slopes of Mt. Olympus at 1100 m.s.m.
Agrostemma githago	317-69-02577*	Portugal	Collected from wild plants at Baleia, Coimbra, at 100 m.s.m.
Agrostemma githago	000-69-12133*	Switzerland	Collected from wild plants in Valais, at 1300 m.s.m.

* Collections of seeds stored in the Seed Bank at Royal Botanic Gardens, Kew
† Metres supra mare

vegetables were almost all obtained from the same supplier and used without any additional selection or cleaning. These were as follows:

Lettuce   Grand Rapids, Lobjoit's Cos, Webbs Wonderful, Kwiek.
Carrot   Autumn King, Early Nantes, Early Horn, Red-cored Chantenay.
Leek   Walton Mammoth, Prizetaker, Titan Osena, Musselburgh.
Brassicae   (Brussels Sprouts) Early Half Tall, (Cabbage) January King, (Calabrese) Green Sprouting, (Cauliflower) Novo.

All seeds were tested on thermo-gradient bars using methods previously described in detail (Thompson, 1970c; Fox and Thompson, 1971) and only briefly recapitulated here. Each bar was divided into 22 sections and set up to cover a span within or inclusive of the range 0°C to 43°C. Replicates of 50 seeds (25 for *Agrostemma githago*) were sown on each section and counts of germinated seeds were made daily for the first seven days of each experiment and then at longer intervals for up to four weeks. Results are expressed as percentage germination (radicle emergence) on successive days after sowing for each section along the temperature range studied. The centre points of successive sections along the thermo-gradient bars, and hence the plotted position for germination responses, varied from one another by 1·5°C to 1·8°C.

The results of germination counts made on successive days were plotted as percentage germination (y axis) against temperature (x axis), and the graphs were used to derive curves, called here 'germination character curves', by interpolating for each day the maximum and minimum temperatures at which 50 per cent of the final maximum number of seeds germinated. In the presentation of the results which follows here, detailed germination curves are presented for one population of each species or cultivar of vegetable examined, followed by germination character curves for three or four species or cultivars.

RESULTS

GERMINATION OF POPULATIONS OF WILD SPECIES

Silene vulgaris (Moench) Garcke

This species is very widely distributed throughout Europe and includes a number of forms which have, by various authorities, been differentiated as separate species or subspecies. This study included two populations within the general morphological range of *Silene vulgaris*

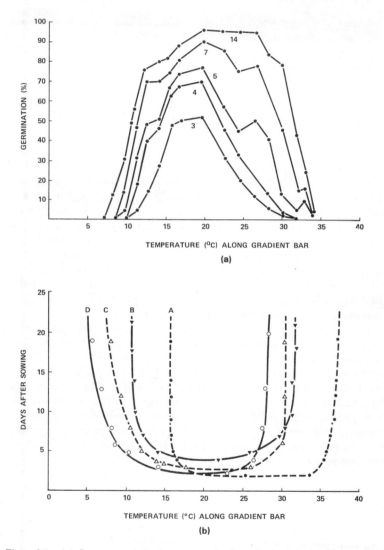

Figure 3.1 (a) Percentage germination on successive days after sowing on a thermo-gradient bar of Silene vulgaris *from Valais, Switzerland; (b) germination character curves of populations of* S. vulgaris *from Rudolstadt, E. Germany (A), Valais, Switzerland (B), Akershus, Norway (C), Pollensa, Mallorca (D)*

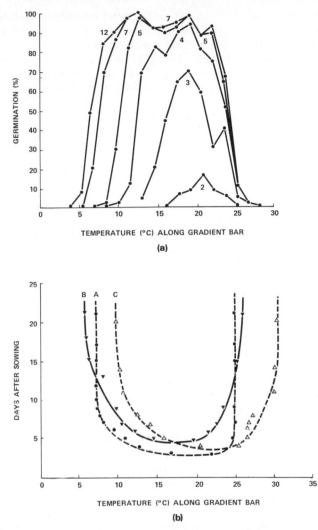

Figure 3.2   (a) Percentage germination on successive days after sowing on a thermo-gradient bar of Silene italica *from Bras, S. France; (b) germination character curves of populations of* S. italica *from Bras, France (A), Western Turkey (B), Pirin Planina, Bulgaria (C)*

subspecies *vulgaris,* one population of *S. v.* subspecies *maritima* (With.) A. and D. Löve and one of *S. v.* subspecies *commutata* (Guss.) Hayek. Results are shown for detailed germination responses of an alpine population of *S. v.* subspecies *vulgaris* in *Figure 3.1(a).* Germination

C

SEED ECOLOGY

of the alpine population was restricted to a median temperature range in which no seeds germinated below 7°C or above 34°C, but a very high proportion germinated over the range 20–27°C. The pattern of the response, with a number of seeds germinating on successive days over broad temperature ranges, and no evidence of a high degree of synchronous germination at particular temperatures, was characteristic of a heterozygous population. Germination character curves (*Figure 3.1(b)*) displayed considerable variation between the populations. *S. v.* subspecies *commutata* had a lower minimum and maximum than any other, as expected of a Mediteranean population (the milder the winter, the lower the minimum germination temperature—see Thompson, 1970a), and the population from Rudolstadt (East Germany) had a higher minimum than any other. The differences between *S. v.* subspecies *maritima* from Norway and *S. v.* subspecies *vulgaris* from alpine meadows in Switzerland were very small throughout the range of temperatures over which germination occurred.

Silene italica L. Pers

This species occurs throughout the northern fringes of the Mediterranean basin, characteristically in those parts with relatively cold winters and hot dry summers, and extends into many parts of southern Europe. Detailed curves are shown for a population collected in Var on the edge of the Maritime Alps in southern France (*Figure 3.2(a)*) and character curves for an ecologically similarly situated population in western Turkey, as well as from the lower slopes of the Pirin Planina close to the Struma Valley in Bulgaria (*Figure 3.2(b)*). Germination of the French population was favoured by moderately low temperatures, but did not occur when these fell below about 5°C, and was inhibited by high temperatures so that practically no seed germinated at or above 25°C. Results of tests on the Turkish population also conformed to this pattern but the more continental origin of the Bulgarian population was reflected by a higher minimum and maximum, resulting in limits for 50 per cent germination of approximately 10°C and 30°C.

Gypsophila spp.

The four *Gypsophila* species used in these experiments have extreme continental ranges, extending across the steppeland of southern Russia towards or into south-eastern Europe to the west, and

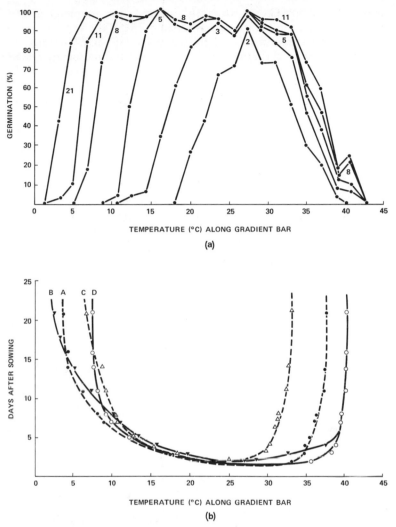

Figure 3.3   (a) Percentage germination on successive days after sowing on a thermo-gradient bar of Gypsophila perfoliata; *(b) germination character curves for seeds of* G. perfoliata *(A)*, G. scorzonerifolia *(B)*, G. paniculata *(C)*, G. acutifolia *(D)*

towards central Asia to the east. Detailed germination curves (*Figure 3.3(a)*) shown for *G. perfoliata* L. display a pattern quite different from responses illustrated previously. Germination occurs rapidly over a very wide range so that within seven days limits are established from about 8°C to 42°C. Subsequently, germination occurred at progressively lower temperatures down to about 1·5°C. The other species (*Figure 3.3(b)*) each displayed this capacity for rapid germination, but the extent of the range varied from one to another. *Gypsophila*

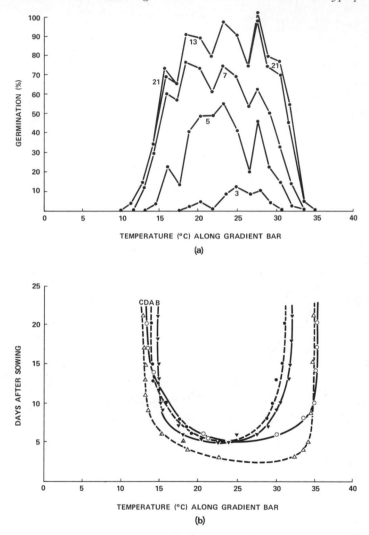

Figure 3.4 (*a*) *Percentage germination on successive days after sowing on a thermo-gradient bar of* Lychnis flos-cuculi *from Brno, Czechoslovakia;* (*b*) *germination character curves of populations of* L. flos-cuculi *from Carpathians, U.S.S.R. (A), Brno, Czechoslovakia (B), Eastern Sweden (C), Coimbra, Portugal (D)*

scorzonerifolia Ser. germinated over practically the entire range of temperatures at which normal metabolic processes occur in plants and the narrowest range was found in *G. paniculata* L. with a maximum for 50 per cent germination of 32°C and a minimum of about 6·5°C.

Lychnis flos-cuculi L.

This species occurs exclusively within deciduous forest zones of Europe in areas of wet meadowland or marshland. It is widely distributed geographically but stringently limited to particular situations by its habitat requirements. Results shown in detail for a population collected near Brno in Czechoslovakia (*Figure 3.4(a)*) display responses characteristic of deciduous forest species: the minimum temperature for germination is relatively high, and the maximum moderate. The seeds germinate relatively slowly, requiring five days to exceed 50 per cent even at the most favourable temperatures. Comparisons between this population and three others (*Figure 3.4(b)*) reveal very close agreement in the positions of temperature minima, and a spread of only about 4°C over the extreme differences between their maxima. Most of the populations shared the character of relatively slow germination, but one, from Sweden, was considerably faster than the rest, reaching 50 per cent within three days of sowing. The day-by-day pattern of development of the responses was characteristic of highly heterozygous populations.

GERMINATION OF POPULATIONS OF WEED SPECIES

Each of the species in this section most probably owes part, at least, of its range to its ability to colonise in the wake of man's activities. Their status as weeds of cultivation, ruderals or wild plants varies widely between populations or species, depending on local circumstances.

Petrorhagia prolifera L. P. W. Ball and Heywood

This is one of a number of species widely distributed in southern, eastern and central Europe, especially in steppeland regions or areas of local grassland. It occurs naturally in southern and eastern Europe but its occurrence as a weed in association with cultivated areas results in its distribution over a much wider range than its natural one. Detailed germination responses, shown in *Figure 3.5(a)*,

42

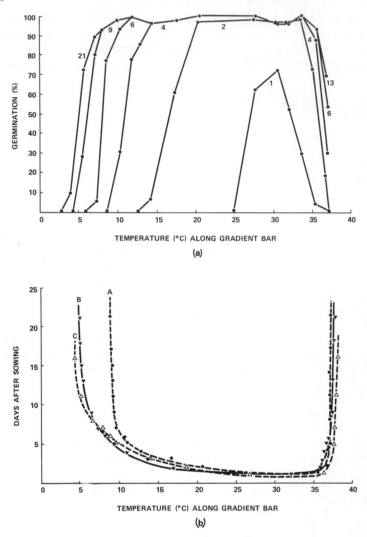

Figure 3.5 (a) Percentage germination on successive days after sowing on a thermo-gradient bar of Petrorhagia prolifera *from eastern France; (b) germination character curves of populations of* P. prolifera *from Budapest, Hungary (A), Art-sur-Meurthe, France (B), Vogtland, Germany (C)*

for a population from eastern France demonstrate rapid germination over the range 25°C to 35°C, and, finally, germination covering a very wide range of temperature, from below 5°C to above 35°C. Germination character curves for this and two other populations

(*Figure 3.5(b)*) demonstrate close similarities in responses over most of the range of temperatures at which seed germination occurs, with a rather higher minimum for the Hungarian population than those from France and Germany.

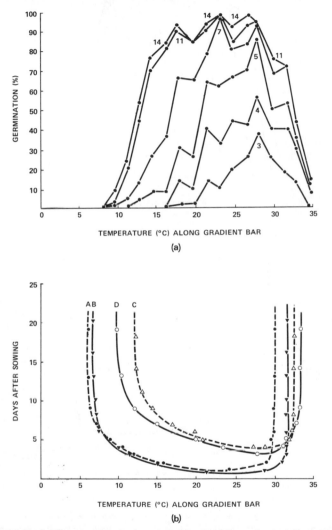

(a)

(b)

Figure 3.6   (a) Percentage germination on successive days after sowing on a thermo-gradient bar of Silene conica *from Godollo, ˙Hungary; (b) germination character curves of populations of* S. conica *from Greifswäld, Germany (A), Budapest, Hungary (B), Godollo, Hungary (C), Bonafles, Portugal (D)*

44

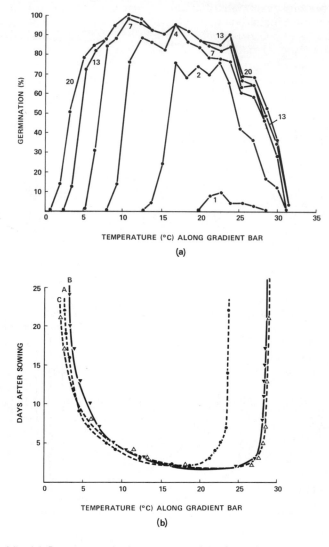

Figure 3.7 (a) Percentage germination on successive days after sowing on a thermo-gradient bar of Silene gallica *from Portugal; (b) germination character curves of populations of* S. gallica *from Gran Canaria, Canary Islands (A), Coimbra, Portugal (B), Pretoria, South Africa (C)*

Silene conica L.

This species is widely distributed in the wild in parts of southern Europe and in association with man over much of continental, eastern and northern Europe, excluding most of the British Isles and Scandinavia. Detailed responses shown for a population from Godollo in Hungary (*Figure 3.6(a)*) demonstrate responses over a median temperature range with limits close to 10°C and 35°C. At least four days were required to reach levels of 50 per cent, and day-by-day development was characteristic of a heterozygous population. Comparisons between different populations (*Figure 3.6(b)*) showed closely similar maxima, but the time course of germination and the minima of the different populations showed more considerable differences. Two populations germinated relatively rapidly even at low temperatures and established minima well below 10°C; the other two germinated more slowly, especially below 20°C, and established minima at or above 10°C.

Silene gallica L.

This species occurs as a wild plant around the Mediterranean basin, and as a weed in association with man's activities over much of southern, central and western Europe, as well as in many other parts of the world where it has been introduced and become established as an alien, usually as a weed. Detailed responses are shown here for a population collected near Coimbra in Portugal (*Figure 3.7(a)*) and these are characteristic of a species of Mediterranean origin, with a very low minimum and relatively low maximum for germination, combined with rapid responses over the median range between about 15°C and 25°C. Comparison of the responses of this population with those of collections from the Canary Islands and South Africa are shown as germination character curves (*Figure 3.7(b)*). Minima were closely similar for all three populations, as were the courses of germination at temperatures below 20°C, and indeed no differences occurred throughout the whole response range for the populations from South Africa and Coimbra. However, temperature maxima for 50 per cent germination of the Canary Island population were much lower.

Agrostemma githago L.

This species most probably originated in the area of the eastern Mediterranean and spread as a weed of cereals and vetches throughout

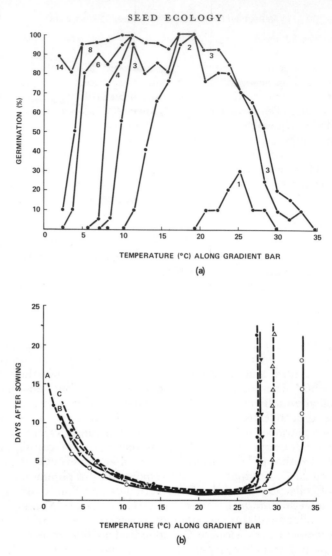

Figure 3.8 *(a) Percentage germination on successive days after sowing on a thermo-gradient bar of* Agrostemma githago *from Greece; (b) germination character curves of populations of* A. githago *from Strykowo, Poland (A), Mt. Olympus, Greece (B), Coimbra, Portugal (C), Valais, Switzerland (D)*

Europe in neolithic times. Detailed curves shown in *Figure 3.8(a)* for a population collected near Mt. Olympus in Greece, growing under wild or nearly wild conditions, display a characteristic Mediterranean response closely similar to that of *Silene gallica* L., but with a lower minimum. This collection also shows the day-by-day response pattern typical of a homozygous population in which a high proportion of the seeds germinate synchronously at any particular temperature. Comparisons between this population and others collected as weeds growing in Poland, Portugal and Switzerland are presented as germination character curves in *Figure 3.8(b)*. Responses below 25°C were closely similar, with uniformly low minima approaching close to 0°C, and germination of the seeds of several populations of this species has been observed at 0°C, after the seeds had become encased in ice. Maxima for the four populations varied, although no obvious similarity of response occurred between populations from areas possessing geobotanical affinities. For example, the relative similarity of responses between populations from Portugal and Greece might have been expected, but the even closer similarity between the Greek population and the one from Poland denies the validity of such correlations.

GERMINATION OF CULTIVARS OF VEGETABLES

Lettuces

Detailed response curves are shown for the cultivar Grand Rapids (*Figure 3.9(a)*), and responses of this and three other cultivars are compared as germination character curves in *Figure 3.9(b)*. These display minima close to freezing point, very rapid germination over the median range up to about 25°C, and maxima for 50 per cent germination between 27·5°C and 32°C. The similarities, in all respects of these lettuce cultivars and the populations of *Agrostemma githago* shown previously are striking. The detailed responses of Grand Rapids display a developmental pattern typical of a markedly homozygous batch of seeds, and also show that at high temperatures values close the maximum are established within a few days of sowing. In this particular batch of Grand Rapids there is a long tail of seedlings germinating at high temperatures up to 35°C, but definition of maxima in lettuce achenes are usually much sharper.

Carrots

Detailed germination responses are shown for the cultivar Autumn King (*Figure 3.10(a)*), and its germination character curve (*Figure 3.10(b)*) is compared with those of three other cultivars. Responses span a wide temperature range from about 5°C to 38°C, and few differences, if any, occur in the curves obtained for different cultivars. Development of the curves is typically heterozygous, and they display features associated with wild species of continental origin. These

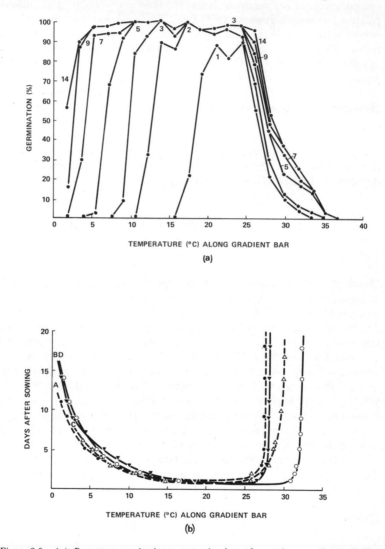

Figure 3.9   (a) Percentage germination on successive days after sowing on a thermo-gradient bar of the lettuce cultivar Grand Rapids; (b) germination character curves of lettuce cultivars Lobjoit's Green Cos (A), Grand Rapids (B), Webbs Wonderful (C), Kwiek (D)

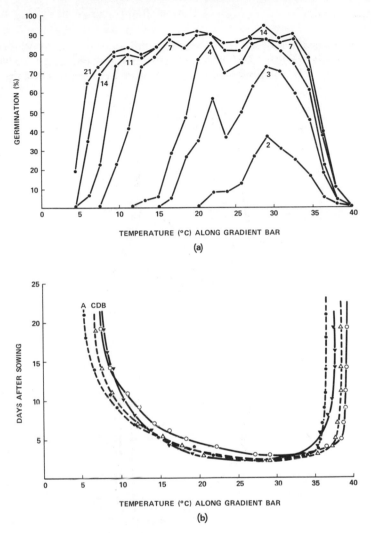

Figure 3.10   (a) Percentage germination on successive days after sowing on a thermo-gradient bar of the carrot cultivar Autumn King; (b) germination character curves of carrot cultivars Autumn King (A), Early Nantes (B), Early Horn (C), Red Cored Chantenay (D)

include minima between 5°C and 10°C and moderately high maxima, in association with rapid germination responses which result in 50 per cent germination over a wide temperature range within three days of sowing.

Leeks

Germination responses are shown in detail for the cultivar Walton Mammoth (*Figure 3.11(a)*), and germination character curves (*Figure 3.11(b)*) are presented for comparisons between this and three

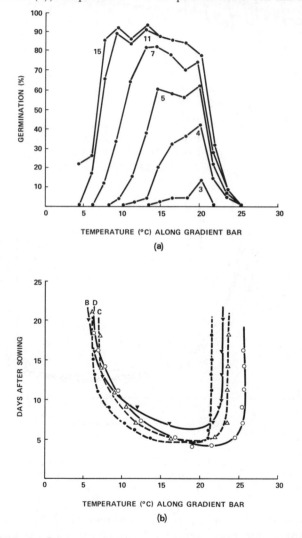

Figure 3.11 (a) Percentage germination on successive days after sowing on a thermo-gradient bar of the leek cultivar Walton Mammoth; (b) germination character curves of leek cultivars Walton Mammoth (A), Titan Osena (B), Musselburgh (C), Prizetaker (D)

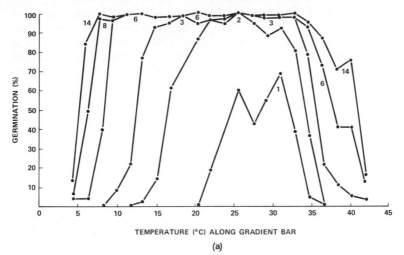

(a)

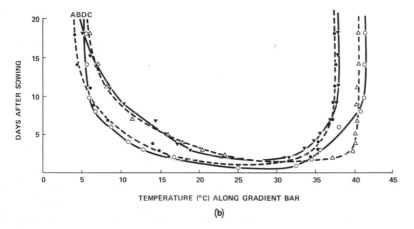

(b)

Figure 3.12 (a) Percentage germination on successive days after sowing on a thermo-gradient bar of the brussels sprout cultivar Early Half Tall; (b) germination character curves of different types of Brassica—*cabbage, January King (A), cauliflower, Novo (B), broccoli (calabrese), Green Sprouting (C), brussels sprout, Early Half Tall (D)*

other cultivars. The responses display characteristically low maxima and moderate minima, producing germination responses only over the range 5–23°C, combined with relatively slow development, reaching 50 per cent four to six days after sowing. These temperature responses are closely similar to those shown earlier for two populations of *Silene italica*. Variations between different cultivars are small below 20°C, but maxima ranged from 21·3°C for 50 per cent germination of Walton Mammoth, to 25·5°C for Prizetaker. The pattern of development is characteristic of a heterozygous batch of seeds.

Brassicas

Detailed responses are shown for the brussels sprout Early Half Tall (*Figure 3.12(a)*), and germination character curves (*Figure 3.12(b)*) are compared for this and for cultivars of cabbage, cauliflower and calabrese. All are characterised by very wide-ranging germination responses, not unlike those shown earlier for *Gysophila* spp., and by very rapid responses at median temperatures over a wide range from about 15°C to 35°C.

It might appear that differences occur in temperature maxima and minima for different types of *Brassica*, and in the rate (speed) of germination over the range 8–20°C in particular. These differences are being examined elsewhere, but at present it seems unlikely that they can be correlated with particular crop forms, since comparisons, for example within different cultivars of cabbage, or within different cultivars of cauliflower, suggest similar variations from one cultivar to another.

DISCUSSION

Examination of the germination responses described in the previous section reveals, usually, some degree of similarity of response within populations of a species or cultivars of a vegetable. This similarity may exist as close agreement in the time course of germination throughout the entire temperature range, or it may be limited to agreement only at particular parts of the response range. A species such as *Silene vulgaris* which occupies ecologically varied habitats over a wide geographical area and which is morphologically extremely variable may, as found here, also possess varied physiological responses, which express themselves in the diversity of the germination behaviour of different populations.

The number of populations discussed here for each species is insufficient to characterise the range of variation occurring within the species, or to provide material for correlation with geographical distribution. However, comparisons between species which are more ecologically limited and less varied morphologically than *S. vulgaris* demonstrate that sets of populations within a particular species, such as *S. italica* or *Lychnis flos-cuculi*, are likely to resemble one another in a number of descriptive parameters and indeed may be markedly similar, even when populations from widely separated collections are compared. Similarly, comparisons within a group of closely related species, with similar ecological preferences and geographical ranges, as exemplified here by species of *Gypsophila*, may reveal affinities which can be used to characterise the group as a whole.

Species of wild plants growing under competitive conditions may be stringently limited in the range of their germination responses, in common with other adaptive responses which permit successful establishment within a particular area. However, it appears that when such plants are transported into alien locations as weeds, particular genetic qualities may enable establishment in areas with climates quite unlike those in which the plant originated, with or without contingent adaptive modification of their responses. A situation where transposition has been accompanied by little, if any, change in germination response is illustrated in this paper by the similarities between the responses of populations of *S. gallica* from Portugal and South Africa, and the point has been examined in more detail for populations of *Agrostemma githago* distributed all over Europe (Thompson, 1972), a selection of which are shown in this paper. However, dispersal of a species as a weed may not necessarily be associated with the retention of a particular form of germination response, as is illustrated in this paper by differences between populations of *Silene conica* and by the much lower maximum temperature for the germination of one population of *S. gallica* compared with the two other populations of the same species.

Clearly much more study is required to identify those factors which are likely to result in stability of germination responses during dispersal over a wide geographical area, as in *A. githago*, rather than in adaptive modifications from one area to another as suggested by results with *S. conica*. Experiments by other workers have produced conflicting evidence on the stability, or otherwise, of germination responses over a plant's range. Thus, Harper (1965), McNaughton (1966) and Stearns and Olson (1958) have published results demonstrating differences between different populations, whereas Lauer (1953) found few differences between populations collected in various parts of Europe. Examination of the results presented above suggests that such modifications as do occur are likely to be limited to relatively small changes in response at particular parts of the temperature range, so that quantitative variations between populations are kept within limits which preserve an essential qualitative resemblance between

populations of a species growing as a weed to much the same degree
as has been found between different populations of individual species
growing as wild plants. Further study is needed of the change-
limiting factors which restrict adaptive capacity and so preserve a
recognisable 'species character' between different populations of a
species.

Examination of the responses of strains of cultivated plants presented here show that these plants do not possess a generalised
germination response, produced, perhaps adaptively, by selection
for particular qualities imposed by the special features of cultivation
by man. Nor is there any suggestion that strains of vegetables
selected for sowing at different seasons in the open ground or in an
artificially heated environment, such as a glasshouse, are likely to
differ from one another in major respects. Thus, the germination
responses found from one cultivar to another of a particular crop
plant were notably similar, though not necessarily the same, and in
contrast the overall comparison between one crop plant and another
shows that each possesses quantitative and qualitative characteristics
which are distinctively different.

This suggests either that cultivation imposes few selective pressures
on the germination character of a species, or that the germination
response is not easily changed by adaptive pressures produced during
cultivation, or that the primeval character of these crop plants was
largely pre-adapted to the particular requirements of cultivation.
These alternatives are not mutually exclusive and indeed the first
could be a consequence of the third; and all could interact to a greater
or lesser degree, depending on the germination responses of the crop.
Thus, all the forms of *Brassica* shown in *Figure 3.12* germinate rapidly
and completely over so wide a temperature range that there is little
scope for further modification to improve their performance over the
range of temperatures relevant to cultivation, complying with the
first alternative proposed above. The responses of these forms may
still be similar to the natural germination character of the primitive
ancestors of the group, a character which accords closely with that
found in plants with an extreme continental distribution, represented
here for wild populations by four species of *Gypsophila* and for weeds
by populations of *Petrorhagia*.

Similarly, the wide temperature tolerances found in strains of
carrots may be compared with the germination responses of species
of weeds of continental origin, and would appear to deny the
probability of any important Mediterranean element in the origin of
this crop. However, scope for selective improvement in germination
at low temperatures, below about 15°C, appears to exist in this crop
(as shown by Hegarty, Chapter 23) and this represents a situation

in which the second alternative, suggesting the presence of certain restrictions in the adaptive capacity of species to selection, may have acted to preserve an ancestral character which appears to conflict to some extent with the requirements for a crop plant.

This situation is exemplified more clearly in the figures illustrating germination responses of lettuces and leeks (*Figures 3.9* and *3.11*). Both these crops, particularly the former, possess temperature maxima which are sufficiently low effectively to limit their geographical distribution as crop plants, and the seasons when they can be sown. Leeks possess in addition, restraints on the rate of germination and on their response to low temperature which produce a relatively restricted germination character, and this conflicts in theory and practice with the third alternative proposal that primeval ancestors of crop plants were pre-adapted to cultivation to the point where little further improvement was possible. The fact that this crop, of very ancient origin indeed, still possesses such restrictions on range and rate of germination provides a good example of the comparative stability of such responses. Under normal methods of cultivation and seed raising in this crop, selective pressures would be exerted on a field scale, after mass cross-pollinations within a heterozygous out-breeding population; it is possible that the stability of these responses would not be so marked under more controlled breeding programmes.

The germination responses of lettuce, although confined by a moderate maximum (about 30°C), are more clearly conformable with the requirements of a cultivated plant at median and low temperatures, and it might have been expected that selective breeding, easily achieved by single plant selections in this almost totally inbreeding crop, would have resulted in marked differences between different strains. This situation is not evident in *Figure 3.9*, and such differences as do occur are almost confined to differences in maxima within a restricted range. Comparisons between results for strains of lettuce and those obtained for populations of *Agrostemma githago* reveal almost complete agreement in the time course of germination at different temperatures and the range of variation between different strains or populations. Thus, cultivation and deliberate selective breeding by man for characters other than those contributing to germination responses, would appear to have had no selective effect on germination beyond that achieved under natural conditions by a weed species of possibly similar natural origin and geographical dispersal.

The results shown here display, repetitively, patterns of variation within restricted ranges, which preserve quantitative and qualitative distinctions between species or types of vegetable. The overall qualitative similarities noted here between populations of a species do not necessarily contradict the findings of other workers who have described differences between populations, or even between individuals of particular species such as *Rumex* spp. (Cavers and Harper, 1966). In most of the examples shown here variations between populations or strains of vegetable would have been considerable if comparisons had been limited to particular, narrowly defined, regions of the temperature range : usually those parts closely approaching the temperature maxima and minima, although occasionally responses over parts of the median temperature range differed as well. Thus, tests restricted to particular *loci* along the temperature scale could have indicated considerable differences between populations. It is only when comparisons are made over the whole of the temperature range, between a number of populations of a number of species, that a scale can be established against which the relative magnitude of changes in response may be measured. When this is done the essential common and distinctive qualities innate within the germination responses to temperature, of different populations of a species, can be recognised and an attempt made to relate them to the geographical pattern of distribution of the species.

If this approach is used it is possible to make generalisations about response patterns of a species as a whole from consideration of the responses of different populations, or about response patterns of a type of vegetable as a whole from the responses of different strains. The results of the tests reported here demonstrate that, for some species at least, differences in germination responses between widely separated populations may be quite small. They also demonstrate that populations of a species distributed as a weed can retain characteristics which preserve an affinity between the populations amounting to a qualitatively recognisable species character. This situation has been discussed in much more detail with reference to *Agrostemma githago* (Thompson, 1972), in which tests on 19 populations distributed throughout Europe produced results which showed that all populations possessed a characteristic Mediterranean germination response, suggesting that changes during its many years of association with man as a weed had been small. Extension of these comparisons to the results obtained with different strains of four types of vegetable demonstrate similarly characteristic patterns occurring among strains which differentiate one kind of vegetable from another. The existence of a species-specific character for each vegetable, and the occurrence of response patterns analogous to those found for wild species from particular geographic locations, or for weed species of a particular geographic origin, strongly suggest that the germination responses of these vegetables also owe their origin to the requirements of particular geographical conditions. It therefore seems most likely that cultivation has had little effect on the germination responses of these vegetables, apart from marginal changes similar to those found in the weed species, and that the individual characteristics of each type still reflect, to a very large degree, the responses developed by their wild progenitors by selective adaptation to the climate of the geographical regions in which they were originally distributed.

ACKNOWLEDGEMENTS

I have pleasure in thanking Dr. K. Ferguson for collections of seeds, and Miss Erica Dodd for assistance with these experiments. I should also like to acknowledge gratefully gifts of seed collected by staff of the following institutions: Jardin Alpin de la Fondation J.-M. Aubert, Valais, Switzerland; Botanic Garden, Martin-Luther University, Halle, East Germany; Botanic Garden, University of Oslo, Norway; Botanic Garden, Geneva, Switzerland; Botanic Garden, University of Dijon, France; Botanic Garden, Technical University, Dresden, East Germany; Botanic Garden, V. L. Komarov Institute of Botany, Leningrad, U.S.S.R.; Agricultural and Silvicultural University, Brno, Czechoslovakia; Botanic Garden, University of Uppsala, Sweden; Botanic Garden, University of Coimbra, Portugal; Botanic Garden, Hungarian University, Budapest, Hungary; Botanic Garden, Nancy, France; Cambridge University Botanic Garden, England; Institute Przemyslu Zielarskiego, Poznan, Poland; Bucharest Botanic Garden, Romania.

REFERENCES

CAVERS, P. B. and HARPER, J. L. (1966). 'Germination polymorphism in *Rumex crispus* and *Rumex obtusifolius*', *J. Ecol.*, **54**, 367–382

EVENARI, M. (1949). 'Germination inhibitors', *Bot. Rev.*, **15**, 153–194

FOX, D. J. C. and THOMPSON, P. A. (1971). 'A thermo-gradient bar designed for use in biological studies', *J. exp. Bot.*, **22**, 741–748

HARPER, J. L. (1965). 'Establishment, aggression and co-habitation in weedy species', in *The Genetics of Colonizing Species*, (ed. Baker, H. G. and Stebbins, G. L.) Academic Press, London and New York

KOLLER, D. (1955). 'The regulation of germination in seeds', *Bull. Res. Coun., Israel*, **5D**, 85–108

LANG, A. (1965). 'Effects of some internal and external conditions on seed germination', in *Encyclopedia of Plant Physiology*, (ed. Ruhland, W.) 15 (2), Springer Verlag, Berlin, Heidelberg, New York

LAUER, E. (1953). 'Über die Keimtemperatur von Ackerunkräutern und deren Einfluss auf die Zusammensetzung von Unkrautgesellschaften', *Flora, Jena*, **140**, 551–595

MCNAUGHTON, S. J. (1966). 'Ecotype function in the *Typha* community type', *Ecol. Monogr.*, **36**, 297–325

MCWILLIAMS, E. L., LANDERS, R. Q. and MAHLSTEDE, J. P. (1968). 'Variation in seed weight and germination in populations of *Amaranthus retroflexus* L.', *Ecology*, **49**, 290–296

STEARNS, F. and OLSON, J. (1958). 'Interaction of photoperiod and temperature affecting seed germination in *Tsuga canadensis*', *Am. J. Bot.*, **45**, 53–58

STOKES, P. (1965). 'Temperature and seed dormancy', in *Encyclopedia of Plant Physiology*, (ed. Ruhland, W.), 15 (2), Springer Verlag, Berlin, Heidelberg, New York

THOMPSON, P. A. (1968). 'Germination of Caryophyllaceae at low temperatures in relation to geographical distribution', *Nature, Lond.*, **217**, 1156–1157

THOMPSON, P. A. (1970a). 'Germination of species of Caryophyllaceae in relation to their geographical distribution in Europe', *Ann. Bot.*, **34**, 427–449

THOMPSON, P. A. (1970b). 'A comparison of the germination character of species of Caryophyllaceae collected in central Germany', *J. Ecol.*, **58**, 699–711

THOMPSON, P. A. (1970c). 'Characterisation of the germination responses to temperature of species and ecotypes', *Nature, Lond.*, **225**, 827–831

THOMPSON, P. A. (1972). 'Effects of cultivation on the germination character of the Corn Cockle (*Agrostemma githago* L.)', *Ann. Bot.*, (in press)

VEGIS, A. (1963). 'Climatic control of germination, budbreak and dormancy', in *Environmental Control of Plant Growth*, (ed. Evans, L.), Academic Press, New York, 265–288

WENT, F. W. (1949). 'Ecology of desert plants. II. The effect of rain and temperature on germination and growth', *Ecology*, **30**, 1–13

DISCUSSIONS

Osborne: Can you explain the great stability which you find in the response of seed populations to temperature?

Thompson: The stability over many generations of the germination response to temperature refers to mass selection within whole populations; a small proportion of extreme genotypes will almost certainly have a different range of germination temperatures.

Wareing: Selection for ability to germinate—and therefore perhaps to grow—at low temperatures appears worth attempting to extend the growing season of crop plants; it has been done with maize.

Whittington: Monogerm sugar beet varieties have been selected for field emergence at low temperatures (6°C) by Ellerton.

Thompson: The minimum temperature for commercial onion varieties is usually 8°C–9°C, but some F_1 varieties germinate at 4°C–6°C.

Rorison: To what extent are alternating temperatures beneficial?

Thompson: The benefit from alternating temperatures varies from species to species: high temperatures may be inhibitory or neutral; if the high temperature is inhibitory, the combination will also be inhibitory, but a combination of a favourable low and an unfavourable but neutral high temperature may promote germination. Some species have an absolute requirement for temperature fluctuations of 5°C or 10°C to be able to germinate.

4

DIFFERENCES IN THE PROGENY DUE TO DAYLENGTH AND HORMONE TREATMENT OF THE MOTHER PLANT

Y. GUTTERMAN

Department of Botany, Hebrew University of Jerusalem, Israel

This paper deals with four groups of experiments:

1. Effect of short day (SD) or long day (LD) conditions on germinability of *Ononis sicula* Guss.; *Lactuca sattiva* L. var. Grand Rapids; *Portulaca oleracea* L. and *Carrichtera annua* (L.) Asch.
2. The influence of daylength and red and far-red light on the germinability at different temperatures of *Portulaca oleracea* L.
3. The influences of daylength and hormones on the germinability of the seeds produced and on the performance of the resulting generation of plants of *Lactuca scariola* L.
4. The influence of the position of the caryopses in the dispersal unit on the germinability of *Aegilops ovata*.

INTRODUCTION

Interest in the influence of daylength on the progeny was aroused during several years' field observations on annual plants in the Negev desert of Israel.

Large differences were found in the appearance of individual plants from year to year and within each year, depending on the distribution of the rains and the date of germination which ranged from September (13 h days) through December (11 h days) to March (13 h days) (*Figure 4.1*). Large differences were also found in the numbers of plants of different species from year to year. When 22 species of these annual desert plants were grown under permanent SD (8 h days) or

SEED ECOLOGY

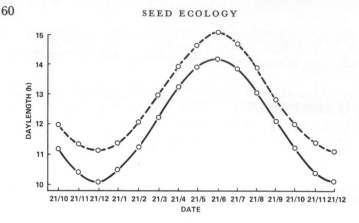

Figure 4.1 (a) ———— = *daylength in Jerusalem from sunrise to sunset (daylength in Jerusalem is nearly identical with that of Avdat)* ————— = *photoperiodic daylength (from morning light intensity of five lux to the same intensity in the evening)*

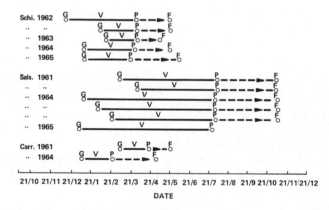

Figure 4.1 (b) Date of germination (G), *first appearance of flowers* (P), *first appearance of ripe fruit* (F) *and length of vegetative period* (V) *for* Schismus arabicus *(Schi)*, Salsola inermis *(Sals.) and* Carrichtera annua *(Carr.) as observed in Avdat in various years (Evenari and Gutterman, 1966; Gutterman, 1966).*

LD (20 h days), differences were again found: in the number of leaves before the appearance of flower buds, in leaf shape and in the pattern of branching (Evenari and Gutterman, 1966; Gutterman, 1966) (*Figures 4.2 and 4.3*).

Then the question arose: are the seeds matured under SD or LD also different, as the mother plants differ?

Figure 4.2 Sclerocaryopsis spinocarpos. *Greenhouse plants 60 days after germination, grown under long (20 h) days (= LD) or under short (8 h) days (= SD) (Gutterman, 1966)*

Figure 4.3 Ononis sicula *plants, 100 days after germination, grown in the greenhouse in Jerusalem under long (20 h) days (= LD) or under short (8 h) days (= SD) (Gutterman, 1966)*

RESULTS AND DISCUSSION

Germinability affected by SD or LD conditions

To test this possibility checks were made on the seeds of the desert annual *Ononis sicula*, matured under SD or LD conditions, and the following was found.

In LD the maturing seeds become yellow, are relatively large and heavy and imbibe only after a long time in contact with water (45–90 per cent after 80 days) (*Figure 4.4*). In SD seeds remain green

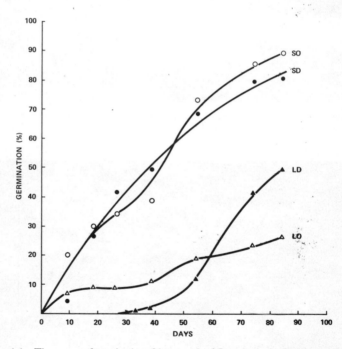

Figure 4.4   Time course of germination of intact seeds of Ononis sicula *plants grown in the greenhouse under long (20 h) days (= LD) or short (8 h) days (= SD) and outdoors under LD (= LO) or SD (= SO). (Gutterman, 1966; Evenari, Koller and Gutterman, 1966)*

or brown. The embryos are ripe before the yellow stage is reached. The green seeds imbibe sooner than the yellow ones. The brown seeds are capable of immediate imbibition on contact with water, but in storage they lose their viability within one or two years (Evenari,

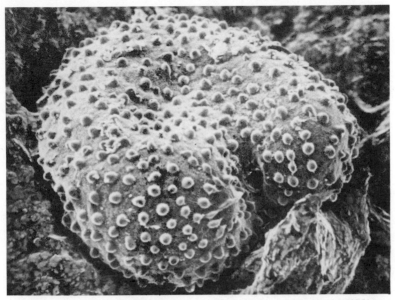

Figure 4.5   Whole seeds of Ononis sicula × *81 (Gutterman and Heydecker, 1973)*

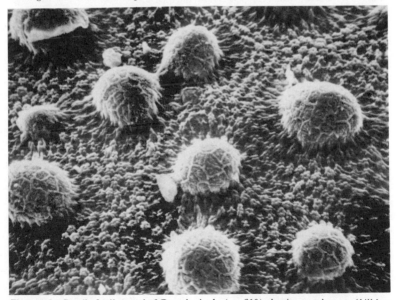

Figure 4.6   Detail of yellow seed of Ononis sicula *(× 610) showing protuberances 'hills' on surface of seed. The three following enlargements (Figures 4.7–4.9) show sections of the seed surface between these 'hills' (Gutterman and Heydecker, 1973)*

64

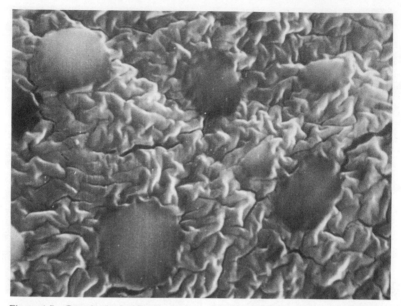

Figure 4.7 Ononis sicula. *Detail of brown seed (× 5250). Relatively thin and smooth surface covering an undeveloped structure (Gutterman and Heydecker, 1973)*

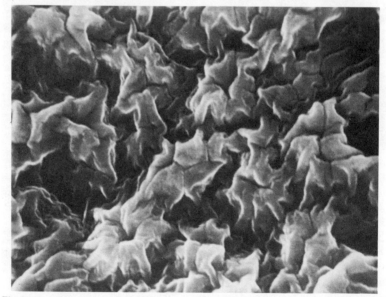

Figure 4.8 Ononis sicula. *Detail of green seed (× 5600). Surface structure well developed (Gutterman and Heydecker, 1973)*

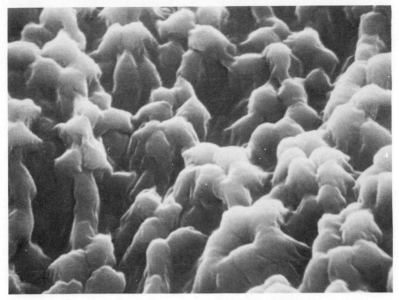

Figure 4.9   Ononis sicula. *Detail of yellow seed (× 5800). Well-developed and thickened surface structure (Gutterman and Heydecker, 1973)*

Koller and Gutterman, 1966; Gutterman, 1966), rather earlier than green or yellow seeds.

The difference in the germinability is caused by the degree of permeability of the seed coat to water. A little scratch on the seed coat causes immediate imbibition by all seeds on contact with water. The difference in the cuticle surface structure of these three kinds of seed was studied by means of a scanning electron microscope. Part of the seed surface is covered with 'hills' (*Figures 4.5 and 4.6*), but the differences in structure observed are confined to the 'valleys' between the 'hills'. The structure is undeveloped in brown seeds (*Figure 4.7*), well developed in green seeds (*Figure 4.8*) and well developed and thickened in yellow seeds (*Figure 4.9*) (Gutterman and Heydecker, 1973).

A change of daylength for no more than the last eight days of ripening can affect the colour, size, weight and thereby the germination of the seeds (Gutterman, 1969; Gutterman and Evenari, 1972). The ability of one single plant to develop, according to the daylength during maturation, three kinds of seeds that differ in their rate of germination is very important for survival under the difficult

desert conditions. Changes in daylength also have effects on seed
germinability in other species:

1. A change of daylength during the last 12 days of ripening can
   affect the germinability in the lettuce (*Lactuca sativa*) cultivar
   Grand Rapids (*Table 4.1*). Seeds from SD conditions germinate
   faster than seeds from LD.

2. In *Portulaca oleracea* (characterised below) seeds from 8 h
   days germinate faster than those from 11 h days and from
   13 h days (*Table 4.2*). A change of daylength during the last

Table 4.1 TREATMENTS GIVEN TO THE PLANTS OF *LACTUCA SATIVA* GRAND RAPIDS 517 AND
THE GERMINABILITY OF THEIR SEEDS, AT 26°C IN THE DARK WITH SHORT (5 MIN) ILLUMINA-
TIONS OF WHITE LIGHT—200 (4 × 50) SEEDS IN EACH TREATMENT

Photoperiodic conditions during growth of mother plants	Germination (%) after 48 h (light: 1 × 5 min)*	Germination (%) after 11 days (light: 11 × 5 min)*
SD, then LD (80)†	0	4 ± 1·4
LD, then SD (80)	29·5 ± 2·6‡	32 ± 3·2
SD, then LD (12)	5 ± 1·2	8 ± 1·4
LD, then SD (12)	13·5 ± 3·4	18 ± 3·4
continuous LD	5 ± 0·6	6 ± 0·8
continuous SD	16·5 ± 2·2	24 ± 2·5

* Number of short (5 min) illuminations given to the seeds. First illumination was given 1·5 h after wetting.
† Number of days under the photoperiodic conditions before harvest. The germination experiment began
immediately after harvest.
‡ ±S.E.

Table 4.2 INFLUENCE OF THE PHOTOPERIODIC TREATMENTS GIVEN TO *PORTULACA
OLERACEA* PLANTS ON THE NUMBER OF LEAVES AND AGE OF PLANTS AT THE TIME OF
APPEARANCE OF THE FIRST FLOWER BUD AND ON THE GERMINATION OF THEIR SEEDS AT 35°C

Daylength (h) in which parent plants were grown	Number of leaves at the time of appearance of first flower bud	Age of plants at the time of appearance of first flower bud (days)	Germination conditions	
			Continuous light: % germination at 24 h	Dark, with short illuminations every 24 h: % germination at 144 h
8	4·4 ± 0·5*	12·0 ± 0·7	61	85
11	3·2 ± 0·4	9·7 ± 0·2	40	65
13	7·0 ± 0	14·0 ± 0	13	45
15	10·5 ± 0·2	19·0 ± 0	—	—

* ±S.E.

Table 4.3 THE INFLUENCE OF CHANGE OF DAYLENGTH DURING THE LAST 8 DAYS OF
MATURATION OF *PORTULACA OLERACEA* FROM 16 H DAYS TO EITHER 13 H OR 8 H DAYS ON
GERMINATION AT 40°C IN THE DARK WITH 5 MIN OF WHITE LIGHT EVERY 24 H. FIRST
ILLUMINATION WAS GIVEN 1·5 H AFTER WETTING (GUTTERMAN, 1969)

Photoperiodic treatments of the mother plants (h) (last 8 days)	Germination (%) * time (h)				
	24	96	120	144	312
16	0	0	8·7 ± 4·3	34·7 ± 2·9	88·0 ± 2·4
13	0·7 ± 0·8†	7·3 ± 2·9	28·0 ± 2·4	38·7 ± 1·6	90·0 ± 1·4
8	1·3 ± 1·6	27·3 ± 4·3	48·0 ± 5·1	51·3 ± 4·1	94·0 ± 1·4

* No. of seeds in each treatment, 200 (4 × 50)
† ±S.E.

8 days from 16 h of light to either 8 h or 13 h, affects the
germinability (*Table 4.3*) (Gutterman, 1969).

3. In *Carrichtera annua* seeds from LD (20 h) germinate much
faster than from SD (8 h) even after storage for 9 years: LD,
51·5 per cent; SD, 7 per cent after 26·5 h in contact with water,
at 25°C and in continuous light.

*The influence of daylength and either red (R) or far-red (FR) light, during
the maturation of the seeds, on the germinability at different temperatures of*
Portulaca oleracea.

Portulaca oleracea, a facultative short day plant, is an important weed in
citrus groves in sandy or loamy soils until the trees become large
enough to shade the whole soil surface, when this species disappears.
The plant is, however, also widely used as an annual summer
vegetable, but it disappears even in relatively mild winters, because it
needs both high temperatures and high light intensity.

Plants of this species were grown in 8, 11, 13 and 15 h days and
their seeds were collected. *Table 4.2* shows that *P. oleracea* is a facultative
SD plant with an optimum daylength for early flowering of 11 h. But
the daylength in which the parent plants were grown was positively
correlated not only to their time to flowering but also to the length
of time to the germination of their seeds (*Table 4.2*) (Gutterman,
1969).

In another experiment all the *P. oleracea* plants were grown at
temperatures of 27°C during the day and 22°C during the night.
The various photoperiodic treatments were:

1. 8 h of daylight and 16 h of darkness (SD).
2. 8 h of daylight, 8 h of white light followed by 8 h of darkness (LD).
3. 8 h of daylight, followed by 2 h of R and then 14 h of darkness.
4. As 3 but FR instead of R.
5. 8 h of daylight and 2 h of R in the middle of the dark period.
6. As 5 but FR instead of R.
7. 8 h of daylight, 14 h of darkness followed by 2 h of R.
8. As 7 but FR instead of R.

The supplementary light given was in the low energy range $(2000–2500 \ \text{ergs cm}^{-2} \ \text{sec}^{-1})$.

The results of the photoperiodic experiments on the mother plants (*Table 4.4*) show that the plants flower under both SD (after 7·9 leaves) and LD (after 11·3 leaves). Plants given 2 h of red or far-red light, either at the beginning or at the end of the dark period, or 2 h of far red at the middle of the dark period, behave more or less like plants under SD, concerning the number of leaves before the

Table 4.4 THE INFLUENCE OF PHOTOPERIODIC CONDITIONS AND LIGHT QUALITY (SD, 8 H; LD, 16 H; 2 H OF RED OR FAR-RED LIGHT BEFORE, AFTER OR IN THE MIDDLE OF THE DARK PERIOD), GIVEN TO PLANTS OF *PORTULACA OLERACEA*, ON THE NUMBER OF LEAVES AT TIME OF APPEARANCE OF FIRST FLOWER BUD AND ON THE GERMINABILITY OF THEIR SEEDS AT 40°C (GUTTERMAN, 1969)

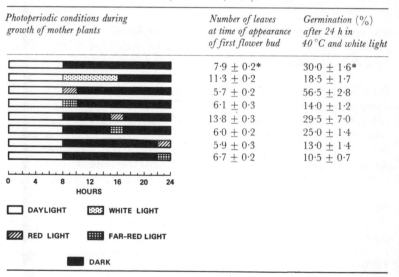

Photoperiodic conditions during growth of mother plants	Number of leaves at time of appearance of first flower bud	Germination (%) after 24 h in 40 °C and white light
	7·9 ± 0·2*	30·0 ± 1·6*
	11·3 ± 0·2	18·5 ± 1·7
	5·7 ± 0·2	56·5 ± 2·8
	6·1 ± 0·3	14·0 ± 1·2
	13·8 ± 0·3	29·5 ± 7·0
	6·0 ± 0·2	25·0 ± 1·4
	5·9 ± 0·3	13·0 ± 1·4
	6·7 ± 0·2	10·5 ± 0·7

0   4   8   12   16   20   24
HOURS

☐ DAYLIGHT   ▦ WHITE LIGHT

▨ RED LIGHT   ▦ FAR-RED LIGHT

■ DARK

Number of plants in each treatment, 65–70.
Number of seeds in each treatment, 200 (4 × 50).
* ±S.E.

appearance of flower buds. But plants given 2 h of red at the middle of the dark period react as if they were grown under LD (*Table 4.4*).

The percentage of germination at 40°C in white light of seeds matured under SD conditions was higher than that of LD seeds. But the biggest difference in the germination was between seeds matured under red or far-red light at the beginning of the dark period.

No difference in germination was observed between treatments of red and far-red light in the middle and at the end of the dark period (*Table 4.4*). However, at the sub-optimal temperature of 25°C (in white light), where the final germination percentages were low, seeds derived from mother plants which had received far-red light germinated to a higher percentage than seeds from plants which had received red light (*Table 4.5*). This difference was especially large for

Table 4.5 THE INFLUENCE OF THE TREATMENTS GIVEN TO THE MOTHER PLANTS OF *PORTULACA OLERACEA* (SEE TABLE 4.4) ON THE GERMINATION OF THEIR SEEDS AT TEMPERATURES OF 25°C, 30°C AND 40°C

Photoperiodic conditions during growth of mother plants	No. leaves at time of appearance of first flower bud	Total germination (%)		
		25°C	30°C	40°C
	7.9±0.2*	11.0±0.7	80.0±2.4	96.0±0.5
	11.3±0.2	1.0±0.3	58.5±2.2	85.0±1.1
	5.7±0.2	2.0±0.0	63.5±5.5	98.5±0.6
	6.1±0.3	26.0±1.8	63.5±3.7	79.0±2.4
	13.8±0.3	2.0±0.5	75.5±3.6	93.0±1.0
	6.0±0.2	14.0±2.2	85.5±1.6	96.0±0.8
	5.9±0.3	14.0±1.1	85.5±2.4	91.0±1.0
	6.7±0.2	16.0±1.7	87.0±1.9	87.5±1.8

```
0    4    8    12   16   20   24
              HOURS
```

☐ DAYLIGHT   ▨ WHITE LIGHT

▨ RED LIGHT   ▨ FAR-RED LIGHT

■ DARK

* ±S.E.

seeds from plants which had received 2 h of far-red light before the start of the dark period.

Transfer of the non-germinated seeds from 25°C to 30°C brought about an increase of the germination percentage. Seeds from plants which had received 2 h of red or far-red light before the beginning of the dark period now reached the same germination percentage, whereas seeds from plants which had received 2 h of far-red light at

the middle of the dark period still germinate to a higher percentage
than their red counterparts.

Transfer to 40°C brought about the same final germination
percentage which would have been obtained if these seeds had

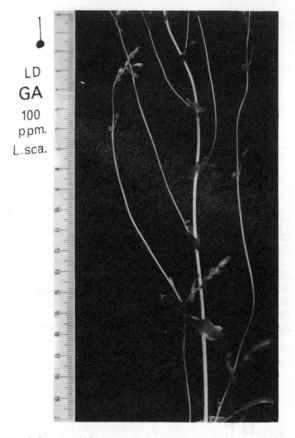

LD
GA
100
ppm.
L.sca.

Figure 4.10   Inflorescence of Lactuca scariola *plant treated with 0·1 ml of 100 ppm of GA₃,
once a week, grown under LD conditions (Gutterman, Thomas and Heydecker, unpublished)*

germinated directly under the same conditions. At all temperatures
seeds from SD germinated better than seeds from LD (Gutterman,
1969).

How can these results be explained? It seems possible that the
changes in germinability are caused by different levels and/or
different combinations of different material factors (hormones,

enzymes, nutrients, etc.) which determine germination. Downs (1964), Goeschel, Pratt and Bonner (1967), Engelsma (1968; 1969), Holland and Vince (1968), Stoddart and Lang (1968), Galston and Davies (1969), Zeevaart (1969a; 1969b; 1971a; 1971b), Teltscherová (1970), Wagner and Comming (1970), Ende and Zeevaart (1971), Loveys and Wareing (1971) and others point out the relation between the synthesis of various substances in plants and such environmental conditions as light, photo-period, red and far-red regime. Zeevaart (1966) found a relationship between hormonal treatment given to the mother plant before or after anthesis and the presence of certain substances in the seeds as a consequence of these treatments (see also

Figure 4.11   Inflorescence of Lactuca scariola *plant treated with 10 ml of 5000 ppm of CCC, once a week, and grown under LD conditions (Gutterman, Thomas and Heydecker, unpublished)*

Felippe and Dale, 1968). According to theoretical schemes proposed by various authors the actively absorbing form of far-red phyto-chrome (P_{fr}) ties up with substrate (S) or with a 'reactant' during the germination process (Mancinelli and Borthwick, 1964; Ikuma and Thimann, 1964; Negbi, Black and Bewley, 1968; Amen, 1968; Gutterman, Evenari and Heydecker, 1972).

Figure 4.12   Inflorescence of Lactuca scariola *plant grown under LD conditions (Gutterman, Thomas and Heydecker, unpublished)*

Combining known facts and suppositions (a) that the environmental conditions affect the material composition of the plants, (b) that substances are transferred from the mother plant to the seeds relative to the amounts present in the mother plant, and (c) that the substrates or reactants transmitted in this way combine with the phytochrome during the germination process, the following concept may be arrived at. Under the influence of the various environmental conditions tested the mother plants form either different substances or different absolute or relative amounts of certain substances which are transferred to their seeds. One, some or all of these substances are either by themselves the substrate or reactant which combines with phytochrome or they affect the substrate or reactant formed in the seed. Since germinability under various germination conditions (light, temperature, etc.) is *inter alia* a function of the substrate–phytochrome or reactant–phytochrome complex it is obvious that germinability is linked to the environmental conditions under which the mother plant grows. The picture is even more complicated because there are changes in the seed also during maturation (Skene and Carr, 1961; Felippe and Dale, 1968) and even after full maturation, depending on the storage conditions (Ross and Bradbeer, 1971).

In order to obtain further information on the relationship between hormone and daylength treatments of the mother plants on the germinability of their seeds and on the development of the progeny, experiments were carried out with the long day plant *Lactuca scariola* L.

The influence of daylength and hormones on the germinability and on the following generation of Lactuca scariola *L.*

The effect of weekly applications of hormones in LD (16 h) were compared with those of transferring plants from LD to SD (8 h) or

Figure 4.13   Leaves of Lactuca scariola *plants grown under LD (middle) and treated with GA₃ (left) or CCC (right) once a week (Gutterman, Thomas and Heydecker, unpublished)*

keeping them untreated in LD. The transfer or the beginning of hormone treatment coincided with the end of the rosette phase of growth when the stems began to elongate and flower buds appeared. Morphological changes in the developing inflorescences were visible within a few days of first applying the hormones. Under the influence of gibberellic acid (GA_3) the inflorescences elongated more, and of chlormequat (CCC) elongated less than those of untreated plants in LD (*Figures 4.10, 4.11 and 4.12*). The leaves of the plants from these

Table 4.6 *LACTUCA SCARIOLA* EFFECTS OF WEEKLY APPLICATIONS OF GROWTH SUBSTANCES AND OF DAYLENGTH ON THE PROGENY OF TREATED PLANTS (POTTED PLANTS IN GROWTH ROOMS)

Treatment	Weight of 100 seeds harvested 16–21 days after start of treatment (mg)	Germination in incandescent light		Seedlings showing flower buds 53 days after emergence (%)
		After 13·5 h	After 96 h	
A. *16 h days*				
No chemical treatment	75	$43 \pm 1·0\ddagger$	$81 \pm 3·0\ddagger$	0
GA_3 100 ppm*	63	$25 \pm 1·0$	$70 \pm 2·0$	0
GA_3 10 ppm*	92	$44 \pm 4·0$	$84 \pm 2·0$	0
CCC 5000 ppm†	83	$38 \pm 4·0$	$77 \pm 3·0$	6·25
B. *8 h days after flower initiation*				
No chemical treatment	96	$58 \pm 6·0$	$87 \pm 6·0$	5·10

* Injected into stem below inflorescence (0.1 ml).
† Watered on to soil in pot (10 ml).
‡ ±S.E.
Seeds from all parents given the chemical treatments described and from parents transferred to short days had a significantly higher gibberellin content than the untreated 'long day' plants. The GA content tended to be higher in seeds from plants treated with 10 ppm than from 100 ppm GA_3, and highest in seeds from the CCC-treated and from the 'short day' parents.

treatments were also different in their shape, spininess and leaf margin (*Figure 4.13*). The first batch of seeds harvested, though not later ones, differed in size according to the treatment of the parent plants (*Table 4.6*).

Seeds from all parents given the chemical treatments described and from parents transferred to short days had a significantly higher gibberellin content than the untreated 'long day' plants. The GA content tended to be higher in seeds from plants treated with 10 ppm than from 100 ppm GA_3, and highest in seeds from the CCC-treated and from the 'short day' parents. Germination of the later-produced main batch of seeds was consistently influenced by the treatment of the mother plant (Gutterman, Thomas and Heydecker, unpublished). Seeds from plants which had been moved from LD to SD contained a higher level of assayable gibberellins, germinated more rapidly and produced seedlings that flowered earlier than those from plants remaining untreated in a LD regime. The assayable gibberellin level in seeds from plants in LD treated with 100 ppm GA_3 was little higher than that of seeds from untreated plants grown in LD, their germination in incandescent light was slow and the flowering of the resulting seedlings was not advanced. By contrast, the assayable gibberellin level of seeds from plants treated with 10 ppm GA_3 was significantly higher than that of the LD control plants though their germination was not advanced. It is particularly interesting that in seeds from plants treated with 5000 ppm CCC the gibberellin level was at least equally high as of seeds borne by plants treated with 10 ppm GA_3, and that progeny from CCC-treated plants flowered as early as progeny of plants transferred from LD to SD when strictly comparable seedlings, i.e. those germinating after equal imbibition periods, were compared (*Table 4.6*).

The influence of the position of the caryopses in the dispersal unit on their germinability

Dispersal units of *Aegilops ovata* L., a facultatively photoperiodic (LD) annual plant (Datta, Gutterman and Evenari, 1972a), contai 1 in their spikelets caryopses differing in their morphology, weight and germinability according to their position in the spikelets. *A. ovata* plants usually produce three spikelets (type III) and rarely two (type II) or four (type IV). In the schematic drawing of two spikes of type III (*Figure 4.14*), on the left-hand spike the two lower spikelets contain two caryopses each (a_1, a_2 and b_1, b_2). The highest spikelet contains only one caryopsis ($c_{(2)}$). The a_1 and b_1 caryopses (the lower of each pair) are larger, heavier, brighter and less hairy than the a_2, b_2 and $c_{(2)}$. On the right-hand spike, each spikelet contains only one caryopsis (a, b, and $c_{(1)}$). All of them are similar to caryopses a_1 and b_1. The various caryopses differ in their germinability (Datta, Evenari and Gutterman, 1970).

 Plants originating from caryopses of the diverse orders were grown under various photoperiodic and thermoperiodic conditions. Plants grown under LD produced heavier caryopses than plants grown under SD and plants grown under low temperature ($15/10°C$) formed heavier caryopses than their counterparts grown under high temperatures ($28/22°C$).

 Under LD and $15/10°C$, mother plants derived from caryopses a_1, b_2 and c produced caryopses possessing different germination

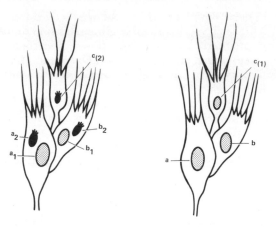

Figure 4.14   Schematic drawing of two spikes with three spikelets each (type III) of Aegilops
ovata *L. On the left-hand spike the two lower spikelets contain two caryopses each (a_1, a_2 and b_1, b_2).
The highest spikelet contains only one caryopsis $c_{(2)}$. The a_1 and b_1 caryopses are larger, heavier,
brighter and less hairy than the a_2, b_2 and $c_{(2)}$. On the right-hand spike each spikelet contains only
one caryopsis (a, b and $c_{(1)}$). All of them are similar to caryopses a_1 and b_1 (Datta, Gutterman
and Evenari, 1972b)*

characteristics not only according to the origin of the dispersal units
but also according to the order of caryopsis from which the mother
plants themselves developed (*Table 4.7*). In this case the influence
of external conditions on the first (grandmother) generation is
transmitted through the second (mother) generation to the following
generation or generations (Datta, Gutterman and Evenari, 1972b).

CONCLUSIONS

From this work and that of other workers (Wieser, 1927; Black and
Naylor, 1959; Koller, 1962; Wentland, 1965; Whalley, McKell and
Green, 1966; Jacques, 1968; Schwemmle, 1969; Thompson, 1970;
Evans and Dunstone, 1970; Karssen, 1970; Junttila, 1971) it
becomes more and more obvious that the external conditions under
which the mother plants grow have a far-reaching influence on the
germinability of the seeds and therefore on the progeny. This
influence is of great ecological significance for the survival mechanisms
of the various species investigated. But the physiological mechanisms
which bring about this effect remain in all cases unknown and merit
further investigation.

Table 4.7 GERMINATION PERCENTAGES AFTER 24, 72 AND 504 h OF VARIOUS ORDER CARYOPSES COLLECTED FROM MOTHER PLANTS DERIVED FROM CARYOPSES a_1, b_2 AND C. THE MOTHER PLANTS WERE GROWN UNDER LD CONDITIONS AT TWO TEMPERATURE REGIMES: 28°C DURjNG THE LIGHT AND 22°C DURING THE DARK PERIOD (28/22); AND 15°C DURING LIGHT AND 10°C DURING THE DARK PERIOD (15/10). GERMINATION CONDITIONS: 15°C AND CONTINUOUS LIGHT (DATTA, GUTTERMAN AND EVENARI, 1972b)

Order of caryopses from which mother plants developed	Order of caryopses collected from mother plants	LD 28/22			LD 15/10		
		24 h	72 h	504 h	24 h	72 h	504 h
a_1	a_1	100			84·4	96·1	96·1
	a_2	81·1	98·1	98·1	7·5	84·4	88·4
	b_1	100			95	100	
	b_2	60	100		10	90	95
	$c_{(1)}$	100			100		
	$c_{(2)}$	63·1	100		8·5	82·6	95·6
b_2	a_1	100			55	95	98·3
	a_2	97·1	100		1·5	83·8	83·8
	b_1	100			51·2	80·4	87·7
	b_2	85·3	100		2·3	59·5	73·8
	$c_{(1)}$	100			0	75	100
	$c_{(2)}$	90·3	96·7	96·7	0	37·2	66·6
c	a_1	100			21·2	100	
	a_2	96·4	98·2	98·2	0	91·2	96·2
	b_1	100			65·6	96·2	100
	b_2	76	100		0	68	78·4
	$c_{(1)}$	100			—	—	—
	$c_{(2)}$	86·9	100		0	28	51·2

REFERENCES

AMEN, R. D. (1968). 'A model of seed dormancy', *The Bot. Rev.*, **34**, No. 1, 1–31
BLACK, M. and NAYLOR, J. M. (1959). 'Prevention of the onset of seed dormancy by gibberellic acid', *Nature*, **184**, 468–469
DATTA, S. C., EVENARI, M. and GUTTERMAN, Y. (1970). 'The heteroblasty of *Aegilops ovata* L., *Israel J. Bot.*, **19**, 463–483
DATTA, S. C., GUTTERMAN, Y. and EVENARI, M. (1972a). 'The photoperiodic and temperature responses of plants derived from the various heteroblastic caryopses of *Aegilops ovata* L.', *Indian J. Bot.*, (in press)
DATTA, S. C., GUTTERMAN, Y. and EVENARI, M. (1972b). 'The influence of the origin of the mother plant on yield and germination of their caryopses in *Aegilops ovata* L., *Planta*, **105**, 155–164
DOWNS, A. J. (1964). 'Photocontrol of anthocyanin synthesis', *J. Wash. Acad. Sci.*, **54**, 112–120
ENDE, H. VAN DEN and ZEEVAART, J. A. D. (1971). 'Influence of day-length on gibberellin metabolism and stem growth in *Silene armeria*', *Planta*, **98**, No. 1, 164–176

ENGELSMA, G. (1968). 'The influence of light of different spectral regions on the synthesis of phenolic compounds in gherkin seedlings, in relation to photomorphogenesis. IV. Mechanism of far-red action, *Acta Bot. Neerl.*, **17** (2), 85–89

ENGELSMA, G. (1969). 'The influence of light of different spectral regions on the synthesis of phenolic compounds in gherkin seedlings, in relation to photomorphogenesis. VI. Phenol synthesis and photoperiodism', *Acta Bot. Neerl.*, **18**, 347–352

EVANS, L. T. and DUNSTONE, R. L. (1970). 'Some physiological aspects of evolution in wheat', *Aust. J. Biol. Sci.*, **23**, 725–741

EVENARI, M. and GUTTERMAN, Y. (1966). 'The photoperiodic response of some desert plants', *Z. Pflanzenphysiol.*, **54**, 7–27

EVENARI, M., KOLLER, D. and GUTTERMAN, Y. (1966). 'Effects of the environment of the mother plants on the germination by control of seed-coat permeability to water in *Ononis sicula* Guss.', *Aust. J. biol. Sci.*, **19**, 1007–1016

FELIPPE, G. M. and DALE, J. E. (1968). 'Effects of CCC and gibberellic acid on the progeny of treated plants', *Planta*, **80**, 344–348

GALSTON, A. W. and DAVIES, P. J. (1969). 'Hormonal regulation in higher plants', *Science*, **163**, 1288–1297

GOESCHEL, J. D., PRATT, H. K. and BONNER, B. A. (1967). 'An effect of light on the production of ethylene and the growth of the plumular portion of etiolated pea seedlings', *Pl. Physiol.*, **42**, 1077–1080

GUTTERMAN, Y. (1966). 'The photoperiodic response of some desert plants and the effect of the environment of the mother plants on the germination by control of seed-coat permeability to water in *Ononis sicula* Guss.' (Hebrew), *M.Sc. Thesis*, Jerusalem, The Hebrew Univ.

GUTTERMAN, Y. (1969). 'The photoperiodic response of some plants and the effect of the environment of the mother plants on the germination of their seeds' (Hebrew with English summary), *Ph.D. Thesis*, Jerusalem, The Hebrew Univ.

GUTTERMAN, Y. and EVENARI, M. (1972). 'The influence of daylength on seed coat colour, an index of water permeability of the desert annual *Ononis sicula* Guss', *J. Ecol.*, **60**, No. 3, 713–719

GUTTERMAN, Y., EVENARI, M. and HEYDECKER, W. (1972). Phytochrome and temperature relations in *Lactuca sativa* L. Grand Rapids seed germination after thermo-dormancy: *Nature New Biology*, **235**, No. 57, 144–145

GUTTERMAN, Y. and HEYDECKER, W. (1973). 'Studies of the surfaces of desert plant seeds. Effect of daylength during maturation on the seed coat of *Ononis sicula* Guss.' *Ann. Bot.*, (in press)

HOLLAND, R. W. K. and VINCE, D. (1968). 'Photoperiodic control of flowering and anthocyanin formation in *Fuchsia*', *Nature*, **219**, No. 5153, 511–513

IKUMA, H. and THIMANN, K. V. (1964). 'Analysis of germination processes of lettuce seeds by means of temperature and anaerobiosis', *Plant Physiol.*, **39**, No. 5, 756–767

JACQUES, R. (1968). 'Action de la lumière par l'intermédiaire du phytochrome sur la germination, la croissance et le développement de *Chenopodium polyspermum* L.', *Physiol. Vég.*, **6**, No. 2, 137–164

JUNTTILA, O. (1971). 'Effect of mother plant temperature on seed development and germination in *Syringa reflexa* Schineid,' *Meldinger fra Norges Landbrukshogskole*, **50**, 1–16

KARSSEN, G. M. (1970). 'The light promoted germination of the seeds of *Chenopodium album* L. III. Effect of the photoperiod during growth and development of the plants on the dormancy of the produced seeds', *Acta Bot. Neerl.*, **19**, No. 1, 81–94

KOLLER, D. (1962). 'Preconditioning of germination in lettuce at time of fruit ripening', *Am. J. Bot.*, **49**, No. 8, 841–844

LOVEYS, B. R. and WAREING, P. F. (1971). 'The red light controlled production of gibberellin in etiolated wheat leaves', *Planta*, **98**, 109–116

MANCINELLI, A. L. and BORTHWICK, H. A. (1964). 'Photocontrol of germination and phytochrome reaction in dark germinating seeds of *Lactuca sativa* L,', *Ann. di. Bot.*, **28**, 9–24

NEGBI, M., BLACK, M. and BEWLEY, J. D. (1968). 'Far-red sensitive dark processes essential for light- and gibberellin-induced germination of lettuce seed', *Plant Physiol.*, **43**, 35–40

ROSS, J. D. and BRADBEER, J. W. (1971). 'Studies in seed dormancy V. The content of endogenous gibberellins in seeds of *Corylus avellana* L.', *Planta*, **100**, 288–302

SCHWEMMLE, J. (1969). 'Influence of the habitat on the germination of *Oenothera* seeds', *Biol. Zbl.*, **88**, 37–46

SKENE, K. G. M. and CARR, D. J. (1961). 'A quantitative study of the gibberellin content of seeds of *Phaseolus vulgaris* at different stages in their development', *Aust. J. biol. Sci.*, **14**, 13–25

STODDART, J. L. and LANG, A. (1968). 'The effect of daylength on gibberellin synthesis in leaves of red clover (*Trifolium pratense* L.), in *Biochemistry and physiology of plant growth substances* (ed. Wightman, F. and Setterfield, G.) 1371–1381, Runge Press, Ottawa

TELTSCHEROVA, L. (1970). 'Changes in the level of endogenous cytokinins in apical buds of *Chenopodium rubrum* L.', *Biologia Plant.*, **12**, No. 2, 134–138

THOMPSON, P. A. (1970). 'Changes in germination responses of *Silene secundiflora* in relation to the climate of its habitat,' *Physiologia. Pl.*, **23**, 739–746

WAGNER, E. and CUMMING, B. G. (1970). 'Betacyanin accumulation, chlorophyll content, and flower initiation in *Chenopodium rubrum* as related to endogenous rhythmicity and phytochrome action', *Can. J. Bot.*, **48** (1), 1–18

WENTLAND, M. J. (1965). 'The effect of photoperiod on the seed dormancy of *Chenopodium album*'. Thesis, University of Wisconsin, U.S.A.

WHALLEY, R. D. B., MCKELL, C. M. and GREEN, L. R. (1966). 'Effect of environmental conditions during the parent generation on seedling vigor of the subsequent seedlings of *Oryzopsis miliacea* (L.) Benth & Hook', *Crop Sci.*, **6**, 510–512

WIESER, G. (1927). 'Der Einfluss des Sauerstoffs auf die Lichtwirkung bei der Keimung lichtempfindlicher Samen', *Planta*, **4**, 526–572

ZEEVART, J. A. D. (1966). 'Reduction of the gibberellin content of *Pharbitis* seeds by CCC and after-effects in the progeny', *Plant Physiol.*, **41**, No. 5, 856–862

ZEEVAART, J. A. D. (1969a). 'Gibberellin-like substances in *Bryophyllum daigremontianum* and the distribution and persistence of applied gibberellin GA$_3$, *Planta*, **86**, 124–133

ZEEVAART, J. A. D. (1969b). 'Changes in gibberellin content of *Bryophyllum daigremontianum* in connection with floral induction', *Neth. J. agric. Sci.*, **17**, 215–220

ZEEVAART, J. A. D. (1971a). 'Effects of photoperiod on growth rate and endogenous gibberellins in the long-day rosette plant Spinach', *Pl. Physiol.*, **47**, 821–827

ZEEVAART, J. A. D. (1971b).'(+) – Abscisic acid content of spinach in relation to photoperiods and water stress', *Plant Physiol.*, **48**, No. 1, 86–90

DISCUSSIONS

Karssen: We have found that the influence of the photoperiodic conditions during seed production on the germination of seeds can last for years, whereas the regulation by phytochrome may disappear after a few weeks.

Gutterman: Seeds of desert plants would hardly benefit by such a transient mechanism because full germination after one single rainfall might destroy the whole population. In such plants, seeds produced on the same branch may have different germinability and still show large differences in their germination after many years.

Cavers: We can confirm that the germination pattern in one environment gives no idea of the pattern in another.

A Questioner: What is the relationship between seed size and dormancy?

Gutterman: Seeds may or may not differ much in size according to their nutrition on the parent plant, but no clear relationship between seed size and germinability has been established in *Portulaca* seeds.

Côme: In apple and some other seeds failure to germinate is due to the presence of phenolic compounds in the seed coats. Do such compounds play a part in your seeds?

Gutterman: The coat-imposed differences in germinability of the different coloured seeds of *Ononis sicula* are not associated with phenolic inhibitors: a slight scratch of the seed coat, after many days' ineffective contact with water, will cause immediate germination.

5

THE PRODUCTION OF HIGH-QUALITY SEEDS

R. K. SCOTT

Department of Agriculture and Horticulture,
University of Nottingham School of Agriculture

P. C. LONGDEN

Broom's Barn Experimental Station, Higham, Bury St. Edmunds

This paper deals with sugar beet seed, chiefly that grown in England, and includes a brief account of seed supply and crop development. It also examines factors which affect seed yield and quality, and discusses the value of seed for growing a root crop.

SEED SUPPLY

Figure 5.1 shows that most of the seed used in the U.K. is home produced. Since 1965, the amount of seed used has declined; farmers using precision drills and monogerm varieties sow less seed per acre: in 1971, 969 tonnes of seed were used, compared with 1881 tonnes in 1966. Whilst the quantity required diminishes, the quality becomes more important.

Figure 5.2 shows where seed crops grew in England in 1969. Most are in South Lincolnshire (A) and the Isle of Ely (B), where many root crops are also grown. Intermingling of biennial seed and annual root crops can cause severe localised outbreaks of downy mildew (Byford and Hull, 1967) and other airborne diseases. Seed crops have, therefore, recently been grown where root crops are comparatively few, especially in Oxfordshire and Gloucestershire

SEED ECOLOGY

where 186 ha of seed crops were grown in 1969 compared with only 12 in 1964.

The crop is now grown by one of three methods. In 1970, 937 ha were direct drilled and 31 ha of steckling beds were grown for

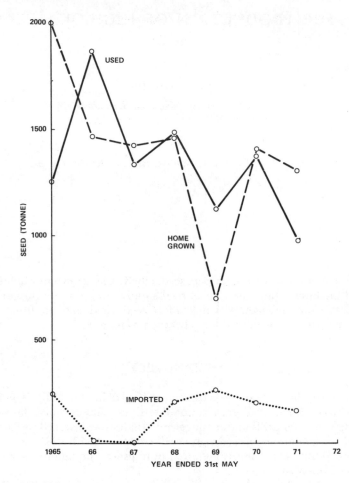

Figure 5.1   U.K. sugar beet seed production, import and use (M.A.F.F. Seedsmen's Stock Returns)

transplanting. Of the direct-drilled crop, 71 per cent was sown in spring, under cereal cover, mainly in the drier east of England, and the remainder was sown in summer, without cover, in areas remote from root crops. Some of the direct-drilled crop is not harvested and a

Figure 5.2   Locations and sizes of sugar beet seed crops, 1969

proportion of stecklings are not used. In recent years about 1000 ha
have been harvested. A few hectares are still transplanted but almost
the entire crop is now grown without hand labour.

CROP DEVELOPMENT

The ability of biennial plants to produce seed depends on many
factors that affect first the vegetative and then the reproductive
phases of development. The amount of vegetative growth before
'bolting' (production of flowering shoots), the degree of vernalisation,
and the conditions under which the bolting plant develops,
particularly during flowering and ripening, can all affect the yield
of seed.

   Plants which overwinter are called 'stecklings'. The roots are
2–3 cm in diameter and 10–20 cm long, restricted in size by late

sowing and a dense stand to increase their chance of surviving the winter. Stecklings will survive unprotected in the field on free-draining soils in the maritime climate of England. Only in very severe winters (such as 1962/63) are many plants lost.

After the winter, stecklings produce a rosette of new leaves, then bolt in late May or early June—earlier in early sowings and direct-drilled crops than in transplanted crops. Stems elongate very rapidly, as much as 15 cm per day. The leaves on the shoot are smaller and more widely spaced than the rosette leaves. The root hardens and continues to increase in weight and total sugar content. Lateral shoots develop in the axils of the leaves on the main stem, grow rapidly and produce flower-bearing branches. The inconspicuous flowers are in the axils of bracts along the upper parts of the multi-branched stem. Each flower usually produces one seed.

In the traditional multigerm cultivars, the cluster is formed from two or more flowers grown together at their bases and becomes hard and irregular, but in monogerm cultivars each 'cluster' is formed from one flower. Seeds develop within perianth remains which contain germination inhibitors. Although inhibitors may be beneficial in preventing premature germination before and during seed crop harvest, too much inhibition causes difficulties after sowing when the root crop is to be established.

INFLUENCE OF PLANT BREEDING ON
SEED PRODUCTION

The primary concern of plant breeders has been to achieve the correct blend of desirable characteristics in the root crop, e.g. sugar yield, juice purity and 'bolting resistance'. Except for incorporating the monogerm character, seed-bearing characteristics are of secondary consideration. The growth characteristics of sugar beet seed crops challenge seed producers because it is a bulky, leafy crop with juicy stems that do not dry readily. Furthermore, the outbreeding nature of the plant and its indeterminate growth habit cause individual seeds on a plant to mature at widely different times.

In recent years, techniques for producing hybrid seed have further complicated seed-production systems. Anisoploid cultivars are produced from a mixture of fully fertile, open pollinating, diploid and tetraploid plants. Triploid cultivars are produced on diploid male-sterile plants pollinated by fully fertile tetraploid multigerm plants. Incorporation of the monogerm character into these diploid male-sterile plants has given triploid monogerm cultivars.

The proportion and arrangement of pollinators and male-sterile plants must be decided to ensure that sufficient pollen is available to give the maximum yield of hybrid seed. Ideally, such a decision would be based on knowledge of the competitive abilities of the various components of the mixture, but in practice the basis is more empirical. The seed grower may be supplied with pre-mixed elite seed or be instructed to grow the crop in separate rows of pollinator and male-sterile plants. When mother seed of pollinators and male-steriles is pre-mixed the desired ratio and random distribution of pollinators is achieved only if both components are sown and establish themselves in proportion to the number of viable seeds in the mixture. Some breeders consider that the survival of one component differs unpredictably from that of the other, so they separate the two lots of seed in different rows. Male or female parents may also be grown in separate rows in steckling beds and stecklings mixed before planting.

Segregation in rows allows pollinators to be removed after flowering, and for this purpose growers leave access paths between strips of pollinator and male-sterile plants. Breeders choose pollinators that flower at the same time as the male-steriles; nevertheless, a grower may sometimes need to remove the tops of the main stems to delay flowering of the earlier component.

FACTORS AFFECTING YIELD AND QUALITY

The factors affecting yield and quality fall into two categories: those over which little or no control is possible, e.g. the weather, particularly at flowering and ripening, and those that can be determined by the grower.

In a recent survey, Scott (1968) found widespread interest in, but lack of information on, the influence of weather during pollination on seed quality. In Ireland, England, Poland and Austria, too much rain during flowering was considered a common cause of small yields and poor germination. Although moderately dry winds promote dehiscence of anthers and pollen release, many contributors, particularly in hot climates, consider that drought following flowering often causes low yields and poor germination. In W. Germany, Turkey, Spain, parts of France, and Arizona, hot dry winds are reported to decrease yield and germination because they interfere with fertilisation or kill developing embryos. Thus, extremes—either wet and cold or dry and hot—during flowering and seed set may be harmful. Experiments are in progress on the problem in Oregon, France and England.

Scott (1970) described the influence of weather in England on pollen release in transplanted open-pollinated diploid crops. Pollen release began at the end of June, increased until mid-July and decreased, little being released at the end of July in each of three years. In 1967, the weather was sunny and warm with low humidity during the day, and more pollen was released than in 1965 and 1966 when the weather was unsettled. Pollen is liberated with a characteristic diurnal periodicity as the atmosphere within the crop becomes drier, giving maximum concentration in the air in the morning—earlier with diploid than with tetraploid varieties. Pollen grains from tetraploids are larger $(21-28 \mu)$ than from diploids $(18-21 \mu)$ and in the present work the amount of pollen trapped within a tetraploid cultivar was only 66 per cent of that from a diploid cultivar (Scott and Longden, 1970). Pollen catches decreased at a similar rate with increasing distance from diploid and tetraploid pollinators, yet the germination of seeds harvested from male-sterile plants 8 m from the pollinators was as good as from male steriles growing adjacent to them. Despite considerable fluctuations in pollen density in different crops, seasons, days and times of day during this three-year project, yield and germination of seed from crops or from individual male-sterile plants exposed to different concentrations of pollen have not been clearly related to the pollen catch. Perhaps during periods of adverse weather (e.g. high humidity) the development of stigmas is also arrested, so compensating for short periods with little pollen.

Now that we have established the influence of weather on pollen release and concentration in the atmosphere, we must study pollen deposition on stigmas and investigate if weather which favours pollen release also favours pollen germination, pollen tube penetration and fertilisation. Successful fertilisation does not, of course, ensure the production of viable seeds. In all samples some seeds fail to develop fully and are not viable (Tekrony and Hardin, 1969). It is not known to what extent unfavourable weather, e.g. drought, poor mineral nutrition, or *Lygus* bug feeding (Hills, 1963) contribute.

THE WEATHER AT HARVEST

Most contributors to Scott's survey (Scott, 1968) considered that harvesting too early was the most frequent cause of poor germination. In seasons when crops ripen late, growers may cut before the crop is fully mature, to avoid the risk of losing the crop if weather becomes impossible for harvest.

Experiments from 1965 to 1971, in the Cotswolds, included similar agronomic treatments but experienced weather which varied at flowering and harvest (*Table 5.1*). Because of changes in genetic

Table 5.1 YEAR-TO-YEAR VARIATION IN YIELD, GERMINATION AND WEATHER, 1965–71

	Multigerm 1965	1966	1967	Mixture† 1968	Monogerm 1969	Mixture† 1970	Monogerm 1971
Optimum commercial yield of seed* (kg/ha)	2472	3351	2384	4003	4443	4681	5447
Optimum commercial seed germination (%)	78	78	79	59	54	65	64
Germination (%) at 3rd Sept.	44	74	78	36	51	60	64
Rainfall (mm)							
July	89·9	80·8	76·5	141·5	49·0	40·6	37·3
August	61·5	78·7	43·9	91·4	87·9	51·8	133·1
Sept.	75·4	52·3	59·2	108·2	32·8	40·4	13·5
Days when rain fell							
July	17	17	12	11	6	15	8
August	14	14	15	16	15	11	21
Sept.	15	9	18	20	10	11	5
Sunshine (h)							
July	123	150	234	136	244	192	259
August	181	189	167	124	158	186	134
Sept.	123	168	116	115	97	157	166
Monthly means of daily mean max./min. temperature (°C)							
July	18·5/11·0	19·8/11·7	22·7/13·1	19·7/11·9	23·3/13·1	20·4/11·8	23·2/13·1
August	20·2/11·0	19·9/11·1	20·7/12·3	19·5/12·3	21·3/13·2	21·0/12·2	20·3/12·7
Sept.	16·8/9·1	18·6/10·3	17·7/10·9	18·3/11·0	17·7/10·8	19·5/11·2	19·5/9·8

* Optimum commercial yield is that where the crop is most fit to cut, taking into consideration the combination of yield and germination giving the maximum gross return to the seed grower.
† Monogerm male sterile, multigerm pollinators harvested together.

material, only crops in the years 1965 and 1967, 1968 and 1971, and 1969 and 1970 are strictly comparable. Despite very different weather during flowering and maturation—cool and damp in 1965 and hot and dry in 1967—similar yields and, eventually, germination potential were obtained, but much later in 1965. The weather sequence in 1968 was expected to cause poor yields and germination: high July rainfall with very heavy rain when the crop was in full bloom, followed by very wet, dull and cool conditions in August and September. The crop was slow to mature but eventually the yield and germination were satisfactory. The seed harvested in 1969, when plants wilted frequently during June and July, had the poorest germination. The 1971 season had perhaps the ideal sequence of weather from flowering onwards: a sunny July, when much pollen was produced, a wet dull August with unrestricted water supply and a dry sunny September ideal for complete maturation and for harvest. Yields were large and high-quality seed developed early. In each year of the series the weather which occurred at flowering also occurred at harvest, and it is therefore impossible to distinguish any one stage which may be particularly sensitive to climatic factors. The essential difference between these years was the earliest date on which seed of high quality could be harvested.

AGRONOMY

Agronomic experiments on seed production try to discover treatments which give the best product for sowing the root crop as economically as possible. Good yields of seed should be combined with high quality, since this combination determines gross returns. Generally, laboratory germination tests in sand or on paper, according to the rules of the International Seed Testing Association, indicate the quality of the seed and predict its performance in the field (*Table 5.2*), but the contribution by Perry (Chapter 17) is relevant here.

Table 5.2 RELATIONSHIPS BETWEEN LABORATORY GERMINATION AND SEEDLING EMERGENCE IN THE FIELD, 1966–71

Year sown	Pairs of data	Slope	Intercept	Correlation coefficient
1966	14	0.40 ± 0.19	15.1 ± 11.1	0.51
1967	20	0.61 ± 0.14	11.7 ± 9.4	0.73
1969	25	0.48 ± 0.08	17.5 ± 3.4	0.80
1970	48	0.94 ± 0.15	-8.1 ± 9.0	0.69
1971	59	0.68 ± 0.09	1.6 ± 5.7	0.70

Methods of growing and plant population

Transplanted crops have about 20 000 plants per hectare, contrasting with 750 000/ha in direct-drilled crops. This leads to a marked difference in the growth form of seed bearers, which are 'bushy' plants in transplanted crops but have single tall stems with seeds clustered at the top in direct-drilled crops. Some of the characteristics of seeds are a consequence of the growth habit; the restricted branching of plants grown close together results in earlier, more uniform ripening of seeds of more uniform but smaller size. However, the different

Table 5.3 METHODS OF GROWING SUGAR BEET SEED, 1964–67

| | Sown in situ | | | | Transplanted | |
| | Under barley | | In open | | | |
	early	late	early	late	Autumn	Spring
Yield (t/ha)	2·4	2·5	2·8	2·3	2·3	2·4 ± 0·10
Germination (%)	67	66	67	63	60	65
Germination (angles)	55	54	55	53	51	54 ± 1·10

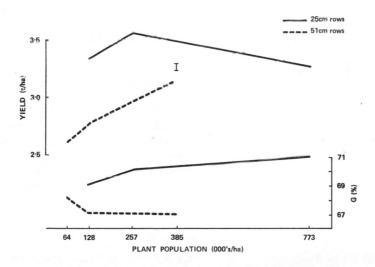

Figure 5.3   Plant population effects on sugar beet seed yield and germination, 1964–69

plant forms gave similar seed yields and germination percentages (*Table 5.3*). July sowings without cover gave the most consistent yields of seed with high germination.

Figure 5.3 shows the effects of differences in plant population of crops open-sown in July, on yield and germination. Seed yields of both multigerm and monogerm varieties were greater in narrow than in wide rows and yields declined as plants were grown further apart in widely spaced rows. The more nearly square the arrangement the better was the yield: 128 000 plants/ha produced 565 kg/ha more seed when grown 30 cm apart in rows 25 cm apart, than 15 cm apart in rows 50 cm apart. The germination percentage was generally slightly greater in seed from crops grown in narrow rows. Crops with high plant populations ripened earlier, possibly by limiting the flowering periods and by preventing secondary growth. This is an advantage, since it is this late growth that produces immature seed in widely spaced crops, especially when they lodge. On the other hand, narrow rows and close spacing tended to produce a smaller proportion of large seeds.

Wood (private communication) has shown that fruits on different locations on monogerm plants give seeds of different sizes and stages of development. On average, individual fruits on the main stem and bases of laterals weighed 12·7 mg, those on the lower laterals 10·1 mg, and those on the upper laterals 9·1 mg. Bell and Bauer (1942) showed that the position of the flower in the cluster of multigerm plants also determines earliness and size because central flowers develop first. Wood has also shown, by the radiographic technique of Longden, Johnson and Love (1971), that the developmental state of a seed at harvest is affected by its position on the plant. On main stems, lower laterals and upper laterals, the proportion of fruits containing fully developed seeds was 67, 52 and 56 per cent, under-developed seeds 25, 32 and 29 per cent, and no seed 7, 13 and 11 per cent, respectively.

High plant populations in which individual plants are much less branched, give the impression of enforced uniformity. In these crops, the plant-to-plant variation in average fruit weight (5·3–13·4 mg) and percentage germination (52–95 per cent) is great, but it is not yet known how density affects either these or the extent of variation in seed weight and development on an individual plant.

Nutrition and manuring

Austin and Longden (1965) have demonstrated that the chemical composition of seeds of pea and carrot influenced the growth of the resulting plants, and although in 1966 they reported that nitrogen fertilisers increased the nitrogen and decreased the phosphorus and potassium contents of red beet seed, those elements may still have been present in non-limiting amounts, because there were no effects on seedling emergence or size. Nitrogen applications of 188, 250 or 312 kg/ha to sugar beet seed crops (Longden, 1970), gave seeds with 2·04, 2·13 and 2·26 per cent N (dry weight basis). There was no effect of N fertiliser on P and K concentration, nor did the application of 125 kg/ha P_2O_5 affect N, P or K concentration. When sown in the field these seeds gave no significant differences in emergence percentages or seedling weight. It thus seems likely that, with normal manuring, variations in the major nutrient content of seeds are insufficient to affect seedling growth.

Fertiliser experiments on two contrasting soils, stony oolitic limestone 'brash' in the Cotswolds (Chipping Norton) and deep silt loam in South Lincolnshire (Sutton St. James), showed striking

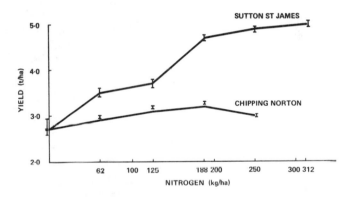

Figure 5.4   Nitrogen effects on sugar beet seed yield

differences in response to top dressings of nitrogen fertiliser in the second year of growth (*Figure 5.4*). The response to N in South Lincolnshire was much greater than in the Cotswolds. The percentage germination of seed produced in the two localities was generally similar; only in 1965, a late season, on a poorly drained site did heavy dressings of nitrogen delay maturity enough to depress germination. Phosphorus, potassium and sodium fertilisers gave small but consistent increases in yield and did not affect germination. In general, fertilisers increase yield because of their effect on the number rather than the size of seeds. Nitrogen had no consistent effect on seed size, but moderate applications of phosphorus and potassium increased it slightly.

Sowing and harvest dates

The times of sowing and harvesting have had the most marked and consistent effects on seed germination. Although crops continue to flower until cut, after a certain date flowers fail to develop into fully mature clusters and seeds.

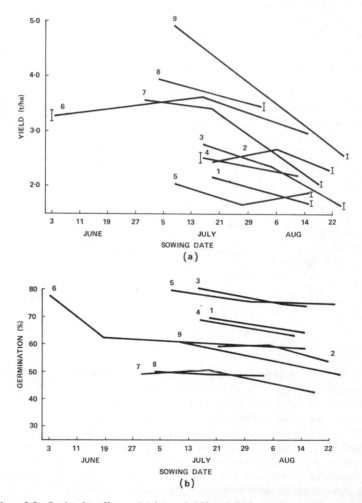

Figure 5.5   Sowing date effects on (a) beet seed yield, and (b) beet seed germination—9 experiments. Multigerm crops: 1, 1964–65; 2, 1965–66; 3, 1966–67; 5, 1967–68. Monogerm crops: 4, 1966–67; 6, 1967–68; 7, 1968–69; 8, 1969–70; 9, 1970–71

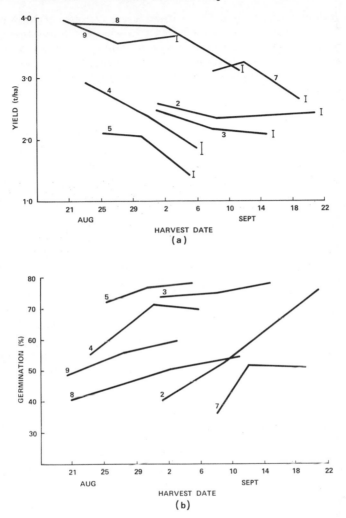

Figure 5.6 Harvest date effects on (a) beet seed yield, and (b) beet seed germination—9 experiments. Multigerm crops: 1, 1964–65; 2, 1965–66; 3, 1966–67; 5, 1967–68. Monogerm crops: 4, 1966–67; 6, 1967–68; 7, 1968–69; 8, 1969–70; 9, 1970–71

Farming practices that influence the initiation and duration of seed development can determine germination percentages. Plants that bolt early (April) also flower early (early July) and ripen early (early September). Late-flowering sugar beet crops do not ripen until late September and harvesting may then continue into the short days and unfavourable weather of late October or November.

Figures 5.5 and *5.6* show the effects of sowing and harvesting dates on seed yield and germination of four multigerm and five monogerm crops (germination is expressed as the percentage of fruits from which one or more seedlings emerge; multigerm crops are therefore liable to show higher figures than monogerm ones). In order to calculate average effects, the data were analysed by linear regression and also by the procedure used by Boyd (1952). Both methods gave similar results. Yield declined by about 20 kg/ha/day and germination by 0·15 per cent/day, with delay in sowing from early July until late August. There was a small but consistent effect of sowing date on seed size, with mid-July sowings producing 1–2 per cent more of the largest seed (retained by a 5·5 or 4·0 mm round-hole sieve for multi- and monogerm, respectively) than mid-August sowings.

Changes in yield and germination with delayed harvesting show that the development of seed yield and germination are out of phase; by the time there are appreciable numbers of germinable seeds, yield is declining because of shedding and bird damage. In our experiments, maximum germination occurred on average about two weeks after maximum yield. Between 21st August and 22nd September yield declined by between 30 and 40 kg/ha/day and germination increased by 1 per cent/day. The decision when to harvest a commercial crop must be a compromise between increasing germination and decreasing yield. Several criteria are used to assess optimum cutting time: the colour of the crop, the texture of cluster (or fruit) and of the true seed, and the amount of shedding. Experience shows that the condition of the true seed may be the most reliable indication of fitness to cut. The best guide available at present is to take seed-bearing branches originating from about the centre of the primary shoots of 20 representative plants. Clusters halfway along these racemes are cut to determine whether the perisperm is mealy rather than milky and the testa is turning from red to brown. More quantitative determinations are needed on this important aspect of seed crop husbandry.

Delay in harvesting increased the proportion of large seeds with multigerm varieties but had no consistent effect with monogerm varieties, possibly because in the monogerm crops bird damage was not so extensive. Early sowing and late harvesting usually increased the proportion of fruits containing fully developed seeds (*Table 5.4*).

Generally, the changes in yield and germination of seed from crops sown on different dates were parallel throughout the harvesting period, i.e. a month's delay in sowing resulted in both lower yield and about 4·5 per cent poorer germination, and there was no evidence that given more time late-sown crops produced as much seed or seed

Table 5.4 EFFECTS OF SOWING AND HARVESTING DATES ON SEED CLASSIFICATION BY
RADIOGRAPH ASSESSMENT

Sown 1969	Harvested 1970	Per cent of seed			Per cent emergence in the field
		Empty	Under-developed	Full	
8 July	20 Aug.	20	9	71	46
8 July	27 Aug.	26	6	68	43
8 July	3 Sept.	22	7	71	43
25 Aug.	20 Aug.	28	10	62	30
25 Aug.	27 Aug.	27	8	65	41
25 Aug.	3 Sept.	27	7	66	38

of such high germination as those sown early. Late sowings always gave plants which in the second year were small, pale green and apparently under-nourished, perhaps because they became vernalised and had their energies diverted to reproductive growth before a large enough root and shoot system had been built.

All our experimental plants were hand harvested, placed in hessian bags and dried on a platform with unheated air before threshing. In three experiments, very similar yields, seed size distributions and germination percentages were obtained from this method (barn drying) as from tripod stacking, mechanised swathing and chemical killing (desiccation) prior to direct combining.

THE VALUE OF SEED FOR GROWING ON

Seed from the thresher can be diameter and density graded to retain the better fruits. The larger and denser the fruit the better is the germination percentage, but beyond a certain point monogermity (the percentage of germinating fruits producing only one seedling) decreases, so the portion retained after grading represents a compromise; in one experiment, for instance, 319 g/kg of threshed seed were recovered at 80 per cent germination and 90 per cent monogermity (*Table 5.5*). Most small fruits contain no embryo and they can be removed by aspiration.

Harper (1971) showed that in eight seed lots fruit and seed weights were related (correlation coefficient $r = 0.65$ to 0.92). In fruits as harvested from mother plants, the range of fruit and seed weights were $3.0–30.0$ mg and $0.5–6.0$ mg, respectively. Germination in the laboratory, emergence in the field and average seedling weights of monogerm cultivars increased with increasing fruit diameter (*Table 5.6*). At equivalent plant populations, crops from large seed out-

Table 5.5 EFFECTS OF GRADING MONOGERM SUGAR BEET SEED*

Aspiration, water gauge (in)	Diameter (mm)			
	<3	3–3·75	3·75–4·5	>4·5
	Proportion (g/kg)			
<2	24	34	14	3
2–3	34	168	110	24
3–4	18	112	125	48
>4	3	32	77	67
	Germination (%)			
<2	22	42	84	92
2–3	63	81	86	86
3–4	59	80	88	88
>4	56	78	86	92
	Monogermity (%)			
<2	100	100	100	100
2–3	99	98	96	90
3–4	99	85	59	48
>4	86	51	30	25

* Produced from a 2:1 mixture of monogerm male-sterile and multigerm pollinator plants.

Table 5.6 EFFECTS OF FRUIT DIAMETER, 1967–71

Diameter (mm)	2·7–3·5	3·5–4·3	4·3–5·5
Emergence (%)	22·1	35·9	54·2
Shoot dry weight (mg)	25·5	37·1	45·3
Sugar yield (t/ha)	6·44	not available	7·44
Seed classified on radiographs as:			
Full (%)	40	63	79
Underdeveloped (%)	33	35	25
Empty (%)	27	5	2

yielded those from small by 20, 10 and 10 per cent, respectively, in 1969, 1970 and 1971. Throughout the growing season, plants from large seed had a slightly greater proportion of dry matter in the storage root. Two hypotheses are now being investigated: (a) that restricting the number of fruits that develop and supplying plentiful water and nutrients can increase seed size and improve quality, and (b) that different seed sizes represent different genotypes, in which case it might be possible to select for large seed size.

REFERENCES

AUSTIN, R. B. and LONGDEN, P. C. (1965). 'Effects of nutritional treatments of seed bearing plants on the performance of their progeny', *Nature Lond.*, **205**, 819–820

AUSTIN, R. B. and LONGDEN, P. C. (1966). 'Factors affecting seed performance: parent plant nutrition: red beet', *Rep. natn. Veg. Res. Stn for 1965*, 47

BELL, G. D. H. and BAUER, A. B. (1942). 'Experiments on growing sugar-beet under continuous illumination', *J. agric. Sci., Camb.*, **32**, 112–141

BOYD, D. A. (1952). 'Effect of seed-rate on yield of cereals', *Emp. J. exp. Agric.*, **20**, 115–122

BYFORD, W. J. and HULL, R. (1967). 'Some observations on the economic importance of sugar-beet downy mildew in England', *Ann. appl. Biol.*, **60**, 281–296

HARPER, F. (1971). 'The relationship between early growth and yield of sugar beet', *Ph.D. Thesis*, University of Nottingham

HILLS, O. A. (1963). 'Insects affecting sugar beets grown for seed', *Agr. Handbook 253*, ARS, U.S.D.A.

LONGDEN, P. C. (1970). 'Manuring the beet seed crop grown in England', *N.A.A.S. quart. Rev.*, No. 87, 112–118

LONGDEN, P. C., JOHNSON, M. G. and LOVE, B. (1971). 'Sugar beet seedling emergence prediction from radiographs', *J. int. Inst. Sugar Beet Res.*, **5**, 160–168

SCOTT, R. K. (1968). 'Sugar-beet seed growing in Europe and North America', *J. int. Inst. Sugar Beet Res.*, **3**, 54–84

SCOTT, R. K. (1970). 'The effect of weather on the concentration of pollen within sugar-beet seed crops', *Ann. appl. Biol.*, **66**, 119–127

SCOTT, R. K. and LONGDEN, P. C. (1970). Pollen release by diploid and tetraploid sugar-beet plants', *Ann. appl. Biol.*, **66**, 129–135

TEKRONY, D. M. and HARDIN, E. E. (1969). 'The problem of under-developed seeds occurring in monogerm sugar beets', *J. Am. Soc. Sugar Beet Technol.*, **15**, 625–639

DISCUSSIONS

Thomson: Your germination percentages are surprisingly low.

Scott: For research purposes we recover and examine all fruits that might contain seeds. This is why we frequently get only 60–70 per cent germination. To get seed of 80–90 per cent germination as sold by the British Sugar Corporation it is necessary to discard a very large fraction of the material; in a typical case we recovered some 320 g per kg harvested.

Town: How do you dry your seeds?

Scott: We put seed-bearing plants in hessian bags and dry them in a stream of unheated air until they contain 10–15 per cent water. Mature crops do not differ much in their water content before drying, but immature ones can differ greatly.

Heydecker: What is the reason for choosing 3 September as your reference date?

Scott: Under English conditions harvesting during the first week in September usually gives an acceptable combination of yield and quality, and we have chosen 3 September as a standard date for comparison between years (*Table 5.1, Figure 5.6(b)*. Later some of the larger seeds drop off and birds eat some of the smaller ones at the tips of the branches.

Hawkins: Germination data give the number of fruits that produce at least one seedling, expressed as a percentage. It must be remembered that diploid multigerm fruits each produce at least 1·8 seedlings, tetraploid ones slightly fewer, but monogerm seeds can produce only one.

Scott: Yes, this is why germination percentages for multigerm seed are higher than for monogerm seed (e.g. *Table 5.1*). There is likely to be a genetic component of seed size, as shown by pronounced differences between individual male sterile plants in the average fruit weight, mean time to germination, overall germination percentage and seedling size of their progeny, but there is also an environmental component, due to the position on, and the growing conditions of, the parent plant. Nevertheless, within all seed lots we have examined, large seeds invariably have resulted in higher germination and larger seedlings, irrespective of the cause of the size. The potential gain from breeding for larger seed size is worth investigating.

Miles: Would more irrigation be desirable?

Scott: We have not experimented with overhead irrigation, but a world-wide survey has shown that poor seed set and germination result both from long periods of cold and wet (e.g. in Ireland, Germany, Poland) and from long hot periods during which irrigation is omitted (e.g. Turkey, Arizona).

6

PROTEIN SYNTHESIS AND VIABILITY
IN RYE GRAINS

B. E. ROBERTS and DAPHNE J. OSBORNE

Agricultural Research Council Unit of Developmental Botany,
University of Cambridge

INTRODUCTION

One of the earliest experiments on the life-span of seeds was carried out by Haberlandt (1873), who stored dry samples of many kinds of European cereals in his laboratory and followed their ability to germinate. Later, Burgerstein (1896) and Robertson, Lute and Kroeger (1943) showed that the period of seed viability was not finite and could be extended or reduced by altering the relative humidity and temperature of storage. Rye (*Secale cereale* L.) was found to lose viability more rapidly than most other cereals, and has therefore been used in the present studies.

Over the years, many workers have described structural and biochemical changes in dry seeds which they have associated with a loss of viability (reviewed in Roberts, 1972) and, based on scant information, many theories have been outlined to explain this phenomenon of seed senescence.

The changes that lead to loss of viability represent either an impairment of processes that occur within the dry seed and are vital to the continuance of the viable state, or an impairment of stored components which are necessary for the synthetic processes that take place during the very early stages of germination.

Marcus and Feeley (1964; 1965) and Weeks and Marcus (1971) showed that viable wheat embryos contain all the components necessary for protein synthesis within the dry embryo. In fact, Marcus, Feeley and Volcani (1966) showed that protein synthesis was activated after 30 min imbibition, and only lagged behind water uptake by 10 min. Subsequent work showed that this initial protein synthesis occurred in the absence of discernible deoxyribonucleic acid (DNA) and ribonucleic acid (RNA) synthesis (Dure and Waters, 1965; Marcus and Feeley, 1964; Chen, Sarid and Katchalski, 1968; Walton and Soofi, 1969). Chen *et al.* (1968) showed that DNA replication and RNA transcription in wheat embryos are initiated after some 15 h germination.

Thus, the early events within germination can conveniently be divided into two phases. *Phase one* (up to 6 h), in which protein synthesis occurs in the absence of discernible DNA and RNA synthesis, and *phase two* (after 6 h), in which active synthesis of proteins and nucleic acids occurs simultaneously.

Changes associated with loss of viability in these two phases of germination in rye have been studied by comparisons of components extracted from embryos of different percentage viability. These include a study of the integrity of the DNA required for the second phase of germination, and the activity in cell-free systems of the components necessary for the synthesis of proteins in the first phase.

GRAIN STOCKS

Rye grain of known percentage viability was obtained from D. B. MacKay and R. J. Flood of the National Institute of Agricultural Botany, Cambridge. In 1964, MacKay and Flood (1968) set up a grain storage experiment in which samples of rye from the same initial stock were stored for five years under similar temperature conditions but at different moisture contents (9–16 per cent). At the termination of their experiment in 1969 the percentage viability of these stocks were then 64, 48, 45, 15 and 0 per cent) (0 per cent = non-viable). These stocks were made available for the present work. A further sample of 95 per cent viable rye was obtained from the grain harvested in 1969 and is called 'viable' in the text.

PLANT MATERIAL

Embryos were extracted from the different percentage viability stocks by the mass isolation technique of Johnston and Stern (1957), and stored over calcium chloride at 2°C.

IN VIVO EXPERIMENTS

In vivo experiments showed that both viable and non-viable embryos

Table 6.1 *IN VIVO* INCORPORATION OF RADIOACTIVE PRECURSORS INTO RYE EMBRYOS AT 6 h GERMINATION

Radioactive precursor	No. embryos in experiment	Values per embryo (counts/min)	
		Total uptake	TCA-insoluble material
^{14}C-Amino acid mixture			
Viable embryos	(100)	71 840	3 597
Non-viable embryos	(100)	90 540	240
Adsorption background	(100)	—	186
^{3}H-Thymidine			
Viable embryos	(100)	370 500	245
Non-viable embryos	—	—	—
Adsorption background	(100)	—	180
^{14}C-Uridine			
Viable embryos	(50)	75 220	171
Non-viable embryos	(50)	54 240	94
Adsorption background	(50)	—	195

Note—Hand-dissected embryos were incubated at 25°C for 60 min in a solution 1 mm deep containing 1 µC/ml of the appropriate radioactive precursor. Embryos were then rinsed and homogenised in 150-fold excess of non-radioactive precursor and aliquots evaporated to dryness and counted in liquid scintillant for determinations of total uptake of radioactive precursor. Other aliquots were mixed with equal volumes of 10 per cent TCA, and the TCA-insoluble material collected on filters, dried, and radioactivity determined by liquid scintillation counting. Full details are given in Hallam, Roberts and Osborne (1972).

take up ^{14}C-labelled thymidine, uridine or amino-acid mixtures. In the early hours of germination (phase one), viable embryos incorporate amino-acids into trichloroacetic acid (TCA) insoluble material, but do not incorporate uridine or thymidine (*Table 6.1*).

Table 6.2 *IN VIVO* INCORPORATION OF RADIOACTIVE PRECURSORS INTO RYE EMBRYOS AT 24 h GERMINATION*

Radioactive precursor	No. embryos in experiment	Values per embryo (counts/min)	
		Total uptake	TCA-insoluble material
^{14}C-Amino acid mixture	*		
Viable embryos	(100)	158 100	19 950
Non-viable embryos	(100)	66 720	119
Adsorption background	(100)	—	186
^{14}C-Thymidine			
Viable embryos	(100)	104 000	70
Non-viable embryos	(100)	126 250	236
Adsorption background	(100)	—	186
^{14}C-Uridine			
Viable embryos	(50)	75 220	171
Non-viable embryos	(50)	54 240	94
Adsorption background	(50)	—	195

* Procedure is given in the footnote to Table 6.1.

E

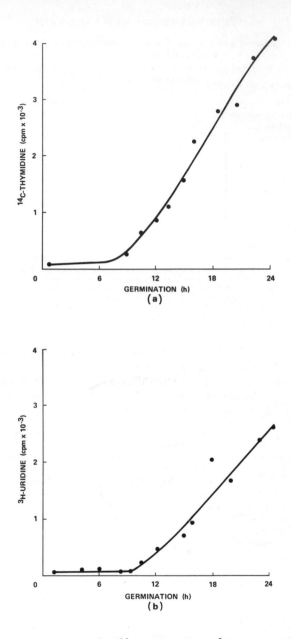

Figure 6.1 The incorporation of (a) ^{14}C-thymidine and (b) ^{3}H-uridine into TCA-insoluble material by individual viable embryos. Incubation at $25°C$ for 60 min in solutions containing 10 $\mu C/ml$ of ^{14}C-thymidine or $100\mu C/ml$ of ^{3}H-uridine

At 24 h of germination, viable embryos incorporate *all* precursors into TCA-insoluble material (*Table 6.2*), synthesis of RNA and DNA both starting at approximately 8–10 h (*Figure 6.1*). However, non-viable embryos neither synthesise proteins in the first phase of germination (*Table 6.1*) nor do they synthesise RNA or DNA in the second phase (*Table 6.2*).

IN VITRO EXPERIMENTS

PHASE TWO: INTEGRITY OF THE DNA

The total DNA present in viable and non-viable embryos was estimated by the diphenylamine method (Burton, 1968) by optical density at 260 nm (Osborne, 1962), and also by microdensitometric measurements of the Feulgen stain of the nuclei of cells from the tip of the secondary root. The results (*Table 6.3*) show that the DNA

Table 6.3   DNA CONTENT OF VIABLE AND NON-VIABLE EMBRYOS

Method of estimation of DNA	$mg\ DNA\ g^{-1}$ dry weight	
	V	NV
Total extractable deoxyribonucleotides		
(a) Diphenylamine*	9·6	10·8
(b) OD at 260 nm†	10·2	9·0
Microdensiometry‡		
Feulgen stain per nucleus	8·4	8·1
Phenol extractable DNA§	10·2	3·1

* Burton (1968).
† Osborne (1962).
‡ Barr and Stroud microdensitometer (GN2) with ×100 objective and an aperture field of 10 mμ. Measured at 540 nm.
§ Marmur (1961).

content of both viable and non-viable embryos is the same. However, when DNA is extracted by the Marmur phenol method (Marmur, 1961), only one-third of the amount spoolable from viable extracts can be obtained from the non-viable embryos, suggesting that much of the DNA is present as low molecular weight fragments. Further study showed that the total spoolable DNA declined progressively with a reduction in the viability of the embryo stocks. This indicated an increase in nuclease activity with loss of viability and the following evidence suggests that in non-viable embryos enhanced nuclease (DNase) acvitity is (1) associated with the DNA and (2) present in the supernatant extract of the embryos.

Table 6.4 DNase activity associated with (a) partially purified DNA and (b) post-ribosomal supernatant extracts of viable and non-viable embryos

Source of DNA	Supernatant	% Increase in TCA-soluble nucleotides (OD 260 nm)
(a) Spoolable V*	—	68
Spoolable NV†	—	112
(b) Enzyme purified V	—	0
Enzyme purified V	V	145
Enzyme purified V	NV	416

* V = viable (95 per cent)
† NV = non-viable (0 per cent).
Note—The solutions were incubated at 25°C for 18 h, then the DNA precipitated with TCA, and the optical activity at 260 nm of the filtrate recorded.

1. If spoolable DNA is isolated from viable and non-viable embryos by the phenol extraction method of Marmur (1961), incubated in buffer at 25°C for 18 h, and residual undegraded DNA then precipitated with TCA, the filtrate from DNA of non-viable embryos then contains more soluble nucleotides than that from viable embryos (*Table 6.4(a)*). Degradation of DNA from viable embryos does not occur if the spoolable DNA is first deproteinised with pronase (*Table 6.4(b)*), a treatment that should remove DNA-associated DNase.

2. DNase activity of soluble protein extracts is shown if deproteinised DNA from viable embryos is incubated with supernatant material soluble after centrifuging extracts for 60 min at 269 000 × g at 1°C from either viable or non-viable embryos. The non-viable supernatant causes more degradation of the DNA than the viable supernatant (*Table 6.4(b)*).

    The progressive degradation of DNA with loss of embryo viability can be correlated with the increases in chromosomal aberrations observed in aged seeds (Roberts, Abdalla and Owen, 1967) and might account for the failure of cell division and growth in non-viable stocks. However, since replication and transcription of DNA do not start till 8 h after water imbibition, it does not account for the inability of non-viable embryos to fulfil the earlier stages of germination (phase one).

PHASE ONE: COMPONENTS OF THE PROTEIN-SYNTHESISING SYSTEM

As in wheat (Marcus, Feeley and Volcani, 1966), the synthesis of protein in viable rye embryos is activated very soon after water is added. There is also an enhancement in protein synthesis at the time of initiation of DNA and RNA synthesis (8 h). The reason for the complete inactivation of the initial protein synthesis in non-viable embryos *in vivo* could be ascribed to the failure of an energy-generating system within the cells. Certainly, respiratory activity is negligible in non-viable embryos (Roberts, Hallam and Osborne, in press) and ultra-structural studies show the mitochondria to be impaired. This leaves open the question of whether a non-viable embryo contains a fully functional protein-synthesising system that requires only an external energy source. This aspect of protein synthesis with loss of viability was therefore studied further using a polyuridylic acid (poly U) dependent *in vitro* system. The presence of adenosine triphosphate (ATP) and guanosine triphosphate (GTP) in a mixture of ribosomes and post-ribosomal supernatant from non-viable embryos, did not overcome the inactivity of the system, indicating that loss of viability is not simply due to an impairment of the energy-coupling system. Mixtures of both components from viable and

Table 6.5 POLYPHENYLALANINE SYNTHESIS BY MIXTURES OF RIBOSOMES AND POST-RIBOSOMAL SUPERNATANTS FROM VIABLE AND NON-VIABLE EMBRYOS

Ribosomes	Supernatant	Incorporation of ^{14}C-Phenylalanine (p moles mg^{-1} rRNA)* (a)	(b)
Viable	Viable	69·4 (100%)	70·1 (100%)
Non-viable	Viable	19·1 (28%)	5·4 (8%)
Viable	Non-viable	18·1 (26%)	55·0 (79%)
Non-viable	Non-viable	2·7 (4%)	0·7 (1%)

* p = pico (10^{-12}). (a) complete system (*Table 6.6*), (b) complete system plus 430 μg wheat germ supernatant fraction. For details of methods see Roberts, Payne and Osborne, *Biochemical Journal* (in press).

non-viable embryos indicate clearly that there is a loss in activity in both the ribosomes and in the post-ribosomal supernatant extracted from the non-viable embryos (*Table 6.5(a)*).

The Ribosomes

Ribosomes from viable and non-viable embryos sediment at the same position in a discontinuous sucrose gradient and also, when mixed, they both sediment at the same position in the gradient. This indicates that there are no gross structural differences between the ribosome types. However, gel electrophoretic analysis of the RNA extracted from ribosome pellets of 95, 64, 48, 15 per cent and

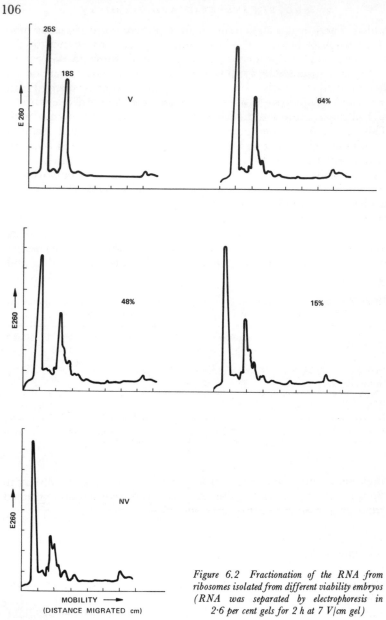

Figure 6.2 Fractionation of the RNA from ribosomes isolated from different viability embryos (RNA was separated by electrophoresis in 2·6 per cent gels for 2 h at 7 V/cm gel)

non-viable embryos show a progressive increase in the degradation of 18S RNA (*Figure 6.2*) (S = Svedberg unit). From these separations it appears that the 25S RNA is intact. However, due to hydrogen bonding in the double-stranded regions, 25S RNA contains considerable secondary structure which tends to hold any fragmented regions together. These double-stranded regions of 25S RNA can be broken and the fragments released by heating the RNA at 70°C for 10 min and rapidly cooling in liquid nitrogen. Subsequent electrophoresis of this RNA indicates a progressive increase in the degree of fragmentation with a decline in embryo viability, the extent of degradation being even greater than that in the 18S RNA.

When ribosomal RNA from non-viable embryos is isolated and subjected to electrophoresis in the presence of Mg^{2+}, which prevents dissociation of the fragments, two distinct bands only, at 25S and 18S, are obtained (P. I. Payne, personal communication). This suggests that no part of the RNA is lost, and that the highly fragmented RNA is stabilised within the ribosome structure. This is further confirmed by the absence of fragments of RNA larger than 4S in any of the post-ribosomal supernatants.

Clearly there is a progressive increase in the fragmentation of both ribosomal RNAs when extracted from ribosomes of lower viability embryos. Furthermore, the same degradative patterns are found in RNAs extracted directly from the dry embryos, indicating that the changes have occurred *in situ* during storage.

Table 6.6 DEPENDENCIES OF THE CELL-FREE PROTEIN-SYNTHESISING SYSTEM FROM VIABLE RYE EMBRYOS FOR *IN VITRO* POLY U DIRECTED POLYPHENYLALANINE SYNTHESIS

	Incorporation of ^{14}C-phenylalanine (p moles mg^{-1} rRNA)
Complete system	868·2
− ATP	110·6
− ATP, energy-generating system	0·06
− Mg^{2+}	0·03
− tRNA	2·4
− Synthetase	0·3
− Poly U	3·8
− Ribosomes	0
+ RNase (50 μg)	0·12

Note—Each value is the average of duplicate determinations.

The activity of the ribosomes (80–100 μg ribosomal RNA (rRNA)) has been assessed in an *in vitro* system, saturated with deacylated transfer RNA (tRNA) (100 μg), supernatant fraction (450 μg protein), poly U (100 μg) in the presence of optimum levels of Mg^{2+} (10 mM) and K$^+$ (25 mM). In this *in vitro* system the synthesis of polyphenylalanine is linear for 80 min at 25°C. The dependencies of this poly U dependent *in vitro* system are summarised in *Table 6.6*.

When ribosomes are pelleted twice during extraction they show a decline in activity which reflects their percentage viabilities. However, when the extraction period is reduced by pelleting the ribosomes only once and rinsing the surface of the pellet three times with the extraction buffer, this decline in activity with loss of embryo viability is much less pronounced. In fact, non-viable ribosomes are then capable of more than 60 per cent of the incorporation of viable ribosomes. This suggests that a considerable amount of ribosome inactivation occurs during extraction, and this is more pronounced in extracts of lower viability embryos. The fact that twice-pelleted ribosomes from 64, 48, 15 per cent and non-viable embryos show a slight decline in their capacity to bind ^{14}C-phenylalanyl-tRNA enzymically, as assessed by the method of Nirenberg and Leder

Table 6.7   THE CAPACITY OF RIBOSOMES FROM DIFFERENT VIABILITY EMBRYOS TO BIND ^{14}C-PHENYLALANYL tRNA, IN PRESENCE OF POST-RIBOSOMAL SUPERNATANT FROM 64 PER CENT VIABLE EMBRYOS

Source of ribosomes (embryo viability)	^{14}C-phenylalanyl-tRNA bound (p moles mg^{-1} rRNA)
64%	62·1
48%	57·5
15%	47·7
0%	43·7

(1964), is one indication of the nature of the ribosomal inactivation (*Table 6.7*).

Transfer RNA

The major lesion associated with loss of protein synthetic capacity as the embryo viability declines, clearly occurs in the supernatant fraction. From the results shown in *Table 6.8* it is apparent also that this supernatant lesion increases progressively with a decline in the percentage viability of the embryo. The addition of deacylated tRNA alone to mixtures of ribosomes and supernatants (as in *Table 6.5(a)*), increases the total activity of the mixtures but does not modify the percentage differences observed between the viable and non-viable supernatants. However, the addition of a complete supernatant fraction from wheat germ enhances the activity of the non-viable supernatant to 79 per cent of that of the viable (*Table 6.5(b)*). This wheat supernatant fraction contains not only tRNA but

Table 6.8   THE CAPACITY OF POST-RIBOSOMAL SUPERNATANTS FROM EQUAL WEIGHTS OF DIFFERENT EMBRYOS TO ACTIVATE POLYPHENYLALANINE SYNTHESIS IN VIABLE RIBOSOMES USING ^{14}C-PHENYLALANINE

Source of supernatants (embryo viability)	^{14}C-polyphenylalanine synthesised (p moles mg^{-1} rRNA) A	B
64%	7·5	34·9
48%	5·8	16
15%	2·8	9
0%	0·5	2·6

Notes
A = 1 g dry weight of embryos.
B = 3 g dry weight of embryos.

also aminoacyl tRNA synthetases and transfer enzymes, all of which are essential to poly U directed *in vitro* protein synthesis.

All the tRNA fractions isolated from 64, 45 and 15 per cent viability embryos have very similar kinetics of aminoacylation of ^{14}C-labelled phenylalanine in the presence of a supernatant from viable embryos, which suggests that the tRNA's capacity for amino-acylation does not decline with loss of embryo viability. Degradation of tRNA is not, therefore, the lesion of non-viability.

Aminoacyl tRNA synthetases

No difference in the capacity of supernatants from 64, 48, 45, 15 per cent and non-viable embryos to aminoacylate tRNA with ^{14}C-labelled amino acids (*Table 6.9*) or ^{14}C-phenylalanine could be

Table 6.9   THE AMINOACYLATION OF tRNA BY ^{14}C-AMINO ACID MIXTURE USING POST-RIBOSOMAL SUPERNATANTS FROM VIABLE AND NON-VIABLE EMBRYOS

Source of supernatant (embryo viability)	Production of ^{14}C-aminoacyl-tRNA (disintegrations per minute)
95%	30 450
0%	31 320
Minus supernatant background	174

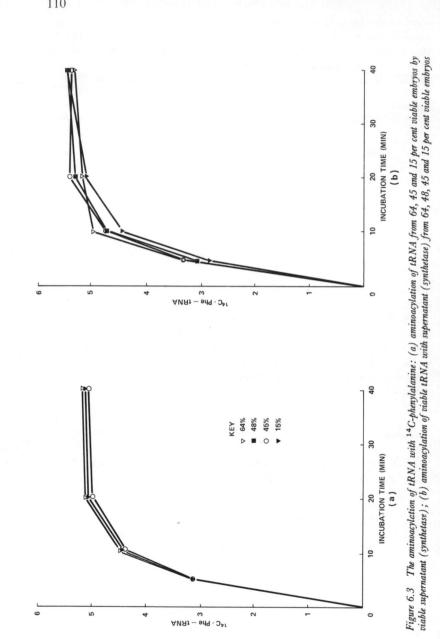

Figure 6.3 The aminoacylation of tRNA with ¹⁴C-phenylalanine: (a) aminoacylation of tRNA from 64, 45 and 15 per cent viable embryos by viable supernatant (synthetase); (b) aminoacylation of viable tRNA with supernatant (synthetase) from 64, 48, 45 and 15 per cent viable embryos

detected (*Figure 6.3*), so it appears that the aminoacyl synthetase activity remains unimpaired as embryos lose viability. The production of aminoacyl tRNAs also is not the limiting factor for the failure of protein synthesis in non-viable embryos. This was confirmed when ^{14}C-phenylalanyl tRNA was used instead of ^{14}C-phenylalanine in

Table 6.10 THE CAPACITY OF POST-RIBOSOMAL SUPERNATANTS FROM DIFFERENT VIABILITY EMBRYOS TO ACTIVATE POLYPHENYL-ALANINE SYNTHESIS ON 95 PER CENT VIABLE RIBOSOMES USING ^{14}C-PHENYLALANYL tRNA

Source of supernatant (embryo viability)	^{14}C-polyphenylalanine synthesised (p moles mg^{-1} rRNA)
64%	305
48%	261
45%	239
15%	156
0%	73

the poly U dependent *in vitro* system (*Table 6.10*): the loss of activity was still apparent in the mixtures containing supernatants from the stocks of lower viabilities.

The transfer enzymes

Using the *in vitro* transfer system of Legocki and Marcus (1970) there was found to be a very slight reduction in the activity of the transferase II component (or fraction II—the fraction associated with the transfer of the aminoacyl tRNA from the initial binding site to the peptidyl tRNA site on the ribosome) with loss of viability (*Table 6.11*). In supernatants treated with a sulphydryl alkylating reagent

Table 6.11 THE CAPACITY OF TRANSFERASE II ENZYMES IN POST-RIBOSOMAL SUPERNATANTS FROM DIFFERENT VIABILITY EMBRYOS TO TRANSFER ^{14}C-PHENYLALANYL tRNA TO THE PEPTIDYL SITES IN 64 PER CENT VIABLE RIBOSOMES

Source of supernatant (embryo viability)	^{14}C-Polyphenylalanine synthesised (p moles mg^{-1} rRNA)
95%	283
64%	272
45%	258
15%	248

N'ethylmaleimide (NEM), the transferase II fraction is inactivated without impairing the activity of transferase I (or fraction I—the fraction that mediates the initial binding of aminoacyl tRNA to the ribosome) (Yarwood *et al.*, 1971). The capacity of NEM-treated supernatants to facilitate the binding of ^{14}C-phenylalanyl tRNA to twice-pelleted ribosomes from 64 per cent viable embryos declined progressively with the loss of embryo viability (*Table 6.12*). This

Table 6.12 THE CAPACITY OF TRANSFERASE I ENZYMES IN POST-RIBOSOMAL SUPERNATANTS FROM DIFFERENT VIABILITY EMBRYOS TO BIND ^{14}C-PHENYLALANYL tRNA TO 64 PER CENT VIABLE RIBOSOMES

Source of supernatant (embryo viability)	^{14}C-Phenylalanyl tRNA bound (p moles mg^{-1} rRNA)
64%	81·2
48%	79·1
45%	70·6
15%	40·9
0%	5

indicates an age-related inactivation of transferase I in embryos during storage.

CONCLUSIONS

From the many studies of loss of viability in seeds it is clear that no *one* lesion is associated with the ageing process. Instead, the whole molecular structure of the cells passes from one of an organised living system to one of molecular disorganisation of the non-living state. This progression is borne out in the study of the integrity of the DNA and of the activity of protein-synthesising systems from embryos of different percentage viabilities when such systems are studied in the cell-free state.

The DNA shows increasing numbers of breaks with loss of viability, and the high molecular weight RNA, when separated from the ribosomes of samples of different viabilities, also shows increasing cleavage to small molecular weight fragments.

A feature of most interest is the finding that the tRNAs and synthetases suffer little loss of function with loss of viability and can therefore be considered as very stable components of the protein-synthesising system. The transferase I enzymes, concerned with binding of aminoacylated tRNAs to ribosome, are particularly labile, however, and are the only components so far studied whose loss of function parallels loss of viability. This transferase I activity could therefore be the limiting link in maintaining viability in a system that is generally declining. This leaves many questions still to be answered, but conditions that afford protection to these particularly labile biochemical systems may permit extension of the storage life of rye.

Since the degradative changes that lead to the non-viable state must continue beyond the stage of loss of function of the most labile component, the concept of 'deader than dead' must be introduced, and many changes that have been considered in the past could actually be reflections of lost viability rather than indications of an initial lesion.

ACKNOWLEDGEMENTS

The authors are grateful to their colleagues Dr. P. I. Payne and Dr. T. A. Dyer for their collaboration.

REFERENCES

BURGERSTEIN, A. (1896). 'Beobachtungen über die Keimkraftdauer von ein bis zehnjährigen Getreidesamen', *Verh. Zool.-Bot. Ges. Wien*, **45**, 414–421

BURTON, K. (1968). 'Determination of DNA concentration with diphenylamine', in *Methods in Enzymology*, Vol. XII, Part B, pp. 163–166 (ed. Grossman and Moldave), Academic Press, N.Y.

CHEN, D., SARID, S. and KATCHALSKI, E. (1968). 'Studies on the nature of messenger RNA in germinating wheat embryos', *Proc. natn. Acad. Sci. USA*, **60**, 902–909

DURE, L. and WATERS, L. (1965). 'Long-lived mRNA: evidence from cotton seed germination', *Science N.Y.*, **147**, 410–412

HABERLANDT, F. (1873). 'Die Keimfähigkeit unserer Getreidekörner, ihre Dauer und die Mittel ihrer Erhaltung', *Wiener landw. Ztg.*, **22**, 126

HALLAM, N. D., ROBERTS, B. E. and OSBORNE, D. J. (1972), 'Embryogenesis and germination in rye (*Secale cereale* L.). II. Biochemical and fine structural changes during germination', *Planta*, **105**, 293–309

JOHNSTON, F. B. and STERN, H. (1957). 'Mass isolation of viable wheat embryos', *Nature Lond.*, **179**, 160–161

LEGOCKI, A. B. and MARCUS A. (1970). 'Polypeptide synthesis in extracts of wheat germ. Resolution and partial purification of the soluble transfer fractions', *J. biol. Chem.*, **245**, 2814–2818

MACKAY, D. B. and FLOOD, R. J. (1968). 'Investigations in crop seed longevity. II. The viability of cereal seed stored in permeable and impermeable containers', *J. natn Inst agric. Bot*, **11**, 378–403

MARCUS, A. and FEELEY, J. (1964). 'Activation of protein synthesis in the imbibition phase of seed germination', *Proc. natn. Acad. Sci. USA*, **51**, 1075–1079

MARCUS, A. and FEELEY, J. (1965). 'Protein synthesis in imbibed seeds. II. Polysome formation during imbibition', *J. biol. Chem.*, **240**, 1675–1680

MARCUS, A., FEELEY, J. and VOLCANI, T. (1966). 'Protein synthesis in imbibed seeds.

III. Kinetics of amino acid incorporation, ribosome activation and polysome formation', *Pl. Physiol.*, **41**, 1167–1172

MARMUR, J. (1961). 'A procedure for the isolation of deoxyribonucleic acid from micro-organisms', *J. mol. Biol.*, **3**, 208–218

NIRENBERG, M. and LEDER, P. (1964). 'RNA codewords and protein synthesis. The effect of trinucleotides upon the binding of sRNA to ribosomes', *Science N.Y.*, **145**, 1399–1407

OSBORNE, D. J. (1962). 'Effect of kinetin on protein and nucleic acid metabolism in *Xanthium* leaves during senescence', *Pl. Physiol.*, **37**, 595–602

ROBERTS, E. H. (1972). *The Viability of Seeds*, Ch. 9, Chapman and Hall, London

ROBERTS, E. H., ABDALLA, F. H. and OWEN, R. J. (1967). 'Nuclear damage and the ageing of seeds with a model for seed survival curves', *Symp. Soc. exp. Biol.*, **21**, 65–100

ROBERTS, B. E., HALLAM, N. D. and OSBORNE, D. J. *Planta* (in press)

ROBERTSON, D. W., LUTE, A. M. and KROEGER, H. (1943). 'Germination of 20 year old wheat, oats, barley, corn, rye, sorghum and soybeans', *J. Am. Soc. Agron.*, **35**, 786–795

WALTON, D. C. and SOOFI, G. S. (1969). 'Germination of *Phaseolus vulgaris*. III. The role of nucleic acid and protein synthesis in the initiation of axis elongation', *Pl. Cell Physiol*, **10**, 307–315

WEEKS, D. P. and MARCUS, A. (1971). 'Preformed messenger of quiescent wheat embryos', *Biochim. biophys. Acta.*, **232**, 671–684

YARWOOD, A., PAYNE, E. S., YARWOOD, J. S. and BOULTER, D. (1971). 'Aminoacyl-tRNA binding and peptide chain elongation on 80S plant ribosomes', *Phytochem.*, **10**, 2305–2311

DISCUSSIONS

The discussions for Chapter 6 are combined with those at the end of Chapter 7.

7

FINE STRUCTURE OF VIABLE AND NON-VIABLE RYE AND OTHER EMBRYOS

N. D. HALLAM

Agricultural Research Council Unit of Developmental Botany, University of Cambridge and Monash University, Victoria, Australia

INTRODUCTION

Although there have been numerous investigations on the fine structure of seeds during germination, not surprisingly, few studies have been carried out on non-viable material. By contrast, problems associated with seed dormancy and its breaking by physical and chemical means are reasonably well documented. The early stages of embryogenesis in dicotyledons such as *Capsella* and *Gossypium* (Jensen, 1965; Jensen and Fisher, 1967; Schultz and Jensen, 1968) and *Pisum* (Bain and Mercer, 1966; Marinos, 1970; Mollenhauer and Totten, 1971) have been investigated, but the maturation and ripening of the seed is one aspect of embryogenesis poorly represented in the literature. This paper summarises some of the changes taking place during late embryogenesis, the fine structure of the mature embryo and the changes that occur within it during germination, as well as the fine structure of non-viable embryos.

In order to study the unimbibed embryo of *Pisum*, Perner in 1965 used osmium vapour to fix unimbibed dehydrated embryos. In his preparations, membranes were narrower than in tissue fixed in the hydrated condition and mitochondria appeared as stellate structures with no obvious cristae. Such a technique has been criticised on the grounds that, compared with more modern techniques, the slowness of fixation by vapour would lead to poor results. However, further work on *Pisum* by Yoo (1970), who used aqueous aldehyde fixatives, gives the impression that mitochondria of the unimbibed embryo are of a normal round shape, and this suggests that the aqueous fixation alone is enough to hydrate the tissue before actual fixation

116

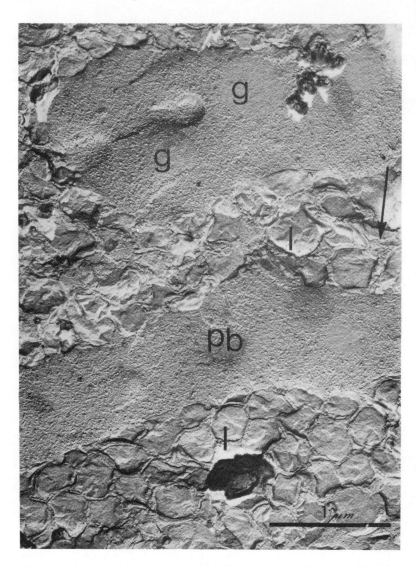

Figure 7.1 Wheat scutellum parenchyma, freeze-etched preparation. Tissue placed in a drop of 100 per cent glycerol frozen within 8 s in liquid Freon 22 cooled by liquid nitrogen. pb— protein body, g—phytin globoid, l—lipid droplet. Arrow indicates direction of shadowing. (Courtesy: J. G. Swift, C.S.I.R.O. Division of Horticultural Research, Glen Osmond, South Australia)

117

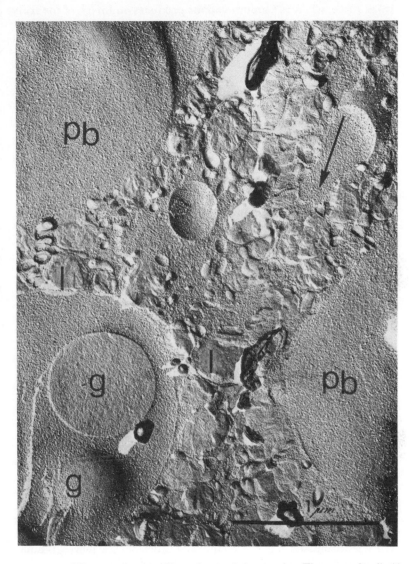

Figure 7.2   Wheat scutellum parenchyma, freeze-etched preparation. Tissue exposed to liquid water for 8 s before snap freezing in Freon 22. pb—protein body, g—phytin globoid, l—lipid body. Arrow indicates direction of shadowing. (Courtesy: M. S. Buttrose, C.S.I.R.O. Division of Horticultural Research, Glen Osmond, South Australia)

takes place, whereas osmium vapour produces an image of the truly dehydrated organelle.

Recent investigations by Buttrose and Swift (1972) have yielded some fascinating results using the technique of freeze etching on unimbibed wheat scutellum. Their study indicates that the water content increases from 7 per cent in the air-dry condition to 26 per cent within the first 10 sec of exposure to water. Gross morphological changes occur during this short time in the fine structure of the tissue. This suggests that water is very rapidly distributed throughout the cellular contents with consequent changes in volume (see *Figures 7.1* and *7.2*). It seems, therefore, that with aqueous fixatives, hydration occurs before fixation and we therefore see subcellular organelles in a clearly recognisable form rather than in their truly dehydrated condition. This must be borne in mind when considering the fine structure of unimbibed tissues.

Specimen preparation for electron microscopy of material described in this paper was essentially as described by Hallam (1972).

EMBRYOGENESIS

In the earlier stages of embryogenesis (e.g., one week after anthesis) many of the cells of the embryo are undergoing mitosis and the embryo tissue is separated from the vacuolated cells of the ovary wall by a band of cells abundant in ribosomes and microtubules, and bounded by a cuticle-like layer.

Cell organelles are surrounded by ribosomes in polysomal aggregations but as development proceeds the number of apparently free ribosomes increases until, at two weeks after anthesis, so many ribosomes are present that polysomes are no longer obvious. At the two-week stage numerous starch-containing plastid-like bodies are present as well as the normal complement of organelles, but within three weeks an obvious condensation of structure has occurred. Endoplasmic reticulum apparently accumulates around electron lucent bodies which are of similar size to mitochondria, and the cytoplasm becomes more electron dense. Dictyosomes appear less frequently in sections, chromatin masses appear clumped and retracted from the nuclear membrane, nucleoli are more granular and mitochondria circular in cross-section.

By four weeks after anthesis, circlets of endoplasmic reticulum completely surround electron lucent bodies and the reduction (apparent simplification) of subcellular organisation is almost complete (*Figure 7.3*). At this stage the endosperm has a milky viscous consistency.

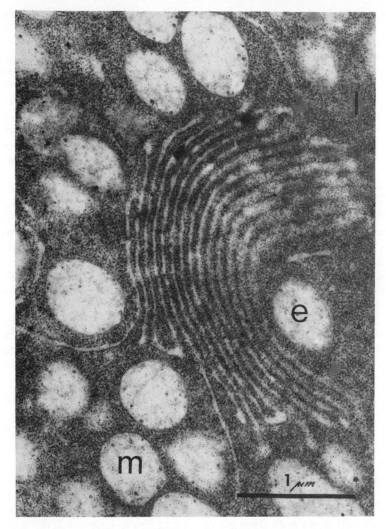

Figure 7.3   Embryo tissue approximately 4 weeks after anthesis. Endoplasmic reticulum aggregates around electron lucent bodies (e). The dense cytoplasm contains numerous ribosomes, mitochondria (m) and randomly distributed, small lipid bodies (l)

THE MATURE EMBRYO

At approximately five weeks after anthesis dehydration of the grain is complete and within the cells of the embryo chromatin masses and nucleoli are a prominent feature, mitochondria are still round in cross-section (*Figures 7.4* and *7.5*), dictyosomes are rare, and are reduced to, flattened stacks of vesicles (see *Figures 7.6* and *7.7*) pro-amyloplasts contain small groups of osmiophilic granules and endoplasmic reticulum exists as small crescents around electron lucent bodies (see *Figure 7.5*).

During the last stage of dehydration, lipid bodies apparently move to the plasmalemma of the cells, forming a close order packing array (*Figure 7.8*). This is possibly one of the last actions of cyclosis before complete dehydration sets in. Lipid bodies are a common feature of seed material and those of pea and bean cotyledons have been extensively studied by Mollenhauer and Totten (1971). These bodies are often described as spherosomes. Frey-Wyssling, Grieshaber and Muellethaler (1963) and other workers have followed the development of spherosomes and suggest that granular proteinaceous materials within a unit membrane are replaced by fats and oils to give rise to an organelle containing these compounds and bounded by a normal tripartite membrane. High resolution electron microscopy, however, shows that lipid bodies are bounded by a single osmiophilic layer, and therefore should not be described as spherosomes (*Figure 7.9*).

By the time the grain is fully dehydrated all subcellular organelles responsible for respiration and protein synthesis are present in a rudimentary form. The metabolic retardation and the organisation of cells for an extended dehydrated existence in the mature grain takes, in the case of rye, four to five weeks. The reverse process from simplicity to complexity is accomplished in a matter of hours during the earliest stages of germination.

THE VIABLE EMBRYO

We have seen from freeze etching studies (*Figures 7.1* and *7.2*) that hydration of isolated wheat scutellum tissue takes place very rapidly and that consequent rapid changes in fine structure occur. It must be assumed that similar changes take place in rye embryos. Even with whole grains there is a rapid uptake of water within the first 10 min of imbibition. Viable and non-viable grains show similar imbibition patterns for the first hour. The increase in fresh weight in the first 10 min is probably due to physical wetting and the phase during the next 45–50 min is one of solubilisation. There is little change in the fresh weight of non-viable material after this time, whereas active uptake of water into the viable grain continues up to 24 h (see *Figure 7.10* and *7.11*).

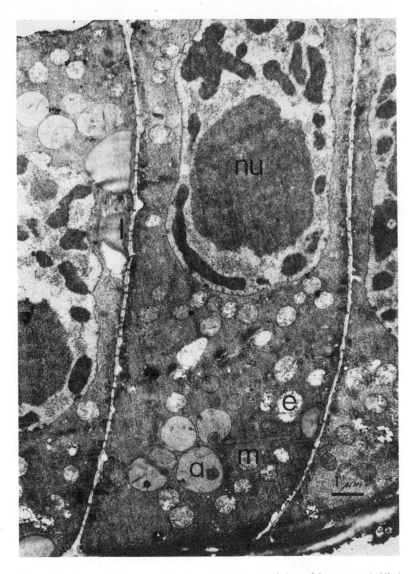

Figure 7.4   Low magnification electron micrograph of pro-cortical tissue of the mature, unimbibed embryo (viable grain). Nuclei contain prominent nucleoli (nu) with electron dense aggregations of chromatin retracted from the nuclear membrane. Pro-amyloplasts are present (a) as well as rounded mitochondria (m) and electron lucent bodies (e). Lipid bodies (l) line the plasmalemma

122

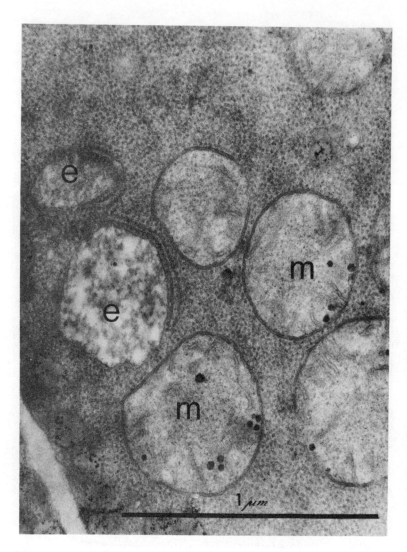

Figure 7.5 In unimbibed tissue mitochondria (m) have few cristae and contain osmiophilic droplets. Electron lucent bodies (e) have small crescents of endoplasmic reticulum associated with them

Figure 7.6   Dictyosomes (d), lipid bodies (l) and mitochondria (m), two weeks after anthesis. Note the numerous dictyosome vesicles (v) and ribosomes

124

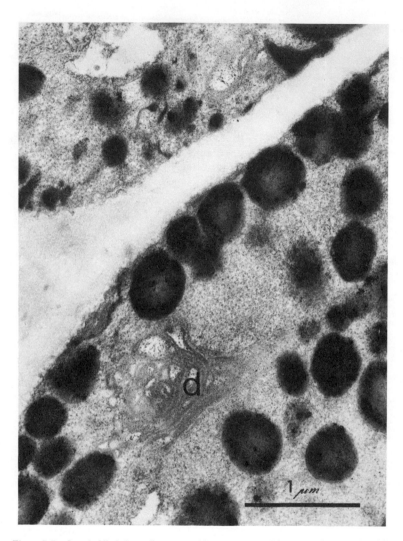

Figure 7.7 In unimbibed tissue dictyosomes (d) are represented as stacks of squashed vesicles usually close to the cell wall. There is a dense packing of ribosomes and lipid bodies are prominent

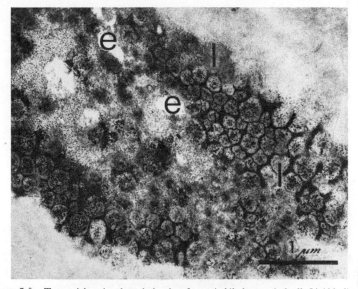

Figure 7.8   Tangential section through the edge of an unimbibed procortical cell. Lipid bodies (l) are seen in a hexagonal pattern on the plasmalemma. The knife has cut through into the ribosomes and electron lucent bodies (e) during sectioning

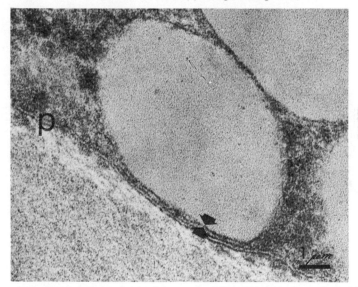

Figure 7.9   The plasmalemma (p) is of the normal tripartite structure, whereas the boundary of lipid bodies (arrowed) is approximately one-third of the width of the plasmalemma. This presumably represents a protein/lipid interface and therefore not a unit membrane

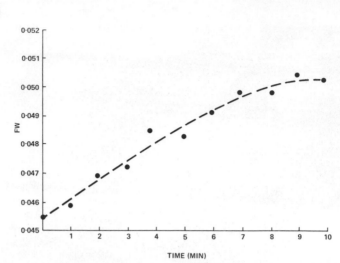

Figure 7.10 Fresh weight of a single rye grain plotted against the time of imbibition. During the initial phase there is a rapid increase in fresh weight complete within the first 10 min. This probably corresponds to physical wetting of the grain

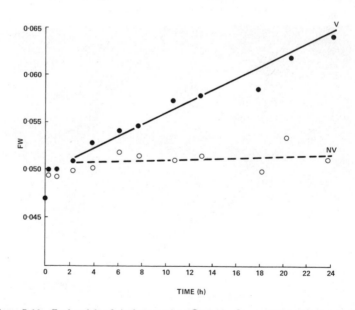

Figure 7.11 Fresh weight of single rye grains (● viable, ○ non-viable) plotted against time of imbibition in hours. From 10 min to 1 h of imbibition the weight remains stable during a solubilisation phase, then an active water uptake occurs only in the viable material

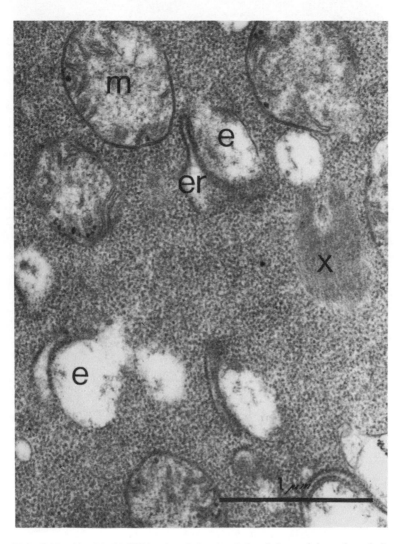

Figure 7.12   After 2 h of imbibition the endoplasmic reticulum (er) around electron lucent bodies (e) is better defined. Ribosomes attached to it stain more heavily than those free in the cytoplasm. x—protein crystaloid, m—mitochondrion

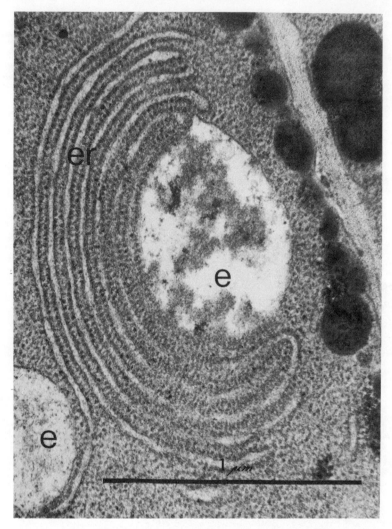

Figure 7.13   By 5 h there is extensive proliferation of rough endoplasmic reticulum (er) around electron lucent bodies (e)

In viable grains no obvious fine structural changes take place until at least 2 h of imbibition. Generally, the membrane systems of mitochondria are better defined at this stage and the distance between the two sheets making up endoplasmic reticulum associated with electron lucent bodies has widened (see *Figure 7.5* and *Figure 7.12*). After 3 h of imbibition, electron lucent bodies in viable material are almost encircled by endoplasmic reticulum, and dictyosomes are present, with numerous vesicles associated with them. By contrast Yoo (1970) found that in *Pisum* embryos, dictyosomes

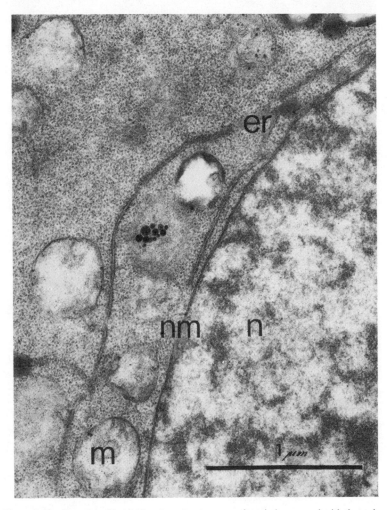

Figure 7.14   From 6 h of imbibition the nucleus is apparently actively concerned with the production of endoplasmic reticulum (er) from the outer nuclear membrane (nm). Note that mitochondria (m) have more cristae than in the unimbibed condition (cf. Figure 7.5) and that chromatin is more diffuse within the nucleus (n) than in unimbibed tissue (cf. Figure 7.4)

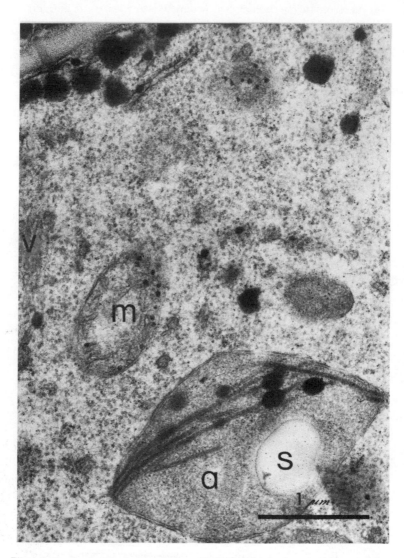

Figure 7.15 Osmiophilic droplets within pro-amyloplasts (a) develop spindle-like attachments by 7 h of imbibition and starch (s) accumulates within the organelle. Note the numerous cristae within mitochondria (m) and dictyosome vesicles (v)

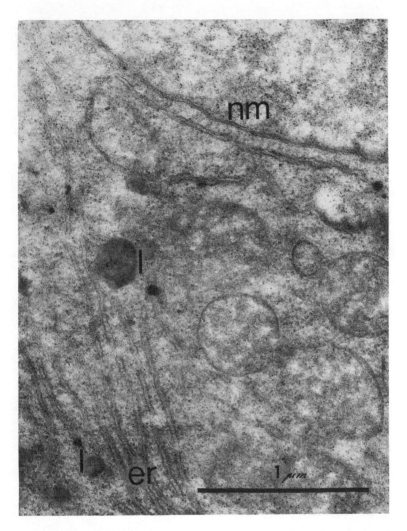

Figure 7.16   At 9 h of imbibition extensive sheets of endoplasmic reticulum (er) are apparently being produced from the nuclear membrane (nm); lipid bodies (l) are often found throughout the cytoplasm from this stage onwards

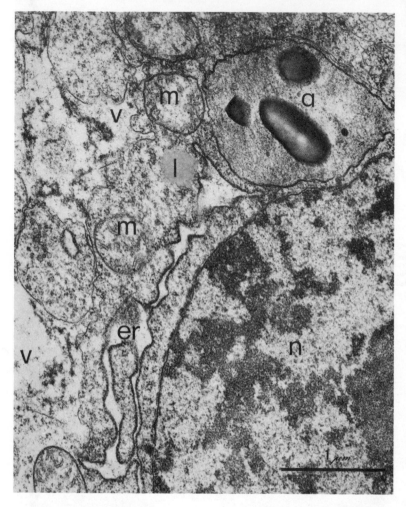

Figure 7.17   By 24 h, cells of the meristem have begun dividing whereas the cells of the pro-cortical and pro-stelar tissues are in an advanced state of development. Numerous mitochondria (m) are present, as are amyloplasts (a) containing starch granules, and developing vacuoles (v), whereas lipid bodies (l) are seen only occasionally. n—nucleus, er—rough endoplasmic reticulum

remained as flat vesicles up to 8 h of imbibition. The proliferation of endoplasmic reticulum associated with electron lucent bodies continues up to about 5 h of imbibition (*Figure 7.13*), where the generation of reticulum appears to be a reversal of the changes taking place during dehydration at late embryogenesis (see *Figure 7.3* and *Figure 7.13*).

By 6 h, however, the site of origin of endoplasmic reticulum changes and appears to be associated with the nuclear membrane—this suggests that cisternae may, in fact, be generated from the nuclear membrane (*Figure 7.14*). From this stage onwards electron lucent bodies are no longer a feature of the cells. There is an increase in the incorporation of ^{14}C–uridine into TCA material from this stage onwards and a consequent increase in protein synthesis (Hallam, Roberts and Osborne, 1972a). By 7 h, mitochondria are rapidly dividing and there are numerous active dictyosomes. Pro-amyloplasts, which at earlier stages appear as organelles containing several osmiophilic particles, begin to accumulate starch. The osmiophilic particles extend spindle-type attachments and starch appears within the organelle (*Figure 7.15*). Metabolic activity increases and apparently cytoplasmic streaming begins as lipid bodies begin to appear in the central areas of the cytoplasm. By 9 h of imbibition (*Figure 7.16*) extensive sheets of endoplasmic reticulum are formed and there is an increase in the incorporation of ^{14}C–thymidine before 12 h of imbibition (Hallam, Roberts and Osborne, 1972a). This precedes division of the meristematic tissue between the 12th and 18th hour of imbibition.

After cell division, the original cells (as opposed to those derived from meristematic divisions) are in an advanced state of differentiation, and vacuolation has taken place by 24 h (*Figure 7.17*). Within some areas, especially coleorhiza and root cap, there are cells which in structure and reaction with acid phosphatase are identical with the lysosomes described from the root cap of *Zea mays* (Berjak, 1968; Berjak and Villiers, 1970). Bodies similar to these are present within the main part of the root primordium but they enlarge during cell expansion and apparently develop into the primary vacuole.

THE NON-VIABLE EMBRYO

The National Institute of Agricultural Botany at Cambridge has provided the recent non-viable rye grains for this study and also samples of wheat harvested in 1870 and oats from 1872. A number of cytochemical tests have been carried out on enzyme systems of viable material as well as recent, old and ancient non-viable grains, and the results are summarised in *Table 7.1*.

For the enzyme acid phosphatase, root primordia were excised and incubated for 24 h at 22°C in 0·01 M sodium glycerophosphate

Table 7.1 ENZYME ACTIVITIES AND FEULGEN REACTION OF EMBRYOS OF DIFFERENT AGES AND VIABILITIES

	Rye Viable		Non-viable		Oats 1872 Non-viable		Wheat 1870 Non-viable		Wheat Non-viable	
	Coleorhiza	Root cap	Coleorhiza	Root cap	Coleorhiza	Root cap	Coleorhiza	Root cap	Thebes 3000–2000 B.C.	Fayum 4441 ± 180 B.C.
Acid phosphatase	+++++	+++++	++++	+++	+++	+++	++	+	Not tested	−
ATPase	+++++	+++++	+					−	Not tested	−
Peroxidase	+++++	+++++	++++		+	+	++	++	−	−
Amylase	+++++	+++++	++++	++++					−	−
Feulgen test	+	+	+				−	−	−	−
I$_2$ test for starch	+	+	+	+	+	+	+	+	+	+

in the presence of lead nitrate. After incubation the tissue was transferred to 1 per cent ammonium sulphide. Black deposits indicate sites of enzyme activity. Organelles containing acid phosphatase in the coleorhiza and root cap are presumably lysosomes (Berjak, 1968). A similar incubation was carried out as a test for ATPase where root primordia were incubated with ATP as a substrate in tris/malate buffer in the presence of magnesium ions and lead nitrate. Again, as with the acid phosphatase test, phosphate liberated from the substrate by the enzyme present within the cells precipitates as lead phosphate and is then converted to lead sulphide by ammonium sulphide. Black deposits form but these are randomly distributed throughout the root primordium. Peroxidase activity in cell walls was tested for by immersing pieces of embryo in an aqueous solution of guaiacol, and then transferring them to 20 vol H_2O_2. The presence of peroxidase is indicated by a yellow coloration in the walls.

To test for the presence of amylase, half-grains were placed on a thin film of 1 per cent starch agar in a petri dish and left for 24 h in an atmosphere saturated with water vapour. The grains were removed and the starch agar irrigated with 1 per cent iodine in potassium iodide. Amylase activity was indicated by areas around and beneath the half-grain that stained a lighter blue with iodine than the surrounding area of agar. Standard Feulgen treatments of a $\frac{1}{4}$ h hydrolysis at 60°C in 1N HCl, followed by 1 h in Schiff's reagent and then bleaching with freshly prepared H_2SO_3, showed the typical purple staining within the nucleus only in viable and recent non-viable material.

Even in the unimbibed condition there are obvious differences in fine structure between viable and non-viable embryos. The outer membrane of many mitochondria of non-viable material is swollen and the inner membrane is lacking in the area of the blister. In addition, the plasmalemma is pulled away from the wall in many places (*Figure 7.18*). We have seen that in viable material no obvious changes occur in fine structure until 2 h of imbibition. At as early as 15 min of imbibition, however, the embryos of non-viable grains are undergoing irreversible degradative changes (*Figure 7.19*). The plasmalemma is pulled further from the wall and some cellular disorganisation is apparent.

In non-viable material, by the sixth hour (*Figure 7.20*), the space between the retracted plasmalemma and the cell wall is filled with fine, granular osmiophilic material suggesting precipitated proteins. Presumably the semipermeable nature of the membrane systems is lost during ageing and the changes seen during imbibition are structural expressions of breakdown at a molecular level. The autolytic processes which are already proceeding slowly in the dehydrated

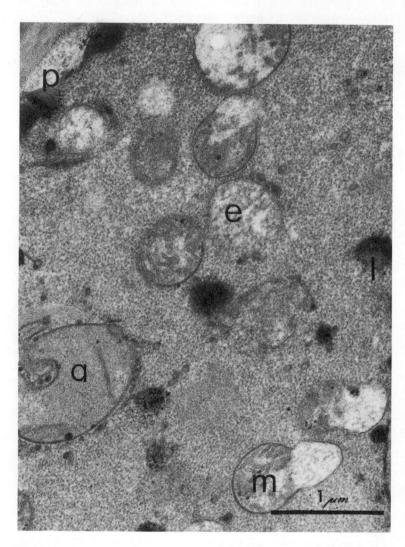

Figure 7.18   Unimbibed non-viable tissue. In many cells the plasmalemma (p) has withdrawn from the cell wall and mitochondria (m) often have a blistered outside membrane. Other organelles, for example pro-amyloplasts (a), electron lucent bodies (e), lipid bodies (l) and the background packing of ribosomes appear normal

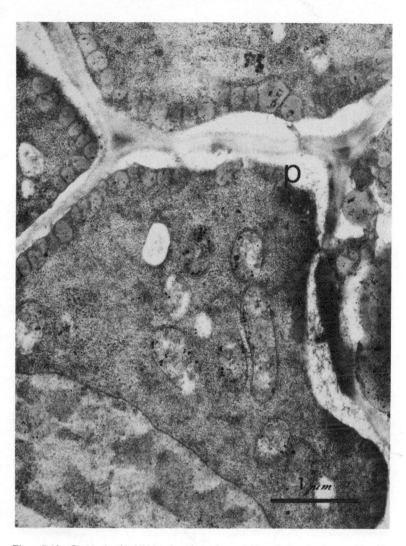

Figure 7.19 By 15 min of imbibition the embryo of non-viable grain has already started breaking down. The plasmalemma (p) has pulled further away from the cell wall. Other organelles appear normal

138

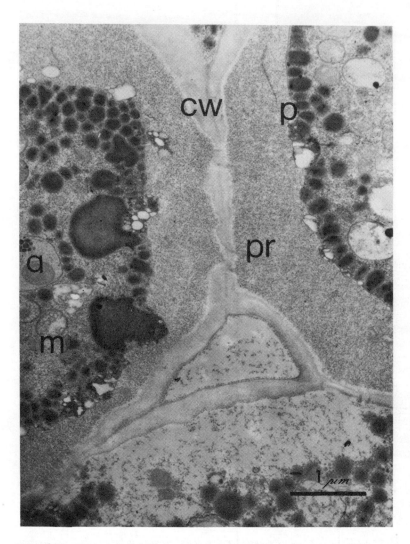

Figure 7.20   The retraction of the plasmalemma is further advanced by 6 h of imbibition and the area between the plasmalemma (p) and cell wall (cw) is filled with an osmiophilic granular material, presumably precipitated proteins (pr). Note the pro-amyloplast (a) and mitochondrion (m)

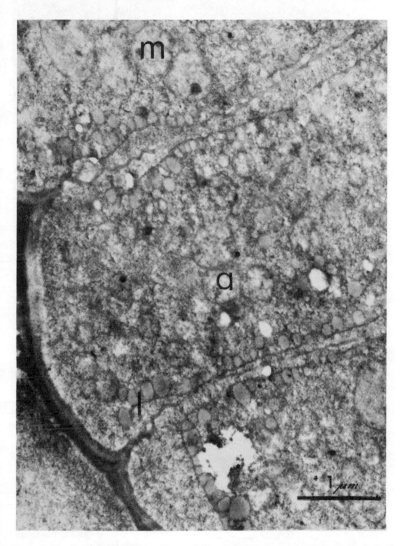

Figure 7.21 By 24 h of imbibition the non-viable material is almost completely autolysed: lipid bodies (l) are still lined up along broken pieces of plasmalemma but few other organelles are recognisable. a—was probably a pro-amyloplast, m—a mitochondrion

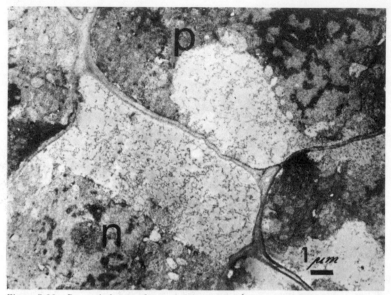

*Figure 7.22   Pro-cortical area of an unimbibed wheat (*Triticum dicoccum *var. Emmer) embryo from Thebes, Egypt (dated 3000–2000* B.C.*). Note that as in recent non-viable material the plasmalemma (p) has pulled away from the cell walls. Nuclei (n) with nucleoli and aggregated chromatin are still recognisable*

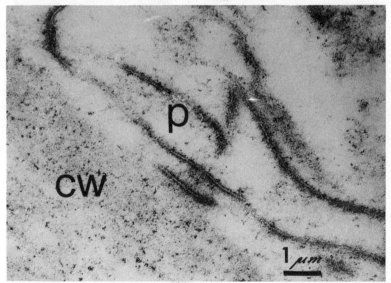

Figure 7.23   Same material as for Figure 7.22. Crumpled plasmalemma (p) between the cell wall (cw) and degraded cytoplasm

embryo are rapidly accelerated on wetting and by 24 h of imbibition the process is almost complete. In the last stages lipid bodies, still close to the broken plasmalemma, are the only clearly recognisable organelles, all others being completely degraded (*Figure 7.21*).

To carry a study of non-viable grains to its logical extreme, we investigated the fine structure of wheat grains from Thebes in Egypt dated at 3000–2000 B.C. from archaeological evidence, and others from the grain silos at Fayum, ^{14}C dated at 4441 ± 180 B.C. (These were kindly made available by the Ashmolean Museum, Oxford.) They showed an amazing amount of fine structure still preserved (*Figures 7.22* and *7.23*). Lipid bodies, nuclei, nucleoli, chromatin and occasionally plasmalemma and bodies suggesting mitochondria were all recognisable. The fine structure of the ancient Egyptian wheat (*Triticum dicoccum* var. Emmer) suggests that changes taking place in an unimbibed condition over a long period of time are similar to those in recent non-viable material where the degradative processes have been accelerated by wetting (see *Figures 7.20* and *7.22*). At 6 h of imbibition, in both recent non-viable rye and in ancient Egyptian wheat, the plasmalemma has pulled away from the cell wall and the space between contains a granular deposit.

However, the history of the Egyptian wheat grains both before storage in tombs or silos by the Egyptians and afterwards must always remain a mystery, and the problem of whether the destruction or breakdown of cellular fine structure is due to preparative treatment before storage or due to time itself must remain an insoluble problem.

CONCLUSIONS

Hydration of the cells of viable grains initiates a rapid return to biochemical and structural complexity. In the case of rye, the early stages of differentiation appear to be a reversal of changes occurring during the organised dehydration and simplification of late embryogenesis. The molecular and structural stability developed within the mature, unimbibed grain is retained throughout the period of viability. At both the biochemical and the structural level it has been shown that the unimbibed rye embryo contains all the necessary components for protein synthesis, whereas lesions exist at the molecular level which prevent this process in non-viable material (Hallam, 1972; Hallam, Roberts and Osborne, 1972b; Roberts, Payne and Osborne, 1972). After 2 h of imbibition active protein synthesis is already measurable in viable material (Hallam, Roberts and Osborne, 1972a), although no dramatic structural change has taken place. Protein synthesis presumably takes place either on polysomes within the dense packing of ribosomes at the early stages, or on the hydrated crescents of rough endoplasmic reticulum associated with electron lucent bodies. It seems that biochemical lesions already present in non-viable unimbibed grains manifest themselves as a breakdown in the molecular integrity of semi-permeable membrane systems within the cells. As non-viable embryos are unable to synthesise proteins it would follow that the slow turnover of compounds responsible for maintenance and repair of membrane systems ceases during long-term dry storage when the embryo loses its viability. From this time onwards the entire cellular system becomes 'leaky' and on wetting the lack of compartmental-isation of enzymes and substrates results in a random mixing and a rapid obliteration of fine structure.

REFERENCES

BAIN, J. M. and MERCER, F. V. (1966). 'Subcellular organisation of the developing cotyledons of *Pisum sativum* L.', *Aust. J. biol. Sci.*, **19**, 45–67

BERJAK, P. (1968). 'A lysosome-like organelle in the root tip of *Zea mays*', *J. Ultrastruc. Res.*, **27**, 233–243

BERJAK, P. and VILLIERS, T. A. (1970). 'Ageing in plant embryos. I. Establishment of the sequence of development and senescence in the root cap during germination', *New Phytol.*, **69**, 929–938

BUTTROSE, M. S. and SWIFT, J. C. (1972). 'Freeze etch studies of mature non-germinated seeds', *Aust. Conf. on Electron Microscopy* (abst.) 81

FREY-WYSSLING, A., GRIESHABER, E. and MUELLETHALER, K. (1963). 'Origins of spherosomes in plant cells', *J. Ultrastruc. Res.*, **8**, 506–516

HALLAM, N. D. (1972). 'Embryogenesis and germination in rye (*Secale cereale* L.) (i) Fine structure of the developing embryo', *Planta*, **104**, 157–166

HALLAM, N. D., ROBERTS, B. E. and OSBORNE, D. J. (1972a). 'Embryogenesis and germination in Rye (*Secale cereale* L.) (ii) Fine structure and biochemistry of viable embryos', *Planta*, **105**, 293–309

HALLAM, N. D., ROBERTS, B. E. and OSBORNE, D. J. (1972b). 'Embryogenesis and germination in Rye (*Secale cereale* L.) (iii) Fine structure and biochemistry of non-viable embryos', *Planta* (in press)

JENSEN, W. A. (1965). 'The ultrastructure and composition of the egg and central cell of cotton', *Am. J. Bot.*, **52**, 781–797

JENSEN, W. A. and FISHER, D. B. (1967). 'Cotton embryogenesis: double fertilisation', *Phytomorph.*, **17**, 261–269

MARINOS, N. G. (1970). 'Embryogenesis of the Pea (*Pisum sativum*). I. The cytological environment of the developing embryo', *Protoplasma*, **70**, 261–279

MOLLENHAUER, H. H. and TOTTEN, C. (1971). 'Studies on seeds. II. The origin and degradation of lipid vesicles in pea and bean cotyledons', *J. Cell Biol.*, **48**, 395–405

PERNER, E. (1965). 'Electronenmikroskopische Untersuchungen an Zellen von Embryonen im Zustand vulliger Samenruhe. I. Mitteilung: Die zellulare Strukturordnung in der Radicula lufttrockener Samen von *Pisum sativum*', *Planta*, **65**, 334–357

ROBERTS, B. E., PAYNE, P. I. and OSBORNE, D. J. (1972). 'Protein synthesis and viability in Rye grain', *Biochem. J.* (in press)

SCHULTZ, R. and JENSEN, W. A. (1968). '*Capsella* and embryogenesis: the early embryo', *J. Ultrastruc. Res.*, 22, 376–392
YOO, B. Y. (1970). 'Ultrastructural changes in cells of pea embryos and radicles during germination', *J. Cell Biol.*, 45, 158–171

DISCUSSIONS

Moore: How do you extract the rye embryos intact?

B. Roberts: We use Johnston and Stern's elegant technique of extracting the embryos, which relies on the plane of weakness along the scutellum. The grains are cooled and mixed with small fragments of solid CO_2 and sheared in a blendor to split off the embryos. These are gravity-separated by floating them in cyclohexane and carbon tetrachloride. Germination of the embryos is, if anything, improved by this treatment.

Street: Has the loss of viability only one cause or many?

B. Roberts: Protein synthesis is activated very early in germination and is therefore a good index of viability, but the lesions in protein synthesis may well not be the first to occur. The first visible changes are in the mitochondria and are reflections of changes in membrane permeability. Varner and others have shown that protein synthesis occurs in dormant grains and that it is necessary for the maintenance of components within embryos; the mitochondria supply the energy for this protein synthesis. Imbibition of water is identical in viable and non-viable seeds for the first 10 min but after 60 min it is an active process only in the viable ones.

Perry: It has been claimed that membrane degradation is a result of lipid autoxidation.

Harrington: Koostra, who worked with me, has shown that in cucurbit seeds, aged for five years, breakdown of lipids—presumably by oxidation—occurs.

B. Roberts: It is dangerous to pin-point any one process as the cause.

Hallam: In wheat grains from 2000 B.C. lipid bodies are still recognisable. Unimbibed non-viable grains show cellular membranes clearly but 15 min after the beginning of imbition these start to disintegrate.

Moore: So-called non-viable seeds are often only dead in part.

Osborne: Yes, but although parts may still have the complete capacity for protein synthesis the nett result is either germination or not. However, it is true that the viable portion of a population gradually deteriorates as more individuals die. In the non-viable seeds which we used in the work described, no protein synthesis at all occurred.

Hanson: Can you trace the deterioration of the seeds before they become non-germinative?

B. Roberts: It would be interesting to investigate a population throughout the period when its germinability is still 100 per cent but its vigour is deteriorating. However, more sensitive criteria than protein synthesis would be necessary for this, e.g. membrane permeability.

Osborne: One could cut embryos in half—one half for germination and one half for protein synthesis experiments, but quantitative interpretation would be difficult.

Maguire: Do the embryo and the endosperm deteriorate simultaneously?

Hallam: Deterioration of the embryo becomes apparent earlier than that of the endosperm and of the aleurone layer, which are thick-walled and more intact even in ancient material. However, fungi attack the endosperm. The cell structure of the endosperm and aleurone is disorganised in 100-year-old material.

B. Roberts: Incorporation experiments have shown that viable aleurone layers can occur together with non-viable embryos.

Thurston: Dormant seeds are capable of surviving in soil despite the soil microflora, but dead seeds disappear within a short time. Evidently the micro-organisms can distinguish between the viable and non-viable seeds. Could this be used to separate out the viable fraction of a seed population?

B. Roberts: Unfortunately, rye seeds are non-dormant.

E. H. Roberts: In Italy, transplantation experiments (viable embryos into non-viable endosperms and vice versa) have shown that the embryo is the essential component in the loss of viability.

Harrington: Have not ionising radiations been shown to cause deterioration of the endosperm as well as of the embryo?

E. H. Roberts: Yes, it has been reported that ageing by ionising radiation causes both.

Moore: Tetrazolium (TTC) tests have shown that seeds age from the tip of the root upwards and the tip of the shoot downwards.

Osborne: A combination of TTC tests with an autoradiographic analysis of results of protein synthesis experiments can pin-point those parts of seeds which are alive and those which are dead.

B. Roberts: TTC tests on our material have been published by MacKay and Flood in the 1968 Report of the National Institute of Agricultural Botany.

Street: At what stage do polysomes appear?

Hallam: Within one week after anthesis the number of ribosomes is so high that polysomes are not visible under the electron microscope. It is not before the 18th hour of imbibition that polysomes appear.

Street: Can protein synthesis take place at extremely low water contents?

Hallam: Buttrose has found that within 8 s of imbibition the water content jumps from 7 to 26 per cent.

B. Roberts: Marcus, in Philadelphia, has found protein synthesis 20 min after the beginning of imbibition, and Varner has shown that even in dry embryos (10–13 per cent H_2O) incorporation of $^{14}CO_2$ is possible.

Osborne: Like hydration, protein synthesis proceeds in succession from the outside inwards.

Street: How do the nucleic acids fit into this sequence of loss of viability?

B. Roberts: We have found an increase in nuclease, but we have not yet found out whether messenger RNA decreases with loss of viability.

8

ENDOGENOUS HORMONES IN THE CONTROL OF SEED DORMANCY

P. F. WAREING, J. VAN STADEN and D. P. WEBB

Department of Botany and Microbiology,
University College of Wales, Aberystwyth

INTRODUCTION

Since there is a mass of evidence that endogenous hormones are intimately involved in the regulation of many aspects of plant growth and development, it would seem likely on *a priori* grounds that they are also involved in the control of seed dormancy. Moreover, it has been known for a considerable time that application of exogenous growth substances, such as gibberellic acid (GA_3) and kinetin, will overcome dormancy in many species of seed showing various types of dormancy (Lang, 1965). In this paper the present state of knowledge regarding the role of hormones in seed dormancy will be reviewed briefly and some new evidence for the involvement of endogenous cytokinins will be presented, especially for certain seeds with chilling or light requirements.

GROWTH PROMOTERS AND CHILLING EFFECTS IN SEEDS

In some species of seed with a chilling requirement only the intact seed is dormant and the embryo itself will germinate if the coat is damaged or removed; in such cases we may speak of *coat-imposed dormancy*. In other species the isolated embryo is dormant, indicating that the cause of dormancy must lie within the embryo itself. However, the distinction between coat-imposed and embryo dormancy is not clear cut, since a number of instances are known in which seeds with dormant embryos pass through a stage of coat-imposed dormancy as the chilling period progresses (Wareing and Saunders, 1971).

A number of studies have been carried out on changes in endogenous gibberellin levels in response to chilling treatment and slightly higher amounts of gibberellins have been shown to be present at the end of the chilling period (e.g., Frankland and Wareing, 1966; Kentzer, 1966). Moreover, Ross and Bradbeer (1968) have shown that if, at the end of the chilling treatment, hazel seeds are transferred to warm conditions for a few days much higher gibberellin levels can be demonstrated. On the basis of this observation they have argued that the primary effect of chilling is to remove a block to gibberellin biosynthesis, and that it is the increase in gibberellin levels which occurs after transfer to warm conditions which is the immediate stimulus to germination. However, other studies on the changes in gibberellin levels in apple seeds at different stages of the chilling treatment have shown that gibberellin levels reach a peak after 30 days and then decline sharply, so that after 60 days of chilling there is no difference in gibberellin levels between chilled and unchilled seeds (Sinska and Lewak, 1970).

There appear to have been no previous studies on changes in endogenous cytokinin levels in seeds in response to chilling, but studies have recently been carried out on the seed of *Acer saccharum* (van Staden and Wareing, 1972) which shows embryo dormancy and requires chilling for 6 weeks (Webb and Dumbroff, 1969). Imbibed fruits were maintained at 5°C or 20°C for 0, 20, 40 and 50 days. They were then extracted and partitioned into fractions soluble in petroleum ether, ethyl acetate, *n*-butanol or water. The extracts were further partitioned by paper chromatography (using isopropanol:ammonia:water (10:1:1) as solvent) and the eluates from the chromatograms were assayed for cytokinin activity using the soya bean callus test (Miller, 1968). The seeds maintained at 20°C for various periods did not germinate and no cytokinin activity could be detected in their extracts. By contrast, seed subjected to periods of chilling showed a steadily increasing percentage germination, reaching 65 per cent after 50 days of chilling. Cytokinin activity could be detected in all three fractions, but most was present in the butanol fraction. This cytokinin activity reached a peak after 20 days of chilling; further extension of the chilling period resulted in a decrease in cytokinins and no activity could be detected in germinated seeds. The chromatographic properties of the active butanol-soluble cytokinin fraction, after further purification by column chromatography on Sephadex LH-20, were found to correspond to those of zeatin.

Parallel studies were made on the changes in gibberellin and abscisic acid (ABA) levels occurring in *A. saccharum* seeds subjected to the same pretreatment periods at 5°C and 20°C (Webb, van Staden and Wareing, 1972). Gibberellins and ABA were extracted and partitioned into 'acidic' and 'basic' ethyl acetate soluble and aqueous fractions. After paper chromatography, gibberellin activity was assayed using the lettuce hypocotyl test (Frankland and Wareing, 1960), and ABA was determined by gas chromatography, using the method of Lenton, Perry and Saunders (1971).

Gibberellin activity was detected in both the acidic and the basic ethyl acetate fractions. The gibberellin activity in the acidic fraction increased during pretreatment at 5°C, reaching a peak after 40 days and then declining to a low level after 50 days of chilling. The corresponding changes in the seed maintained at 20°C (and which failed to germinate) were very much less pronounced. There was a rapid reduction in endogenous ABA levels during chilling, but a much smaller reduction at 20°C.

Diagrammatic presentation of the data illustrates the sequence of changes in growth substances during the chilling period (*Figure 8.1*);

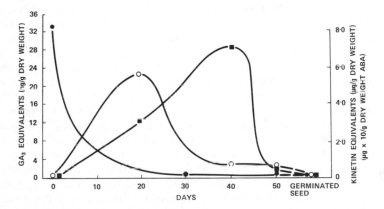

Figure 8.1   The effects of chilling at 5°C on the level of butanol-soluble cytokinins, acidic gibberellin-like substances and endogenous abscisic acid in seeds of Acer saccharum; ■—*acidic gibberellin-like substances measured by the lettuce hypocotyl bioassay;* ○—*butanol-soluble cytokinins measured by the soya bean callus bioassay;* ●—*endogenous abscisic acid measured by GLC (From: Webb, van Staden and Wareing, 1972)*

it is seen that cytokinin activity reached a peak at day 20 and acidic gibberellin activity at day 40, while ABA rapidly fell to a low level during the early part of the chilling period. Application of exogenous GA_3 and kinetin increases the germination of *A. saccharum* seeds and reduces the time required for chilling (Webb and Dumbroff, 1969). Thus, it is possible that the sequence of changes in the levels of growth promoters and ABA during chilling is an essential part of the dormancy-breaking effects of this treatment.

INHIBITORS AND DORMANCY OF SEEDS WITH A CHILLING REQUIREMENT

The occurrence of endogenous germination inhibitors was first postulated by Molisch (1922) to account for the fact that seeds of succulent fruits do not normally germinate in the maternal tissue. The idea that growth inhibitors may be involved in dormancy was first postulated by Hemberg (1949) for buds of potato and *Fraxinus excelsior*. Since there are close analogies between dormancy in buds of temperate woody plants and in seeds with a chilling requirement, it was logical to consider the possible role of inhibitors in such seeds.

The available evidence for the role of inhibitors in seed dormancy is far from conclusive at present (Wareing and Saunders, 1971). While substances which inhibit germination or the growth of *Avena* coleoptiles have frequently been reported from a range of seed species, the occurrence of such substances in seeds does not in itself necessarily imply that they play a functional role in seed dormancy under normal conditions. In order to establish such a role it is necessary to take into account various types of evidence, such as correlations between the levels of inhibitors and the state of dormancy (Wareing, 1965). Using this approach some evidence has been obtained for the role of endogenous inhibitors in seed of *Fraxinus excelsior* and *F. americana* (Kentzer, 1966; Sondheimer, Tzou and Galson, 1968), and of apple (Rudnicki, 1969), in which a decline of inhibitor has been demonstrated during chilling. In these latter cases ABA appears to be a major component of the inhibitory fraction which declines as the seeds emerge from dormancy, and it was seen above that there was a marked decrease in ABA levels in seed of *A. saccharum* during chilling (*Figure 8.1*). On the other hand, the levels of unidentified inhibitory substances did not change during chilling in other species (Frankland and Wareing, 1966).

In peach, it has been argued that the inhibitory effect of the testa can be attributed to the ABA which it contains (Lipe and Crane, 1966). However, in other cases the embryo itself contains inhibitory substances (Webb and Wareing, 1972a), and the effect of the coat appears to be due to the fact that it is impermeable to the inhibitor and hence prevents it from being leached out of the embryo. A good example of this latter situation appears to be provided by the seed of

Acer pseudoplatanus which we have investigated recently (Webb and Wareing, 1972b), and which shows coat-imposed dormancy. The effect of the coat cannot be attributed primarily either to interference with oxygen uptake or to mechanical constraint. On the other hand, pricking or cutting the coat results in germination only if the damaged area is in contact with moist filter paper. These observations can best be explained in terms of the leaching out of an inhibitor (which is not ABA) from the embryo, the level of which is reduced by leaching.

If endogenous inhibitors do, indeed, play a role in the control of seed dormancy, then it would seem likely that they interact with growth promoters, such as gibberellins and cytokinins, in such control (Villiers and Wareing, 1960; Amen, 1968). The effects of both gibberellins and cytokinins are antagonised by ABA in a number of seeds; on the other hand, the effects of ABA can be reversed by cytokinins in some species (Khan and Waters, 1969; Sankhla and Sankhla, 1968), but seldom by GA$_3$. This latter observation has led Khan and Waters (1969) to postulate that endogenous inhibitors, such as ABA, block the gibberellin-mediated germination response, and that cytokinins antagonise the action of the inhibitors, thereby allowing gibberellins to function (see also Khan, 1971).

HORMONES AND LIGHT-REQUIRING SEEDS

It is well known that GA$_3$ and kinetin can replace the light required for germination in a wide range of seed species. Ikuma and Thimann (1963) suggested that the mode of action of these two types of growth substance may be different, GA$_3$ stimulating growth of the radicles and kinetin growth of the cotyledons. Previous studies on changes in endogenous hormones following exposure to red light have not given very decisive results (Wareing and Saunders, 1971). At various periods after exposure increases in both gibberellins (Kohler, 1966; S. B. Taylorson, unpublished) and cytokinins (Barzilai and Mayer, 1964) have been reported, but the changes were so long after exposure to red light that it seemed likely that they were part of the many secondary changes going on during germination. However, very rapid changes in the levels of endogenous cytokinins in seed of *Rumex obtusifolius* have recently been reported following exposure to red light (van Staden and Wareing, 1972). Immediately after exposure of the seeds to 10 min red light (which in other treatments was followed by far red) they were ground in 80 per cent methanol under low intensity green light and left to extract overnight at 4°C. The extracts were then partitioned into butanol-soluble and aqueous fractions. The various fractions (10 g dry weight equivalent)

150

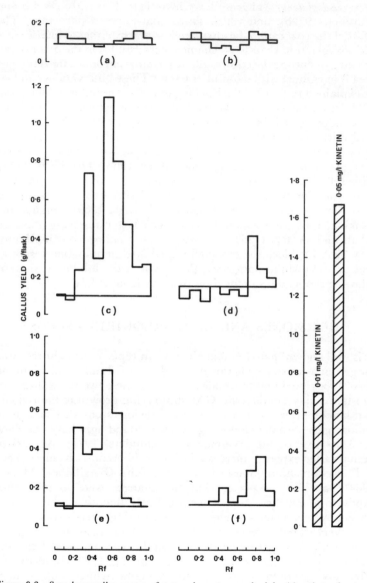

Figure 8.2 Soya bean callus assays of paper chromatograms loaded with n-*butanol extracts from* Rumex *seed. Cultures were grown for 28 days (From: van Staden and Wareing, 1972)*

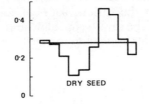

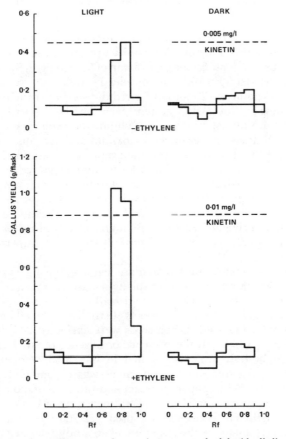

Figure 8.3   Soya bean callus assays of paper chromatograms loaded with alkaline phosphatase treated aqueous extracts from Spergula arvensis *seed (From: van Staden, Olatoye and Hall, 1972)*

were loaded on paper chromatograms (Whatman No. 3MM paper) and developed in water-saturated secondary butanol (WSB). The cytokinin activity in the various zones of the chromatograms was assayed by means of the soya bean callus test.

The light treatments applied to the seeds were as follows:

(a) dark 2 days
(b) dark 4 days
(c) dark 2 days + 10 min red light
(d) dark 2 days + 10 min red light + 2 days dark
(e) dark 2 days + 10 min red light + 5 min far-red light
(f) dark 2 days + 10 min red light + 20 min far-red light

The results (*Figure 8.2*) showed that after exposure to red light for 10 min high levels of cytokinin activity are present in both butanol and aqueous fractions. Exposure to 5 min of far-red light, immediately following the red light did not completely nullify the results of the latter, but exposure to 20 min of far red did so. In germinating seeds only low levels of cytokinin activity were observed. Further purification of the butanol fraction on columns of Sephadex LH-20 gel yielded two fractions showing cytokinin activity, one of which corresponded in its chromatographic properties to zeatin.

The rapid increase in cytokinin levels after exposure to red light suggests that cytokinins may be released from a bound form, and it is tempting to suggest that this may be one of the primary events following exposure to red light. The fact that application of kinetin, benzylaminopurine and zeatin were unable to substitute for the light requirement of *Rumex* seed could possibly be due to impermeability of the seed coats to these substances.

Increases in endogenous cytokinin levels following exposure to red light have also been found in the light-requiring seeds of *Spergula arvensis*. As described by Olatoye and Hall in Chapter 13, there is a marked synergism between the effects of ethylene and red light in the germination of this seed. When seeds were extracted 24 h following exposure to red light, in the presence or absence of ethylene, it was found that both treatments resulted in increased cytokinin levels (*Figure 8.3*). (5 g dry weight equivalent was chromatographed on Whatman No. 3MM paper in water-saturated secondary butanol. Cultures were grown for 24 days. Treatments are as indicated in the figure. The ethylene concentration used was 200 μl/l.) Although no attempts were made in this experiment to study whether there are any rapid effects of red light, the increased cytokinin levels observed preceded any visible signs of germination, and hence in this seed also the hormone changes may be among the essential processes leading to germination.

The role of inhibitors in the dormancy of light-requiring seeds has not been clearly demonstrated, although there is evidence for the involvement of an inhibitor present in the pericarp of the achenes of *Betula pubescens* (Black and Wareing, 1959), which requires light for germination.

CONCLUSIONS

Evidence that the action of chilling treatments on dormant seeds involves changes in endogenous gibberellins, and possibly of inhibitors, has been available for some time. The evidence presented here suggests that changes in endogenous cytokinin levels may also be involved both in seeds with a chilling requirement and in light-sensitive seeds. In the seeds of *Acer saccharum* changes in both gibberellins and cytokinins were observed during the chilling period. It would seem quite probable, however, that in some seeds dormancy-breaking treatments may affect primarily gibberellin levels and in others cytokinin levels. This suggestion is consistent with the fact that in some seeds dormancy can be overcome by application of gibberellins and in others of cytokinins. Further studies are necessary to determine how general are the endogenous cytokinin changes reported here.

Hitherto it has not been possible to demonstrate the direct hormonal control of germination in light-promoted seeds, but the rapid changes in the levels of cytokinins in *Rumex* seeds following exposure to red light suggest that hormone levels may be under rather direct phytochrome control in this species. Further work is necessary to determine how widespread is this effect in other light-requiring seed species.

The position regarding the role of endogenous inhibitors, such as ABA, in seed dormancy is still far from clear. There is increasing evidence for the possible role of ethylene in seed dormancy and germination, especially under soil conditions leading to the generation of this gaseous 'hormone'.

REFERENCES

AMEN, R. D. (1968). 'A model of seed dormancy', *Bot. Rev.*, **34**, 1–30
BARZILAI, E. and MAYER, A. M. (1964). 'Kinins in germinating lettuce seed', *Aust. J. biol. Sci.*, **17**, 798–800
BLACK, M. and WAREING, P. F. (1959). 'The role of germination inhibitors and oxygen in the dormancy of the light-sensitive seed of *Betula* spp.', *J. exp. Bot.*, **10**, 134–145
FRANKLAND, B. and WAREING, P. F. (1960). 'Effect of gibberellic acid on hypocotyl growth of lettuce seedlings', *Nature*, **185**, 255–256

154   SEED ECOLOGY

FRANKLAND, B. and WAREING, P. F. (1966). 'Hormonal regulation of seed dormancy in hazel (*Corylus avellana* L.) and beech (*Fagus sylvatica* L.)', *J. exp. Bot.*, **17**, 596–611

HEMBERG, T. (1949). 'Significance of growth-inhibitory substances and auxins for the rest period of the potato tuber', *Physiologia. Pl.*, **2**, 24–36

IKUMA, H. and THIMANN, K. V. (1963). 'The role of the seed coats in germination of photosensitive lettuce seeds', *Pl. Cell Physiol.*, **4**, 169–185

KENTZER, T. (1966). 'The dynamics of gibberellin-like and growth-inhibitory substances during seed development of *Fraxinus excelsior* L.', *Acta Soc. Bot. Pol.*, **35**, 477–484

KHAN, A. A. (1971). 'Cytokinins: Permissive role in seed germination?' *Science*, **171**, 853–859

KHAN, A. A. and WATERS, E. C. (1969). 'The hormonal control of post harvest dormancy and germination in barley seeds', *Life Sci.*, **8**, 729–736

KOHLER, D. (1966). 'Veränderung des Gibberellingehaltes von Salatsamen nach Belichtung', *Planta*, **70**, 42–45

LANG, A. (1965). 'Effects of some internal and external conditions on seed germination', *Encycl. Plant Physiol.*, **15**, No. 2, 848–893

LENTON, J. R., PERRY, V. M. and SAUNDERS, P. F. (1971). 'The identification and quantitative analysis of abscisic acid in plant extracts by gas-liquid chromatography', *Planta*, **96**, 271–280

LIPE, W. N. and CRANE, J. C. (1966). 'Dormancy regulation in peach seeds', *Science*, **153**, 541–542

MILLER, C. O. (1968). 'Naturally occurring cytokinins', in *Biochemistry and Physiology of Plant Growth Substances*, Runge Press, Ottawa

MOLISCH, H. (1922). *Pflanzenphysiologie als Theorie der Gärtnerei* 5th edn, Jena

ROSS, J. D. and BRADBEER, J. W. (1968). 'Effect of chilling on gibberellin concentrations in hazel seeds', *Nature*, **220**, 85

RUDNICKI, R. (1969). 'Studies on abscisic acid in apple seeds', *Planta*, **86**, 63–68

SANKHLA, N. and SANKHLA, D. (1968). 'Reversal of (+)-Abscisin II induced inhibition of lettuce seed germination and growth by kinetin', *Physiologia Pl.*, **21**, 190–195

SINSKA, I. and LEWAK, S. (1970). 'Apple seed gibberellins', *Physiol. Veg.*, **8**, 661–667

SONDHEIMER, E., TZOU, D. S. and GALSON, E. C. (1968). 'Abscisic acid levels and seed dormancy', *Pl. Physiol.*, **43**, 1443–1447

VAN STADEN, J. and WAREING, P. F. (1972). 'The effect of light on endogenous cytokinin levels in seeds of *Rumex obtusifolius*', *Planta*, **104**, 126–133

VAN STADEN, J., WEBB, D. P. and WAREING, P. F. (1972). 'The effect of stratification on endogenous cytokinin levels in seeds of *Acer saccharum*', *Planta*, **104**, 110–114

VILLIERS, T. A. and WAREING, P. F. (1960). 'Interaction of growth inhibitor and a natural germination stimulator in the dormancy of *Fraxinus excelsior* L.', *Nature*, **185**, 112–114

WAREING, P. F. (1965). 'Endogenous inhibitors in seed germination and dormancy', *Encycl. Plant Physiol.*, **15** No. 2, 909–924

WAREING, P. F. and SAUNDERS, P. F. (1971). 'Hormones and dormancy', *Ann. Rev. Plant Physiol.*, **22**, 261–288

WEBB, D. P. and DUMBROFF, E. B. (1969). 'Factors influencing the stratification process in seeds of *Acer saccharum*', *Can. J. Bot.*, **47**, 1555–1563

WEBB, D. P., VAN STADEN, J. and WAREING, P. F. (1972). 'Seed dormancy in *Acer*: Changes in endogenous cytokinins, gibberellins and germination inhibitors during the breaking of dormancy in *Acer saccharum* Marsh', *J. exp. Bot.*, (in press)

WEBB, D. P. and WAREING, P. F. (1972a). 'Seed dormancy in *Acer*: Endogenous inhibitors and dormancy in *Acer pseudoplatanus*', *Planta*, **104**, 115–125

WEBB, D. P. and WAREING, P. F. (1972b). 'Seed dormancy in *Acer pseudoplatanus*: The role of the covering structures', *J. exp. Bot.*, **23**, 813–829

DISCUSSIONS

Thompson: Professor Wareing has largely explained why we find that GA_3 is not simply a substitute for chilling: GA_3 treated seeds will germinate over a wide range of conditions, whereas chilled seeds retain their specificity.

Gulliver: May not the difference in the effect of chilling and of applied GA_3 on endogenous GA levels be that in the latter case the level stays up?

Wareing: Gibberellin treatment shortens the whole cycle and may stimulate germination and rapid root growth ahead of other changes. But we have no firm information.

Côme: A high concentration of GA_3 stimulates the breaking of dormancy of apple seeds by cold but low concentrations inhibit the effect of cold temperatures; this inhibition cannot be reversed by kinetin.

Wareing: I cannot explain this.

Côme: Apple seeds sown standing on edge at 20°C remain dormant, but at 15°C they will germinate. This suggests an inhibitor which is more inhibiting at a high than at a low temperature.

Wareing: Lettuce seeds too are more inhibited by **coumarin** at high than at low temperatures. But the cause of this is not known.

Osborne: The very rapid changes in cytokinin level in response to red and far-red light suggest enzymic changes. Are the kinetics of the rise and fall temperature dependent?

Wareing: We still need to repeat the experiment at different temperatures.

Smith: If 10 min red light is followed by 20 min far-red light the level of cytokinin falls; does it stay up when red light is followed by 20 min darkness?

Wareing: In wheat leaves, the GA_3 level declines fairly rapidly after such treatment; but we have no information on cytokinins.

Smith: It would be interesting to know whether the mechanism is based on red–far-red reversibility or on initiation by red and removal by far red.

Villiers: Dormant seeds are not inactive; for instance, many organelles are produced during dormancy. Imbibed Grand Rapids lettuce seeds kept in the dark are devoid of polysomes for weeks, but on exposure to red light for 5 min polysomes are produced within two hours.

Wareing: Similarly, Bradbeer has found an increase in membrane-bound ribosomes in hazel seeds during the chilling period.

9

INTERRELATED EFFECTS OF IMBIBITION, TEMPERATURE AND OXYGEN ON SEED GERMINATION

D. CÔME and T. TISSAOUI

Laboratoire de Physiologie des Organes Végétaux,
Centre National de la Recherche Scientifique, Meudon, France

SUMMARY

Provided the embryo is non-dormant very little oxygen is required in the environment of the seed for germination to take place. The lower the temperature, the less oxygen is needed. However, after imbibition, the embryo inevitably receives little oxygen since this gas can only reach it by dissolving in the water which the covering structures ('the coat') of the seed have imbibed. In addition, the coat sometimes contains phenolic constituents which fix part of the dissolved oxygen by oxidation and lower even more the quantity of oxygen available to the embryo. When the temperature rises, the oxygen requirement of the embryo increases but the quantity of oxygen available to it diminishes because oxygen becomes less soluble in the water of the coat, and the phenolic substances, when present, fix more. Germination can thereby be inhibited.

INTRODUCTION

Schematically, one can consider a seed as consisting of an embryo surrounded by covering structures ('a coat') of diverse origins. The embryo represents the essential aspect because it is the part which germinates, provided it is not dormant, can imbibe water, has enough oxygen at its disposal and is placed in a suitable temperature.

Dormancy is a peculiar physiological state which manifests itself by the inability of the imbibed embryo to germinate immediately, irrespective of the environment. The fundamental causes of this non-germination are even now entirely unknown but satisfactory hypotheses exist: embryo dormancy may be the result either of a blockage of phosphorus metabolism (Olney and Pollock, 1960; Bradbeer and Floyd, 1964) or of a repression of genetic material (Khan, 1966; Wood and Bradbeer, 1967). The breaking of dormancy which takes place most frequently through treatment of the moist embryos by cold (4°C or 5°C for several months) may correspond to a derepression of this genetic material, although the exact processes involved are not known.

Even when the embryo is not dormant, or is no longer dormant as a consequence of a suitable treatment, various obstacles to germination can still exist. In this case one speaks of *inhibition* of germination. The coat of the seed is often responsible for this inhibition.

Only the type of inhibition which is connected with insufficient oxygen supply to the embryo is dealt with here, and an analysis is made of how the imbibition of the seed and the temperature affect these phenomena. For this purpose some knowledge of the oxygen requirements of the embryo for germination is necessary.

INFLUENCE OF TEMPERATURE ON THE OXYGEN REQUIREMENTS OF THE EMBRYO

An embryo has germinated when its radicle has visibly elongated, and *germination is therefore a physiological phenomenon which precedes root growth*. With intact seeds, the emergence of the radicle from the covering structures shows that germination has taken place. Morinaga (1926a; 1926b) has shown that the seeds of certain water plants germinate in the absence of oxygen, but in the great majority of cases germination without external oxygen supply is impossible.

Various authors have shown that naked embryos require less oxygen to germinate than do those of intact seeds (Black and Wareing, 1959; Wellington and Durham, 1961). However, it has now been shown that the partial pressure of oxygen required for the germination of apple embryos, whose dormancy has been completely broken by adequate cold treatment, is a function of the ambient temperature (Côme and Tissaoui, 1968; Tissaoui, 1970).

EXPERIMENTAL METHODS

The experimental technique consisted in obtaining, through appropriate capillary tubes, gas mixtures of *exactly determined and constant composition* from compressed nitrogen and from an oxygen/ nitrogen mixture. The atmospheres thus obtained, *often very low in oxygen*, were passed through the germination chamber at a constant flow rate (for details see Côme and Tissaoui, 1968).

The embryos studied, of the cultivars Golden Delicious and Reinette du Mans, had been previously treated with cold for four to five months and were capable of germinating in air at 20°C. They were sown in the germination chambers at 20°C, 15°C and 4°C. After 42 days in these conditions, those seeds which had failed to germinate were placed in petri dishes in order to test whether they had kept their ability to germinate in air at 20°C.

RESULTS

Whatever the temperature, from 4°C to 20°C, the germination of the embryos requires the presence of oxygen in the ambient environment, but *the lower the temperature the less oxygen is necessary*. In *Figure 9.1*, at 20°C curve *N* shows embryos which on germinating produce normal seedlings; curve *T* shows all embryos capable of germinating, including those which produce abnormal seedlings. The shaded zone corresponds to embryos with abnormal germination. At 4°C traces of oxygen are sufficient to permit germination, whereas at 20°C germination is only good if at least 5 per cent oxygen is present. When the oxygen content is lower, a smaller number of embryos

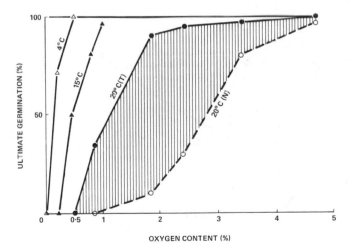

Figure 9.1   Variations in the percentage of non-dormant apple embryos capable of germinating at 4°C, 15°C and 20°C, as a function of the environment

160

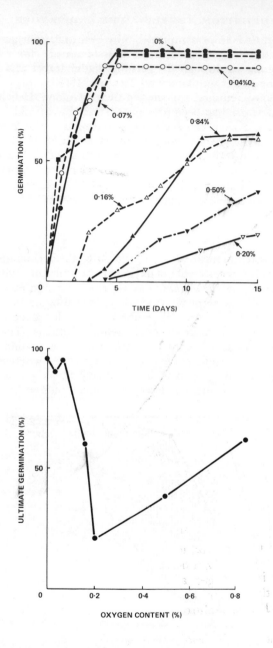

Figure 9.2 Variations with time in the germination percentage of apple embryos which failed to germinate at 20°C

germinate at this temperature, and some of them produce abnormal plants as if they had not yet fully lost their dormancy. The number of embryos which behave in this fashion is the larger and the abnormalities of the resulting plants are the greater, the lower the ambient atmosphere is in oxygen. No such abnormalities ever occur at 4°C or 15°C.

At 20°C embryos placed in an atmosphere totally devoid of oxygen or containing infinitesimal traces of oxygen (less than 0·1 per cent) retain the ability to germinate normally when they are returned to air. *Figure 9.2* shows variations with time in the germination percentage, in air, at 20°C of apple embryos which have failed to germinate at 20°C within 42 days in 0, 0·04, 0·07, 0·16, 0·20, 0·50 and 0·84 per cent of oxygen. It also shows variations of the germination of the same embryos after 15 days, as a function of the oxygen content of the environment in which they had been placed for 42 days before being sown in air.

By contrast, if oxygen is slightly more abundant but nevertheless insufficient to permit germination, the embryos enter into secondary dormancy: they progressively lose their ability to germinate in air. This dormancy becomes most intense in about 0·2 per cent oxygen (*Figure 9.2*). At temperatures which are so low (e.g. 4°C) that they are capable of breaking dormancy, this phenomenon does not occur.

CONCLUSIONS

In order to germinate, non-dormant embryos require more oxygen in the ambient atmosphere the higher the temperature. This phenomenon can be explained as follows. At relatively low temperatures the respiratory metabolism of the embryos is weak, and little oxygen is therefore required to safeguard this degree of metabolism and permit germination to take place. At a higher temperature, the potentially higher metabolic rate of the embryo requires more oxygen. Unless a sufficiently large quantity of this gas is supplied normal germination processes cannot be completed.

It is also likely that at any given oxygen pressure more oxygen penetrates into the embryo when the temperature is lower because the gas is then more soluble. It is therefore understandable that, at a sufficiently low temperature, enough oxygen penetrates into the embryo to permit it to metabolise normally and therefore to germinate. At a higher temperature the embryonic tissues are insufficiently oxygenated to satisfy their metabolic requirements.

All this evidence combines to suggest that germination is a phenomenon which differs from mere growth of the radicle. The experiments described support this suggestion. Thus, though germination is facilitated at lower temperatures, the subsequent growth of the young plant is clearly much retarded under these conditions. Oxygen also plays a different role in the two phenomena: at a given temperature the embryo will germinate equally rapidly as long as the necessary minimum of oxygen is available to it, whereas the growth of the young plant is more rapid, the more abundant is its oxygen supply. In atmospheres very poor in oxygen the embryos may even germinate without any external manifestations of subsequent growth.

Temperature and oxygen also interact in the induction of secondary dormancy. At moderate temperatures, small quantities of oxygen are required for this phenomenon to occur, but at low temperatures no degree of depletion of atmospheric oxygen induces dormancy. Clearly (and *post hoc* obviously) low temperatures which break dormancy also prevent its imposition. Under normal conditions the embryo is surrounded by covering structures (the coat) and oxygen can only reach it by traversing these. But the imbibition of the seeds, which is indispensable for germination, and the temperature greatly influence the quantity of oxygen which traverses these integuments in unit time.

INFLUENCE OF IMBIBITION ON THE PERMEABILITY OF THE SEED-COVERING STRUCTURES TO OXYGEN

Oxygen impermeability of the coats of seeds has often been put forward as an explanation for the germination failure of seeds whose embryos are not dormant. But this mechanism has only been explained satisfactorily since it has been possible to measure the quantity of oxygen which effectively traverses the coats.

THE GENERAL CASE

Depending on its structure, the coat is more or less permeable to oxygen when it is dry (Côme, 1968a; 1968b; 1970), but the imbibition of a seed always results of necessity in a very poor oxygen supply to the embryo. Indeed, the imbibed coat forms a continuous wet layer around the embryo which the oxygen must traverse in solution (Côme, 1962). Since, however, oxygen is sparsely soluble in water, very little can reach the embryo. For any given temperature it is, in fact, the thickness of the wet layer, i.e. the thickness of the coat, which thus determines the rate of oxygen transfer to the embryo.

This phenomenon explains why many seeds do not germinate when they are completely immersed in water, except when the quantity of oxygen dissolved is increased, either through enriching the outer atmosphere (Morinaga, 1926b; Côme, 1962) or by bubbling air or pure oxygen through the water (unpublished work by Madame Thévenot with apple seed embryos). In this connection Heydecker and Orphanos (1968) state that convection currents can transport larger quantities of oxygen than would move by diffusion alone, and that this may favour germination provided there is no barrier which prevents such currents from reaching the embryo.

If other phenomena do not interfere, all the oxygen which dissolves in the imbibed coat is capable of reaching the embryo and may suffice to ensure germination. However, in addition to this purely physical limitation of oxygen transport to the embryo, other mechanisms may interfere which can further reduce this transport. Apple seeds are an example.

THE SPECIAL CASE OF THE APPLE SEED

The integuments of apple seeds are rich in phenolic components (mainly phloridzin, chlorogenic acid and para-coumaryl-quinic acid) which can easily be oxidised when the coats which contain them are moist (Côme, 1967). In the imbibed seeds the embryo is therefore surrounded no longer by a layer of water but by a layer of dissolved phenolic compounds. These then fix at least part of the dissolved oxygen by oxidation. The initially dissolved oxygen is then no longer available to the embryo in its entirety, and the further depletion of oxygen can impair its germination.

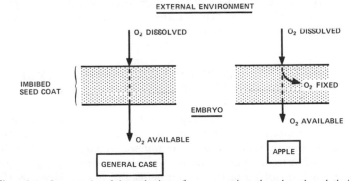

Figure 9.3   Interpretation of the mechanisms of oxygen supply to the embryo through the imbibed coat, for seeds in general and for apple seeds. The length of the arrows represents the volume of oxygen flowing

CONCLUSIONS

Figure 9.3 schematically represents the oxygen supply to the embryo across the imbibed seed coat. In all cases a seed can only germinate if the very small quantities of oxygen which reach the embryo in

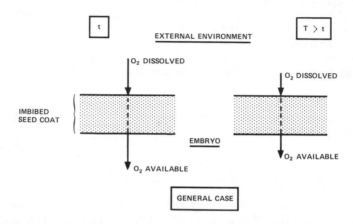

Figure 9.4   Interpretation of the influence of temperature (t, T) on the quantity of oxygen which reaches the embryo through the imbibed seed coat, in general. The length of the arrows represents the volume of oxygen flowing

solution in the imbibed seed coat are sufficient. In addition, germination can be inhibited through the presence of phenolic compounds which form a chemical screen to the passage of oxygen to the embryo. Finally, the temperature greatly influences both the physical phenomenon—the solution of oxygen in water—and the chemical reaction —the oxidation of the phenolic components.

INFLUENCE OF TEMPERATURE ON THE OXYGEN PERMEABILITY OF THE SEED COAT

When the temperature rises, oxygen becomes less soluble in water and a smaller quantity reaches the embryo (*Figure 9.4*). Under the same conditions the oxidation of the phenolic compounds of apple seeds is more intense. The embryo then receives very little oxygen, because less is dissolved in the integuments and more is fixed by the phenolic compounds (*Figure 9.5*). In all cases, the higher the temperature the less oxygen is available to the embryo.

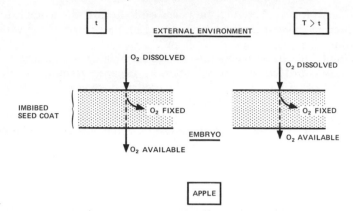

Figure 9.5   Interpretation of the influence of temperature (t, T) on the quantity of oxygen which reaches the embryo through the imbibed coats of apple seeds. The length of the arrows represents the volume of oxygen flowing

CONCLUSION

The imbibition of seeds always places the embryo in very critical conditions regarding oxygen supply. It is probably for this reason that various authors have suggested that a fermentation process accompanies the imbibition of seeds; the respiratory quotient is often high (Stiles, 1960; Brown, 1965), and alcohol (Doireau, 1969) or lactic acid (Phillips, 1947) are products recorded. These conditions of partial anaerobiosis are, however, not necessarily detrimental to germination if they do not persist too long; certain publications (Ballard, 1958; 1961; Ballard and Grant-Lipp, 1967; 1969; Grant-Lipp and Ballard, 1959; Tseng, 1964) even seem to show that such conditions can be stimulating.

The higher the temperature in which the seeds are placed to germinate, the smaller is the quantity of oxygen available to the embryo, whilst its need for oxygen increases. One can therefore understand why germination is often inhibited through relatively high temperatures (for example 25°C and 30°C). The seeds are even more sensitive to thermic conditions when their coats contain substances which, like the phenolic compounds of the apple seed, are capable of fixing dissolved oxygen on its way to the embryo.

From a practical point of view, one should frequently lower the temperature at which the seeds are placed to germinate. Germination may be slower under these cooler conditions but it will take place, and often more uniformly. As soon as the radicle has pierced the

G

seed coat, it may, however, be permissible to raise the temperature without inconvenience in order to speed up the growth of the young seedling.

Finally, it can be noted that excessively high temperatures may have awkward consequences, not only because germination does not take place but also because the seed may progressively lose its ability to germinate subsequently, even at a lower temperature: the oxygen supply falls to a critical low level and the embryo may then enter a state of secondary dormancy. This is the phenomenon known as 'thermodormancy'.

REFERENCES

BALLARD, L. A. T. (1958). 'Studies of dormancy in the seeds of subterranean clover. I, Breaking of dormancy by carbon dioxide and by activated carbon', *Aust. J. biol. Sci.*, 11, 246–260

BALLARD, L. A. T. (1961). 'Studies of dormancy in the seeds of subterranean clover (*Trifolium subterraneum* L.). II. The interaction of time, temperature and carbon dioxide during passage out of dormancy', *Aust. J. biol. Sci.*, 14, 173–186

BALLARD, L. A. T. and GRANT-LIPP, A. E. (1967). 'Seed dormancy. Breaking by un-couplers and inhibitors of oxidative phosphorylation', *Science*, 156, No. 3773, 398–399

BALLARD, L. A. T. and GRANT-LIPP, A. E. (1969). 'Studies of dormancy in the seeds of subterranean clover (*Trifolium subterraneum* L.). III. Dormancy breaking by low concentrations of oxygen', *Aust. J. biol. Sci.*, 22, 279–288

BLACK, M. and WAREING, P. F. (1959). 'The role of germination inhibitors and oxygen in the dormancy of light-sensitive seed of *Betula* spp.', *J. exp. Bot.*, 10, 134–145

BRADBEER, J. W. and FLOYD, V. M. (1964). 'Nucleotide synthesis in hazel seeds during after-ripening', *Nature*, 201, 99–100

BROWN, R. (1965). 'Physiology of seed germination', *Handb. PflPhysiol.*, 15, No. 2, 925–932

CÔME, D. (1962). 'Comment l'oxygène nécessaire à la germination des graines parvient-il à l'embryon? *Revue. gén. Bot.*, 69, 563–573

CÔME, D. (1967). 'L'inhibition de germination des graines de Pommier (*Pirus malus* L.) non dormantes. Rôle possible des phénols tégumentaires', *Ann. Sci. Natur., Bot. & Biol. Veg.*, VIII, 371–478

CÔME, D. (1968a). 'Perméabilité à l'oxygène et au gaz carbonique des enveloppes séminales de diverses espèces, *Pépiniéristes-Horticulteurs-Maraîchers*, No. 90

CÔME, D. (1968b). 'Relations entre l'oxygène et les phénomènes de dormance embryonnaire et d'inhibition tégumentaire', *Bull. Soc. fr. Physiol. vég.*, 14, No. 1, 31–45

CÔME, D. (1970). 'Les obstacles à la germination', *Monographies de Physiologie Végétale*, (ed. Masson and Cie), No. 6, 162

CÔME, D. and TISSAOUI, T. (1968). 'Induction d'une dormance embryonnaire secondaire chez le Pommier (*Pirus malus* L.) par des atmosphères très appauvries en oxygène', *C.r. hebd. Séanc. Acad. Sci.*, 266, 477–479

DOIREAU, P. (1969). Étude de la production d'alcool lors de la germination des différents types de semences', *C.r. hebd. Séanc. Acad. Sci.*, 268, 933–936

GRANT-LIPP, A. E. and BALLARD, L. A. T. (1959). 'The breaking of seed dormancy of some legumes by carbon dioxide', *Aust. J. agric. Res.*, 10, 495–499

HEYDECKER, W. and ORPHANOS, P. I. (1968). 'The effect of excess moisture on germination of *Spinacia oleracea* L.', *Planta*, 83, 237–247

KHAN, A. A. (1966). 'Breaking of dormancy in *Xanthium* seeds by kinetin mediated by light and DNA-dependent RNA synthesis', *Physiologia Pl.*, **19**, No. 4, 869–874

MORINAGA, T. (1926a). 'Germination of seeds under water', *Am. J. Bot.*, **13**, 126–140

MORINAGA, T. (1926b). 'The favorable effects of reduced oxygen supply upon the germination of certain seeds', *Am. J. Bot.*, **13**, 159–167

OLNEY, H. O. and POLLOCK, B. M. (1960). 'Studies of rest period. II. Nitrogen and phosphorus changes in embryonic organs of after-ripening cherry seed', *Pl. Physiol.*, **35**, 970–975

PHILLIPS, J. W. (1947). 'Studies on fermentation in rice and barley', *Am. J. Bot.*, **34**, 62–72

STILES, W. (1960). 'Respiration in seed germination and seedling development', *Handb. PflPhysiol.*, **12**, No. 2, 465–492

TISSAOUI, T. (1970). 'Importance respective de la température, de l'oxygène et de la position des embryons de Pommier (*Pirus malus* L.) dans la germination et l'induction d'une dormance secondaire', *C.r. hebd. Séanc. Acad. Sci.*, **271**, 308–310

TSENG, S. T. (1964). 'Breaking dormancy of rice seed with carbon dioxide', *C.r. Ass. intern. Essais Semences*, **29**, No. 3, 445–450

WELLINGTON, P. S. and DURHAM, V. M. (1961). 'Studies on the germination of cereals. IV. The oxygen requirement for germination of wheat grains during maturation', *Ann. Bot.*, **25**, No. 98, 185–196

WOOD, A. and BRADBEER, J. W. (1967). 'Studies in seed dormancy. II. The nucleic acid metabolism of the cotyledons of *Corylus avellana* L. seeds', *New Phytol.*, **66**, No. 1, 17–26

DISCUSSIONS

Wareing: Were the seeds for the oxygen experiments previously chilled to render the embryos non-dormant?

Côme: Yes, and we then have seeds with non-dormant embryos. But their germination is still inhibited as long as the oxygen diffusion through the seed coat is too limited. There are other complications: if apple seeds are extracted and chilled in moist sand the phenolic inhibitors are leached out and germination is then no longer inhibited by the seed coat; if they are chilled while still in the fruit, no such leaching occurs and then germination remains inhibited. For these reasons we normally use embryos without seed coats in our experiments on oxygen relations. We then find that non-dormant apple embryos take up the same quantity of oxygen during germination at 4°C and 25°C; in low oxygen, at temperatures of 20°C and above, it sometimes happens that germination occurs but subsequently the roots do not grow. Incidentally, dormancy can also be broken without cold, in five to eight days, by keeping moist embryos at 20°C in an atmosphere free of oxygen.

Gordon: Does a microflora develop on apple seeds, as it does on cereal grains?

Côme: No, the phenolics inhibit the development of fungi.

Osborne: Are these phenolics virtually free in solution or are they part of the structure of the cell wall?

Côme: This is not known.

Wareing: If the seeds are washed to leach out the phenolics, does this affect the germination of non-dormant but intact seeds?

Côme: It is a problem of intensity of reaction. Intact seeds germinate at sufficiently low temperatures; the lower the level of phenolics, the higher the temperature at which germination can occur.

Tissaoui: It is noteworthy that there is a clear difference between the most favourable temperatures for germination (low) and for growth (high).

Côme: It is worth noting that phenolics can also be destroyed within the cells by oxidants, by treatment with alcohols and by ultraviolet light, and germination is thereby increased.

Perry: Phenolic compounds in the dry fruit tissues are also thought to be responsible for the inhibition of the germination of sugar beet seeds.

Côme: Phenolics are never direct inhibitors of embryo germination; they always operate by restricting the oxygen supply.

Smith: Are the phenolics in the imbibed apple seed coats oxidised directly or do they act by stimulating an enzyme system?

Côme: They act via an enzymic process.

Heydecker: This may help to explain our experience that beetroot seed clusters placed in a concentrated extract from other, similar seed clusters, can show an enormous rise in oxygen uptake and yet fail to germinate.

Wareing: We have found that in *Xanthium pennsylvanicum* and sycamore (*Acer pseudoplatanus*) seeds, whose embryos are not dormant, the oxygen diffusion through the imbibed seed coat, though restricted, is at a higher rate than the embryo would need for germination. Yet, unless appropriately treated, they will not germinate because they contain an inhibitor.

E. H. Roberts: Barley grains provide indirect evidence that the covering structures are physiologically significant resistances to oxygen transport: (a) removal of the coat causes a substantial rise in gas exchange; (b) the RQ changes from higher than 2 to about 1; (c) an increase in oxygen pressure in the air surrounding an intact grain increases oxygen uptake. Since the oxygen uptake by seeds is chiefly through the cytochrome system which is saturated at low levels of oxygen, the seed coat appears to be a significant barrier.

Wareing: Agreed; still, in *Xanthium* spp. and sycamore seeds at least, enough oxygen gets through to permit germination, and oxygen therefore does not appear to be the limiting factor.

Côme: Air is present inside the dry seeds and in the coat. But much of this air is expelled during imbibition, and we do not know what quantity remains inside.

10

SEED DORMANCY AND SEED ENVIRONMENT— INTERNAL OXYGEN RELATIONSHIPS

MIRIAM M. EDWARDS

*Department of Physiology and Environmental Studies,
University of Nottingham School of Agriculture*

INTRODUCTION

The supply of oxygen to seeds is a function of the external gas concentration, the solubility of oxygen in water, the length and resistance of the diffusion path and the affinity of the enzymes in the seeds for molecular oxygen. Dry seeds have an internal atmosphere containing oxygen (5–7 per cent of the total seed volume in *Vicia faba* L.) in the cavities between the seed structures, in the intercellular spaces and possibly within the cells (Perner, 1965). The respiration rate of dry seeds is extremely low, and the permeability of the dry seed coat is sufficient to allow the internal atmosphere to equilibrate with the air outside. Water moves rapidly into the seed along a steep gradient in water potential from zero at the seed surface ($\psi = 0$) to the low water potential in the tissues ($\psi = < - 100$ J cm^{-3}) (Shaykewich and Williams, 1971). The permeability of the cell membranes may be temporarily increased by the movement of water molecules into them, but the resistance to diffusion of oxygen becomes much greater with the increase in seed water content, due to the increase in thickness of the water films at the cell surfaces and the length of the diffusion path in water relative to air.

Direct measurements have been made of the flow of oxygen through dry and wet seed coats, expressed in arbitrary units for several species, *Phaseolus vulgaris* L., *Pisum sativum* L., *Lactuca scariola* L. var. *sativa*, *Pyrus malus* L. (Côme, 1970), and in absolute units (cm^3 cm^{-2} s^{-1}) for *Cucurbita pepo* L. (Brown, 1940). Assuming linear flow through isolated membranes ($J = -D\,dc/dx$), given the thickness of the seed coat (x) and rate of flow (J), if the diffusion coefficient (D) is known, the oxygen concentration at the embryo surface (c) can be calculated and vice versa. These parameters have been worked out for the intact wet seed coat of charlock (*Sinapis arvensis* L.) (Edwards, 1969), without taking into account that oxygen is taken up along the diffusion path by respiring cells or by terminal oxidases in dead cell layers.

Penetration of oxygen from the embryo surface into the inner tissues has not been measured exactly (Edwards, 1969). Mathematical procedures for the diffusion analysis can be found for a small number of simple geometries with simple boundary conditions (Crank, 1956). These have been used in conjunction with direct experimentation to study the penetration of oxygen into individual cells or aggregates of cells of uniform composition (Goddard and Bonner, 1960), and into more complicated structures, for example, fungal systems (Griffin 1968), roots (Berry, 1949), storage organs (Burton, 1950) and soil crumbs (Currie, 1961; Greenwood and Goodman, 1967). Simplifying assumptions have been made in the diffusion analysis concerning the boundary conditions; the critical surface concentration is assumed to be constant, the diffusion path is restricted to flow along simple geometric co-ordinates, the value of the diffusion coefficient is assumed to be the same as in water and the rate of oxygen uptake is taken at the maximum value throughout the tissue. Modifications of the simplified form of the diffusion analysis have been made by Collis-George and Wallace (1968) and Smith and Griffin (1971) for transient systems with more detailed geometry for the study of small soil organisms. The imbibing seed may be regarded as a transient system, which consists of an aggregate of tissues each with its own limiting boundary conditions, dimensions and diffusion characteristics, and rate of oxygen uptake depending on oxidation processes.

In ideal experimental situations the gas atmosphere is brought to the seed surface, so that the external resistance is reduced to negligible proportions. In imbibing pea seeds (Yemm, 1965) oxygen is taken up increasingly rapidly during the first few hours due to increase in enzyme activity with increase in cell water content. After this, the rate is constant for a period of about 24 h until rupture of the testa by the developing radicle. The restriction to oxygen diffusion can largely be removed by artificial removal of the testa, or by supplying seeds with oxygen instead of air. Increasing oxygen supply also increases the germination of wheat (Durham and Wellington, 1961) and barley (Major and Roberts, 1968), and of naturally dormant seeds of *Xanthium* (Wareing and Foda, 1957), light sensitive seeds of *Lactuca sativa* L. (Ikuma and Thimann, 1963), light hard seeds of *Phacelia tanacetifolia* Benth. (Chen, 1970), and weed seeds, namely *Capsella bursa-pastoris* L. Medik., *Senecio vulgaris* L. (Popay and Roberts, 1970) and *Sinapis arvensis* L. (Edwards, 1969), but not the gemination of seeds of *Trifolium subterraneum* L. (Ballard and Grant-Lipp, 1969). The path of oxygen diffusion into the seed is complicated if the seed coat is perforated by an open micropyle. If the seed coat is artificially perforated to allow entry of oxygen through pores of much greater dimensions than the existing pores in the structure of the seed coat, this introduces artificial diffusion patterns which are not the same as in the seed *in vivo*.

Under non-ideal conditions, an additional external resistance is imposed by the thickness of the film of liquid water surrounding the seed, or if seeds are immersed in water (Orphanos and Heydecker, 1968; Heydecker and Orphanos, 1968; Heydecker, Chetram and Heydecker, 1971). The main conflict seems to be in maintaining an infinitely thin film of liquid water for maximum rate of imbibition, and at the same time minimum resistance to diffusion of oxygen at the seed surface. With thick water films, the external resistance may be altered by convection currents carrying oxygen in solution from the surface of the liquid to the seed.

The complexity of the external resistance in the soil has been described by Currie (1961) and Greenwood and Goodman (1967), where the movement of oxygen to the immediate vicinity of the seed from more remote regions is limiting. The bare platinum electrode has been used to measure oxygen diffusion rates in the soil (Letey and Stolzy, 1964). Dasberg and Mendel (1971) and Wengel (1966) correlated these measurements of maximum soil oxygen diffusion rates with the germination of wheat and maize, and found that a decrease in germination occurred at a critical value which was a function of the water content of the soil. The electrode was modified by Greenwood and Goodman (1967), and their measurements of soil oxygen concentration on a microscale in columns of soil crumbs show regions of high and low concentrations with sharp limiting boundaries in which seeds could be situated at widely different oxygen concentrations, or large seeds on the boundary could have uneven concentrations over the seed surface.

This paper is concerned with a quantitative analysis of the internal oxygen relations of the seed of charlock (*S. arvensis* L.) under different conditions of oxygen supply. The external gas atmosphere was controlled by the use of specially constructed glass gas pipettes, and monitored with a Beckman Model E_2 oxygen analyser. Experimental measurements were made with the standard Warburg procedure, the oxygen microelectrode modified after Greenwood and Goodman (1967), and the standard Rank oxygen electrode.

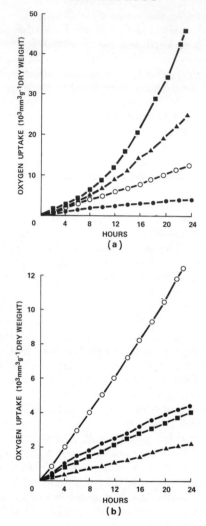

Figure 10.1   *(a) Oxygen uptake with time of excised embryos at various external oxygen concentrations at 25°C (1·0 ■, 0·2 ▲, 0·1 ○, 0·05 ● atm O_2); (b) enlarged scale of oxygen uptake of excised embryos at 0·1 and 0·05 atm O_2, and of seeds at 1·0 ■ and 0·2 ▲ atm O_2 at 25°C*

RESULTS AND DISCUSSION

The seed of charlock (*S. arvensis* L.) is a sphere of approximately 2 mm diameter and 2 mg dry weight. Each seed is surrounded by a

173

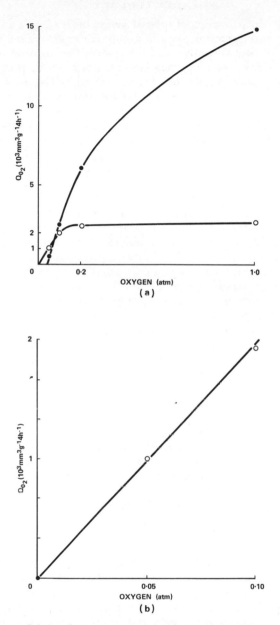

Figure 10.2   (a) Relation of rate of oxygen uptake (Q_{O_2}) and external oxygen concentration (C_0) of excised embryos from 0 to 4 h ○, and from 20 to 24 h ● at 25°C; (b) enlarged scale of (a) 0 to 4 h at 0 to 0·1 atm O_2

seed coat which consists of several layers of dead cells in the testa (mucilaginous epidermis, lignified palisade cells, pigmented cells), inside which is a single layer of living cells derived from the endosperm. The seed coat is permeable to water over the entire surface, but the resistance to diffusion of oxygen is high. The micropyle does not function as an open capillary for movement of water and gases into and out of the seed (Edwards, 1968).

When seeds and embryos which had imbibed at 0 °C were placed at various oxygen concentrations at 25 °C, the rate of oxygen uptake increased depending on the oxygen supply (*Figure 10.1*). Oxygen uptake by the embryos at each of the two higher concentrations increased exponentially with time (*Figure 10.1*(a)), and the rates of oxygen uptake at different times were obtained from linear regressions of the data transformed to logarithms. At the two lower concentrations the rates at different times were obtained from linear regressions on the primary data. The rates of oxygen uptake of seeds in oxygen and in a gas atmosphere equivalent to air were lower than that of excised embryos at 0·05 atm O_2 (*Figure 10.1*(b)). More accurate estimates of the initial and final rates were obtained from the mean hourly rates between 0 and 4 h and between 20 and 24 h, respectively (*Figure 10.2*). The relation between rate of oxygen uptake and external oxygen concentration for the mean rates over 0 to 4 h was a hyperbola (*Figure 10.2*(a)). Using the relationship given in *Figure 10.2*(b), in which the initial rate of oxygen uptake is proportional to external oxygen concentration from 0 to 0·1 atm O_2, the concentration of oxygen at the surfaces of embryos enclosed in seeds in oxygen was estimated to be 0·0385 atm O_2, and in seeds in air 0·0201 atm O_2.

A quantitative analysis of the coefficient of diffusion of oxygen through the seed coat was obtained from these data. Taking the testa as a sphere of external radius (b) 1·105 mm and internal radius (a) 1·068 mm, with concentration at the surface of the seed (C_1) 1·0 atm O_2, and at the surface of the embryo (C_0) 0·0385 atm O_2, and rate of flow (Q_{O_2}) using the equation (Crank, 1956)

$$Q_{O_2} = 4\pi D\ ab\ (C_1 - C_0)/b - a \qquad (1)$$

D, the coefficient of diffusion of oxygen in solution through the wet seed coat was estimated to be $1·125 \times 10^{-7}$ cm^2 s^{-1}. With C_1 equal to 0·2 atm O_2 and C_0 0·0201 atm O_2, the value of D was $6·013 \times 10^{-7}$ cm^2 s^{-1}. The low internal oxygen concentration of the seed was therefore attributed to the resistance to diffusion of oxygen through the seed coat.

From the initial rates of respiration (*Figure 10.2*) and the estimated surface concentrations, an attempt was made to determine the decline in concentration from the surface to the centre of seeds and embryos at various oxygen concentrations. The basic equation for radial diffusion in a sphere, with constant diffusion coefficient K (the coefficient of diffusion of oxygen in water is equal to $1\cdot98 \times 10^{-5}$ cm^2 s^{-1} at 20°C), and with respiration (P) independent of oxygen concentration, is

$$\frac{dc}{dt} = K\left(\frac{d^2c}{dr^2} + \frac{2}{r}\cdot\frac{dc}{dr}\right) - P \tag{2}$$

(Crank, 1956), where c is the concentration of oxygen at any point r along the radius from the surface to the centre of the sphere at time t.

If respiration is dependent on concentration, P is replaced by k^1c, where k^1 is the slope of the line relating rate of oxygen consumption and concentration. In the calculations, the values of c must be expressed in solution (the coefficient of absorption of oxygen gas in water at 25°C (α) equals $2\cdot83 \times 10^{-2}$ cm^3 cm^{-3}). During the first 4 h with excised embryos at high oxygen concentrations, P is independent of concentration (*Figure 10.2(a)*). Assuming $c = C$ where $r = R$ and $dc/dr \to 0$ as $r \to 0$, equation (2) can be solved to obtain values of c at given values of r:

$$c = \frac{P}{6K}(r^2 - R^2) + C_0 \tag{3}$$

With excised embryos at low oxygen concentrations (*Figure 10.2(b)*) P is replaced by k^1c (the value of k^1 was estimated to be $5\cdot91 \times 10^{-4}$ cm^3 cm^{-3} s^{-1} atm^{-1}). The concentration c at any given value of r (equation (3)) becomes

$$c = \frac{C_0 R \sinh r\sqrt{k^1/K}}{r \sinh R\sqrt{k^1/K}} \tag{4}$$

Using equations (3) and (4), and the values for the respiration rates (P) and (k^1C_0) for excised embryos given in *Figure 10.2*, the concentrations at various positions within the embryos was calculated. Similarly, the internal oxygen concentration of seeds at $1\cdot0$ and $0\cdot2$ atm O_2 was calculated using equation (4).

The analyses showed a significant decline in concentration from the surface to the centre of seeds and embryos at all oxygen concentrations (*Table 10.1* and *Figure 10.3*). The embryos in oxygen appeared to be composed of a sphere, in which the respiration rate of all the cells was equal to the maximum value, and concentration declined from $1 \cdot 0$ atm O_2 at the surface to $0 \cdot 768$ atm O_2 near the centre. The lowest surface concentration required to maintain the maximum respiration rate of all the cells was $0 \cdot 332$ atm O_2. With air outside instead of oxygen, there were two concentric zones of tissue merging at a radius of about $0 \cdot 8$ mm. The cells in the outer zone were respiring at the maximum rate, and the concentration in the middle of the meristems ($0 \cdot 83$ mm) was $0 \cdot 108$ atm O_2. The respiration rate of the cells in the inner zone was progressively restricted towards the centre where the concentration was $0 \cdot 0385$ atm O_2 (*Table 10.1* and *Figure 10.3(a)*). In seeds and embryos under low surface concentrations,

Table 10.1 ESTIMATED OXYGEN CONCENTRATIONS (atm) AT THE SURFACE (RADIUS (r) = $1 \cdot 068$ mm), IN THE MERISTEMS (r = $0 \cdot 828$ mm), AND NEAR THE CENTRE (r = $0 \cdot 010$ mm) OF EMBRYOS AT VARIOUS OXYGEN CONCENTRATIONS FROM 0 TO 4 h AT 25°C

| | External oxygen conc (atm) | Internal oxygen concentration (atm) at r (mm) | | |
		1·068	0·828	0·010
Excised	1·000	1·000	0·908	0·768
embryos	0·200	0·200	0·108	0·038
	0·100	0·100	0·059	0·022
	0·050	0·050	0·029	0·011
	0·035	0·035	0·021	0·008
	0·018	0·018	0·011	0·004
	0·010	0·010	0·006	0·002
Seeds	1·000	0·038	0·023	0·008
	0·200	0·020	0·012	0·004

the respiration rate was restricted by oxygen supply throughout the tissue, and the concentration near the centre was effectively above zero (*Table 10.1* and *Figure 10.3(b)*). Allowance for the oxygen taken up by the seed coat did not appreciably alter the values for the internal oxygen concentration of seeds in air or oxygen. Although the evidence was not conclusive, an internal concentration of approximately $0 \cdot 01$ atm O_2 at the embryo surface was found to be sufficient for growth (*Table 10.2*) (Edwards, 1969). Inserting this value into equation (1), the minimum concentration at the seed surface required for growth would be $0 \cdot 045$ atm O_2. Taking $0 \cdot 01$ atm O_2 as a limiting value, in seeds in air more than half the volume of the embryo was above $0 \cdot 01$ atm O_2, and in seeds in oxygen a small proportion of tissue near the centre of the embryo was below the minimum oxygen concentration required for growth (*Figure 10.4*).

Now, one can attempt to relate the oxygen concentrations at

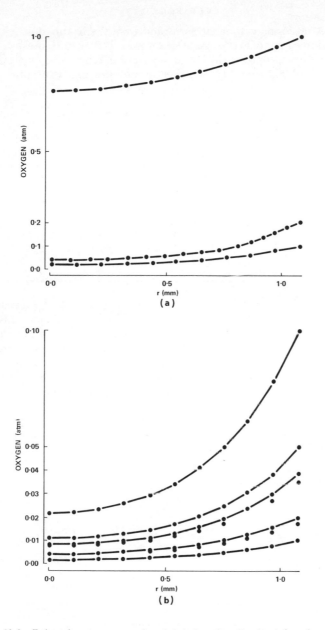

Figure 10.3 Estimated oxygen concentrations (atm) along the radius (mm) from the surface to the centre of excised embryos and of embryos enclosed within seed coats, in atmospheres of various oxygen concentrations from 0 to 4 h at 25°C

various positions within the seeds and embryos to the affinity of the oxidation enzymes in the tissues for oxygen. The overall rate of oxygen uptake was proportional to concentration from 0·05 atm O_2 to less than 0·002 atm O_2, whether or not the gas atmosphere was

Table 10.2 RATES OF OXYGEN UPTAKE AND LENGTHS OF THE EMBRYO AXIS OF EXCISED EMBRYOS MAINTAINED AT CONSTANT LOW OXYGEN CONCENTRATIONS FOR 24 h AT 25°C

External oxygen conc (atm)	Oxygen uptake ($mm^3 \, h^{-1}$ per 100 embryos)	Length (mm)
0·0500	76·15 ± 2·81	6·21
0·0355	59·26 ± 2·24	4·43
0·0250	41·32 ± 1·48	3·76
0·0183	29·90 ± 0·68	2·76
0·0100	14·18 ± 0·27	2·07
0·00004	0·16 ± 0·13	2·08
S.E.		± 0·21

maintained constant ($\pm 3 \times 10^{-5}$ atm O_2) (*Table 10.2, Figures 10.5 and 10.6*). Where the oxygen taken up by the embryos was not replenished, and rate declined with declining concentration, curves were fitted to the data by the method of orthogonal polynomials, from which estimates of the rate at each concentration were calculated. A small amount of oxygen was taken up by embryos in an atmosphere containing only 4×10^{-5} atm O_2, together with air in

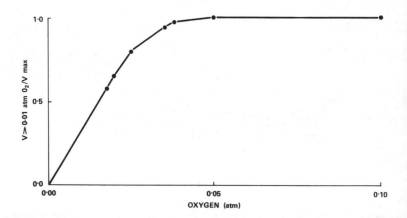

Figure 10.4   Proportion of the volume (V_{max}) of the embryo above 0·01 atm O_2 (V) in seeds in air and in oxygen from 0 to 4 h at 25°C

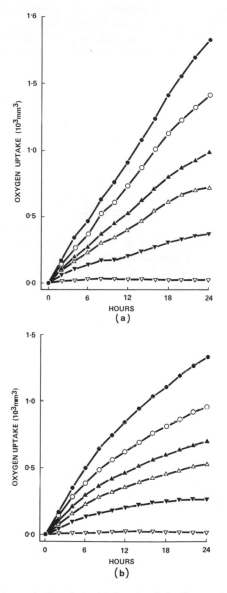

Figure 10.5 Oxygen uptake (atm O_2) with time of excised embryos at (a) constant and (b) declining low oxygen concentrations at 25°C (0.050 ●, 0.035 ○, 0.025 ▲, 0.018 △, 0.010 ▼, 4 × 10⁻⁵ ▽)

the intercellular spaces which comprised about 2 per cent of the tissue volume. In the seeds and embryos at low oxygen concentrations, the activity of a number of terminal oxidases which accept molecular oxygen must be reduced, because the apparent affinity (K_M) of most common oxidases (hydroperoxidases, flavoproteins and auto-oxidisable b-cytochromes) is of the order of 10^{-4} M (Barman, 1969). According to the Michaelis-Menten rule, the apparent affinity (K_M) is that oxygen concentration at which the rate of oxygen uptake becomes half its maximum value. The affinity of pure cytochrome oxidase which is composed of cytochrome a and cytochrome a_3 is 2×10^{-5} M and the rate is maximal at 6×10^{-5} M (0·05 atm O_2) (Okunuki, 1966). An alternative oxidase to cytochrome oxidase is known to have a high affinity for oxygen $(K_M = 10^{-7}$ M), and this is not a cytochrome (Storey and Bahr, 1969; Bendall and Bonner, 1971). In the seeds in air and oxygen the concentration at the embryo surface and in the meristems was similar to the K_M value for cytochrome oxidase (*Table 10.1* and *Figure 10.6*), therefore oxygen uptake through the cytochrome system was probably greater than through any other system (see also Chapter 11).

Many assumptions had to be made in the diffusion analysis; some of these have been examined experimentally.

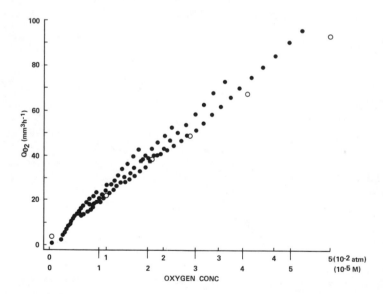

Figure 10.6   Relation between rate of oxygen uptake (Q_{O_2}) and external low oxygen concentration (C_0) expressed in partial pressure (atm) and in moles l^{-1} (M), over a period of 24 h at 25°C

1. In the calculations, the effective surface concentration within the seeds in air was taken to be 0.0201 atm O_2, and the respiration of the seed coat was neglected. Independent estimates of the rates of oxygen uptake of separated seed coats $(15.1\ \mu l\ h^{-1})$ and embryos $(315.9\ \mu l\ h^{-1})$ of 100 seeds in air for 24 h at 25°C, suggested that the error was about 5 per cent. Assuming an error of about 10 per cent, the respiration rates of the separated seed coats and embryos at equivalent low oxygen concentrations $(0.0201$ and 0.0183 atm $O_2)$ were compared with that of intact seeds in air (*Table 10.3* and *Figure 10.7*) (Follows, 1968). Again, estimates of the rates were obtained from linear regressions on the primary data. The value for the intact seeds $(29.41\ \mu l\ h^{-1})$ was not significantly different from the embryos

Table 10.3 RATES OF OXYGEN UPTAKE OF INTACT SEEDS (FOLLOWS, 1968), AND OF SEED PARTS AT EQUIVALENT LOW OXYGEN CONCENTRATIONS, IN WATER AND IN VARIOUS CONCENTRATIONS GIBBERELLIC ACID (GA_3) FOR 24 h AT 25°C

	External oxygen conc (atm)	GA_3 conc (M)	Oxygen uptake per 100 seeds or seed parts $(mm^3\ h^{-1})$
Seeds	0·2000	0	29·41
	0·2000	6×10^{-3}	35·91
S.E.			± 0·86
Excised embryos	0·0183	0	29·40
	0·0183	3×10^{-5}	35·78
	0·0183	6×10^{-3}	39·50
S.E.			± 3·78
Seed coats	0·0201	0	4·72
	0·0201	3×10^{-3}	7·43
	0·0201	6×10^{-3}	4·32
S.E.			± 0·13

at 0.0183 atm O_2 $(29.40\ \mu l\ h^{-1})$, but after addition of the value for the seed coats at 0.0201 atm O_2 $(4.72\ \mu l\ h^{-1})$, the sum of the parts was greater than the whole; therefore the internal oxygen concentration of the seed may have been slightly over-estimated. Depletion of the internal oxygen supply of the seed could not have been due to oxygen uptake by the seed coat, and was presumably due to the physical resistance to diffusion of oxygen through the testa membranes.

2. The initial rates of oxygen uptake, estimated from the mean hourly rates from 0 to 4 h, were assumed to be equal in all cells. This may not be so, because the initial rate of oxygen uptake by the embryo axis tissue may be higher than by the cotyledon tissue.

182

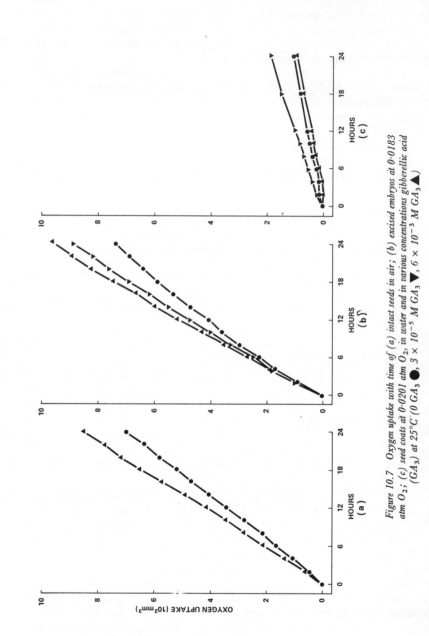

Figure 10.7 Oxygen uptake with time of (a) intact seeds in air; (b) excised embryos at 0·0183 atm O_2; (c) seed coats at 0·0201 atm O_2, in water and in various concentrations gibberellic acid (GA_3) at 25°C (0 GA_3 ●, 3 × 10⁻⁵ M GA_3 ▼, 6 × 10⁻³ M GA_3 ▲)

Measurements of the respiration rates of different parts of the seed were made by means of polarographic techniques, (a) with a modified oxygen microelectrode, after Greenwood and Goodman (1967), and (b) with a standard Rank oxygen electrode (Orton, 1972). Before making the microelectrode measurements the seeds and embryos were embedded 2 mm below the surface of 0·6 per cent agar. A small gold electrode (0·5 mm dia) was covered with a film of 12 per cent cellulose acetyl butyrate containing 0·1 M KCl, which carried the current when a polarising voltage of 0·7 V was applied for 0·3 s at intervals of 10 s. This was protected by a thin polyethylene membrane, which was impermeable to KCl and offered a constant resistance to diffusion of oxygen to the electrode surface. The measured current is dependent on the diffusion of oxygen from the membrane surface to the electrode surface, which is directly proportional to the oxygen concentration at the membrane surface. The electrode can therefore be calibrated to give direct measurements of oxygen concentration at the membrane surface. The electrode was inserted into the agar above the surface of the embryo axis or cotyledon tissue.

Table 10.4 OXYGEN ELECTRODE MEASUREMENTS OF RATES OF OXYGEN UPTAKE BY IMBIBED SEEDS AND SEED PARTS. A. OXYGEN MICROELECTRODE MEASUREMENTS AT $22°C$, B. RANK OXYGEN ELECTRODE MEASUREMENTS AT $22°C$ (ORTON, 1972)

	Oxygen conc (atm)	Oxygen uptake	
		$\mu M\ cm^{-2}\ s^{-1}$ (10^{-6})	$\mu M\ g^{-1}\ dry\ wt\ s^{-1}$ (10^{-2})
A. Intact seed			
Embryo axis	0·1348	4·92	
Cotyledons	0·1438	4·33	
S.E.	±0·0137		
Excised embryo			
Embryo axis	0·0548	10·00	
Cotyledons	0·0924	7·68	
S.E.	±0·0149		
B. Intact seed	0·2000		0·90
Excised embryo	0·2000		3·41
Embryo axis	0·2000		4·89
Cotyledons	0·2000		2·36

The measured oxygen concentration at the surface of the embryo axis was slightly lower than at the surface of the cotyledons of each seed, but the difference was not significant for five different seeds. After removal of the seed coat, the measured oxygen concentration was much lower over all parts of the embryo surface, particularly above the axis tissue. Assuming linear diffusion to the electrode surface, $J = -D \, dc/dx$, where D is the diffusion coefficient of oxygen in 0·6 per cent agar ($1·0 \times 10^{-5}$ cm^2 s^{-1}), c the measured oxygen concentration (M cm^{-3}), and x the depth in agar (0·2 cm), J, the rate of flow (M cm^{-2} s^{-1}) into different parts of the seed and excised embryo, was calculated. Estimates of the integrated flow over the entire surface of the intact seed and excised embryo were compared with the previous data (*Figure 10.2*). Because of the additional external resistance to diffusion through the superficial layer of agar, and the barrier along the diffusion path caused by the presence of the electrode, the values were correspondingly lower. Similar measurements made on the severed parts with the Rank oxygen electrode showed a higher rate of flow, suggesting a higher respiration rate in the embryo axis tissue than in the cotyledon tissue (*Table 10.4*).

3. The data of *Figure 10.2* represent the overall rates of respiration of embryos at different external oxygen concentrations, including the internal gradient, and cannot strictly be extrapolated to find the rate of oxygen uptake at different depths within the embryo. The value of k^1 may be calculated theoretically from the data of *Figure 10.3* (Edwards, 1969, equations (3) and (6); Collis-George, personal communication).

4. The constant diffusion coefficient K was assumed to be the same as the coefficient of diffusion of oxygen in water ($1·98 \times 10^{-5}$ cm^2 s^{-1} at 20°C). Lower values for the diffusion coefficient have been calculated within tissues ($6·2$ to $8·9 \times 10^{-6}$ cm^2 s^{-1}) (Berry, 1949).

The oxygen concentration at the embryo surface of seeds in oxygen (0·0355 atm O_2), in air (0·0183 atm O_2), and at partial pressures greater than 0·05 atm O_2 (0·01 atm O_2) was higher than that required for growth. A significant increase was detected in the length of the embryo axis, associated with increase in cell size and cell number, when the oxygen concentration at the surface of the excised embryo was more than 0·01 atm O_2 (*Table 10.2*) (Edwards, 1969). Virtually all the seeds germinated after treatment with 6×10^{-3}M gibberellic acid (GA_3) for 24 h at 25°C. The rate of oxygen uptake by the treated seeds was significantly higher within 1 h of imbibition with GA_3 ($35·91$ μl h^{-1} per 100 seeds) compared with seeds imbibed in water ($29·41$ μl h^{-1} per 100 seeds) (*Table 10.3* and *Figure 10.7*). The initial rates of water uptake were equal, and assuming constant diffusion coefficient K between 12 and 24 h, the internal oxygen concentration of the GA_3 treated seeds was estimated to be approximately 0·0143 atm O_2, which was sufficient for growth.

It appeared that in charlock seeds and embryos at low oxygen

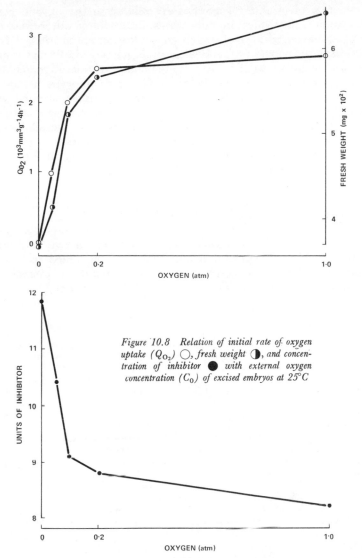

Figure 10.8   Relation of initial rate of oxygen uptake (Q$_{O_2}$) ◯, fresh weight ◑, and concentration of inhibitor ● with external oxygen concentration (C$_0$) of excised embryos at 25°C

concentrations, some other factor was preventing growth when the supply of oxygen was not directly limiting. Estimates of the increase in concentration of growth-inhibiting substances with time and temperature in aqueous solution around seeds and embryos suggested that there was continuous production of inhibitor in the tissues of the embryo which have a low oxygen concentration. The rate of production of inhibitor was inversely related to the oxygen concentration of the tissues; in other words, the inhibitor was formed by reactions favoured at low oxygen concentration (*Figure 10.8*). The inhibitory substances probably arise in all parts of the embryo, but because of the gradient in oxygen, the concentration at the centre may be higher than at the periphery. With excised embryos the rate of diffusion from the surface near where the meristems are situated was rapid, and growth was not prevented except at very low oxygen concentrations. In intact seeds the rate of diffusion of the inhibiting substances was hindered by the presence of the testa, and concentrations were reached which prevented growth when the supply of oxygen was not directly limiting (Edwards, 1969).

ACKNOWLEDGEMENTS

I wish to thank Professor Collis-George for constructive help and correction of the original diffusion analysis. The skilled technical assistance of Mr. J. Blunt and Mr. L. Heathcote, and the mathematical assistance of Mr. K. Gregson, are gratefully acknowledged.

REFERENCES

BALLARD, L. A. T. and GRANT-LIPP, A. E. (1969). 'Studies of dormancy in the seeds of subterranean clover (*Trifolium subterraneum* L.). III. Dormancy breaking at low concentrations of oxygen', *Aust. J. biol. Sci.*, **22**, 279–288

BARMAN, T. E. (1969). *Enzyme Handbook*, Springer-Verlag, Berlin

BENDALL, D. S. and BONNER, W. D. (1971). 'Cyanide—insensitive respiration in plant mitochondria', *Pl. Physiol.*, **47**, 236–245

BERRY, L. J. (1949). 'The influence of oxygen tension on the respiratory rate in different segments of onion roots', *J. cell. comp. Physiol.*, **33**, 41–66

BROWN, R. (1940). 'An experimental study of the permeability to gases of the seed coat membranes of *Cucurbita pepo*', *Ann. Bot.*, **4**, 379–395

BURTON, W. G. (1950). 'Studies on the dormancy and sprouting of potatoes. I. The oxygen content of the potato tuber', *New Phytol.*, **49**, 121–134

CHEN, S. S. C. (1970). 'Influence of factors affecting germination on respiration of *Phacelia tanacetifolia* seeds', *Planta*, **95**, 330–335

COLLIS-GEORGE, N. and WALLACE, H. R. (1968). 'Supply of oxygen during hatching of the nematode *Meloidogyne javanica* under non-competitive conditions', *Aust. J. biol. Sci.*, **21**, 21–35

CÔME, D. (1970). *Les Obstacles à la Germination*, Masson, Paris

CRANK, J. (1956). *Mathematics of Diffusion*. Clarendon Press, Oxford

CURRIE, J. A. (1961). 'Gaseous diffusion in the aeration of aggregated soils', *Soil Sci.*, **92**, 40–45

DASBERG, S. and MENDEL, K. (1971). 'The effect of soil water and aeration on seed germination', *J. exp. Bot.*, **22**, 992–998

DURHAM, V. M. and WELLINGTON, P. S. (1961). 'Studies on the germination of cereals. IV. The oxygen requirement for germination of wheat grains during maturation', *Ann. Bot.*, **25**, 197–205

EDWARDS, M. M. (1968). 'Dormancy in seeds of charlock. II. The influence of the seed coat', *J. exp. Bot.*, **19**, 583–600

EDWARDS, M. M. (1969). 'Dormancy in seeds of charlock. IV. Interrelationships of growth, oxygen supply and concentration of inhibitor', *J. exp. Bot.*, **20**, 876–894

FOLLOWS, M. (1968). 'The effect of gibberellic acid on dormant charlock seeds', *B.Sc. Dissertation*, Univ. of Nottingham

GODDARD, D. R. and BONNER, W. D. (1960). 'Cellular respiration', in *Plant Physiology 1A* (ed. Steward, F. C.), 209–312, Academic Press, New York and London

GREENWOOD, D. J. and GOODMAN, D. (1967). 'Direct measurements of the distribution of oxygen in soil aggregates and in columns of fine soil crumbs', *J. Soil Sci.*, **18**, 182–196

GRIFFIN, D. M. (1968). 'A theoretical study relating the concentration and diffusion of oxygen to the biology of organisms in soil', *New Phytol.*, **67**, 561–577

HEYDECKER, W., CHETRAM, R. S. and HEYDECKER, J. C. (1971). 'Water relations of beet-root seed germination. II. Effects of the ovary cap and of the endogenous inhibitors', *Ann. Bot.*, **35**, 31–42

HEYDECKER, W. and ORPHANOS, P. I. (1968). 'The effect of excess moisture on the germination of *Spinacia oleracea* L.', *Planta*, **83**, 237–247

IKUMA, H. and THIMANN, K. V. (1963). 'The role of the seed-coats in germination of photosensitive lettuce seeds', *Pl. Cell Physiol.*, **4**, 169–185

LETEY, J. and STOLZY, L. H. (1964). 'Measurement of oxygen diffusion rates with the platinum microelectrode. I. Theory and equipment', *Hilgardia*, **35**, 545–554

MAJOR, W. and ROBERTS, E. H. (1968). 'Dormancy in cereal seeds. I. The effects of oxygen and respiratory inhibitors', *J. exp. Bot.*, **19**, 77–89

OKUNUKI, K. (1966). 'Cytochromes and cytochrome oxidase', in *Comprehensive Biochemistry 14 Biological Oxidations* (ed. Florkin, M. and Stotz, E. H.), 232–308, Elsevier, Amsterdam and London

ORPHANOS, P. I. and HEYDECKER, W. (1968). 'On the nature of the soaking injury of *Phaseolus vulgaris* seeds'. *J. exp. Bot.*, **19**, 770–784

ORTON, P. J. (1972). 'The application of polarographic techniques to the measurement of seed respiration', *B.Sc. Dissertation*, Univ. of Nottingham

PERNER, E. (1965). 'Electron-microscopic investigations of cells of embryos in the state of the complete rest-period of the seed. I. The order of cellular structure in the radicula of air-dried seeds of *Pisum sativum*', *Planta*, **65**, 334–357

POPAY, A. I. and ROBERTS, E. H. (1970). 'Factors involved in the dormancy and germination of *Capsella bursa-pastoris* (L.) Medik. and *Senecio vulgaris* L.', *J. Ecol.*, **58**, 103–122

SHAYKEWICH, C. F. and WILLIAMS, J. (1971). 'Resistance to water absorption in germinating rape seed (*Brassica napus* L.)', *J. exp. Bot.*, **22**, 19–24

SMITH, A. M. and GRIFFIN, D. M. (1971). 'Oxygen and the ecology of *Armillariella elegans* Heim', *Aust. J. biol. Sci.*, **24**, 231–262

STOREY, B. T. and BAHR, J. T. (1969). 'The respiratory chain of plant mitochondria. II. Oxidative phosphorylation in skunk cabbage mitochondria', *Pl. Physiol.*, **44**, 126–134

WAREING, P. F. and FODA, H. A. (1957). 'Growth inhibitors and dormancy in *Xanthium* seed', *Physiologia Pl.*, **10**, 266–280

WENGEL, R. W. (1966). 'Emergence of corn in relation to soil oxygen diffusion rates', *Agron. J.*, **58**, 69–72

YEMM, E. W. (1965). 'The respiration of plants and their organs', in *Plant Physiology 4A* (ed. Steward, F. C.) 231–310, Academic Press, New York and London

DISCUSSIONS

Smith: What is the cause of the low oxygen concentration in the interior of the charlock seeds? Is it due to phenolic compounds in the seed coat?

Edwards: No. Absorption of oxygen by phenols in the charlock seed coat does not account for the measured depletion of the oxygen concentration from the atmospheric level to the 2 per cent in the interior of the seed.

Crabb: The respiration rate of barley grains whose pericarp and husk have been removed with sulphuric acid is as high in an atmosphere containing less than 4 per cent oxygen as it is in air. But intact barley grains are often 'water sensitive' and the resulting dormancy appears to be due to an insufficient amount of oxygen reaching the embryo.

Edwards: The resistance to the penetration of oxygen depends partly on the covering structures and partly on the size of the seed and the position of the meristems. Barley grains are larger than charlock seeds and the meristems are situated deeper in the tissues, therefore they may well require an external oxygen concentration of more than 4 per cent to germinate.

Crabb: Does oxygen penetrate at the same rate through the coats of dormant and non-dormant seeds?

Edwards: When dormancy of charlock seeds is broken by gibberellic acid the effective internal oxygen concentration is similar to that in seeds imbibed in water. When dormancy is broken by sulphuric acid treatment, this corrodes the seed coat and increases the internal oxygen supply. The effect of the two substances is different. Corrosion (or removal) of the seed coat removes the coat-imposed dormancy which arises because the slow rate of oxygen supply to the interior of the seed results in the formation of an inhibitor; if it were not for this inhibitor, the oxygen supply through the coat would be sufficient for germination. The gibberellin treatment acts on the embryo within the intact seed coat.

Wareing: The situation is similar to that in sycamore seeds where the oxygen supply through the coat itself would be sufficient for germination but where an inhibitor is present.

E. H. Roberts: It is important to note that removal of the covering structures does increase the oxygen uptake by sycamore and barley seeds. What is an adequate level of oxygen may depend on whether cytochrome oxidase is involved (when a low level of oxygen may be sufficient) or any other oxidase (when the same level may be insufficient).

Wareing: The interference of seed coats with oxygen uptake is proven but one cannot generalise on its role as a limiting factor for germination.

Edwards: After measuring the rate of oxygen uptake by intact and de-coated seeds, the rate of flow through the seed coat and the supply of oxygen to the meristem and relating the oxygen concentration to the activity of enzymes in the seed, it is clear that the seed coat forms a considerable barrier to the movement of oxygen into the seed (the coefficient of diffusion through the coat is 10^{-7} cm^2 s^{-1}); but despite this the evidence shows that sufficient oxygen for growth reaches the embryo surface and the meristems.

Heydecker: Does the carbon dioxide concentration in the seed play any part in the induction of dormancy?

Edwards: Hardly; the carbon dioxide concentration in dormant charlock embryos is lower than that in germinating embryos in full oxygen.

11

OXIDATIVE PROCESSES AND THE CONTROL OF SEED GERMINATION

E. H. ROBERTS

Department of Agriculture, University of Reading

INTRODUCTION

A number of dormancy mechanisms involving oxidation reactions have been postulated from time to time. Since these have been reviewed recently (Roberts, 1969), it is proposed here to concentrate on a hypothesis which we have been developing and which would seem to account for most of the dormancy phenomena in a wide range of species. Although the mechanism involved is only partly elucidated, it helps to explain the many diverse factors which accelerate the loss of dormancy. After discussing all the major categories of factors which are known to break dormancy, this paper will concentrate on those which appear to be most important in the field, and on the ways in which they may interact to control germination in adaptively useful ways in annual weed species.

LOSS OF DORMANCY

This section is a development of the hypothesis that stimulating the activity of the pentose phosphate pathway leads to loss of dormancy. As a result of some investigations on rice concerning the effects of temperature, oxygen pressure and the removal of covering structures, it was postulated that loss of dormancy was due to some oxidation reaction other than conventional respiration (Roberts, 1962). Oxygen can affect the process either in 'dry' storage or after the seeds have been set to germinate. As in many other seeds an increase in temperature during storage increased the rate of loss of dormancy of rice seeds and for all varieties tested the Q_{10} (temperature coefficient) for this process was about 3·1 (Roberts, 1965), at least between 27°C and 47°C. Such a high Q_{10} suggested that the effect of temperature is not chiefly on a simple physical process such as diffusion. It has sometimes been suggested that these after-ripening effects of temperature could not involve enzymic reactions since the seeds are too dry for normal metabolism. However, in the different context of the events occurring in seeds during loss of viability, we know that even at low moisture contents there can be an increase in the products of hydrolytic enzyme activity—e.g. in reducing sugars and fatty acids (Roberts, 1972a; 1973); consequently, the possibility cannot be ruled out that after-ripening effects on 'dry' seeds during storage are the results of metabolic processes, although these may be occurring very slowly.

In order to test further the hypothesis that the oxidation involved in the loss of dormancy is not some aspect of conventional respiration, a number of respiratory inhibitors were tested for their effect on the germination of dormant rice seeds. It was found (Roberts, 1964a; 1964b) that all the common inhibitors of cytochrome oxidase (potassium cyanide, sodium azide, carbon monoxide, hydrogen sulphide and hydroxylamine) stimulated the breaking of dormancy. None of the other seventeen respiratory inhibitors tested had any stimulatory effect.

As with many other seeds it was found that nitrate ions are stimulatory and nitrite is even more effective, but reduced forms of nitrogen —ammonium ions, urea, amino acids—had no effect (Roberts, 1963a; 1963b). These results suggested that the oxidised forms of nitrogen were effective by virtue of their ability to act as hydrogen acceptors rather than as a nitrogen source—a suggestion which was tested by applying methylene blue which, as predicted on this hypothesis, also had some dormancy-breaking activity (Roberts, 1964a). An alternative possibility that nitrate and nitrite act by being converted to hydroxylamine, which in turn inhibits cytochrome oxidase, was mentioned by Roberts (1964b); however, in view of the result with methylene blue and in view of the fact that there is some doubt whether *free* hydroxylamine is a natural intermediate of nitrite reduction in higher plants (Hageman, Cresswell and Hewitt, 1962) this was considered a less likely alternative. It is nevertheless mentioned again here since Hendricks and Taylorson (1972) have recently suggested that nitrate acts as a dormancy-breaking agent in *Amaranthus albus* and lettuce (*Lactuca sativa*) because it is converted to hydroxylamine which inhibits terminal oxidation. At present there is no clear way of distinguishing between these alternative explanations; but this is of no great consequence, for either explanation will fit the hypothesis which has been developed (Roberts, 1969) and will now be elaborated in greater detail.

The early investigations on rice which have just been described were followed up by more detailed work on barley, *Hordeum distichon*, for which, incidentally, it has already been shown that the germination of dormant seeds could be stimulated by cyanide and carbon monoxide (Fischnich, Thielbein and Grahl, 1962) and hydrogen sulphide (Pollock and Kirsop, 1956); furthermore, Ogowara and Hayashi (1964) had shown that cyanide and malonate are stimulatory to dormant seeds of a close relative, *Hordeum spontaneum*. In our investigations (Major and Roberts, 1968a) it was found that the responses to inhibitors varied between cultivars but, in general, the five inhibitors of cytochrome oxidase were still the most potent; however, other respiratory inhibitors were also found to produce significant increases in the germination of dormant seeds—viz. sodium fluoride, iodoacetate, malonate, and monofluoroacetate. Diethyldithiocarbamate (DIECA) had only a small effect and that was equivocal: in one cultivar it increased germination, whereas in another it decreased it. Dinitrophenol (DNP) also sometimes had a small effect. Experiments further showed that in cultivated oats (*Avena sativa*), in addition to the five inhibitors of cytochrome oxidase (KCN, NaN$_3$, CO, H$_2$S and NH$_2$OH), significant stimulation could be obtained with DIECA and malonate (Major and Roberts, 1968a).

It then became clear that respiratory inhibitors other than those which affect cytochrome oxidase could stimulate germination in some species. Measurements of seed respiration showed that the common property of these successful germination stimulators was that they did in fact inhibit oxygen uptake (Major and Roberts, 1968a). This was demonstrated for KCN, NaN$_3$, malonate and fluoroacetate in barley and KCN and NaN$_3$ in rice. But best confirmation of the rule was provided by the results with NaF at 10^{-3} M on rice which did not stimulate germination but, if anything, slightly stimulated oxygen uptake. It is important to mention these results on oxygen uptake since Ballard and Grant-Lipp (1967) have reported that the germination of dormant seeds of *Trifolium subterraneum* is stimulated by respiratory uncouplers. They obtained significant increases in germination using 0·2 mM DNP, 1·0 mM sodium arsenate, and 4·0 mM salicylate. However, much the greatest stimulation was obtained with 0·8 mM sodium azide which, while acting as an uncoupler at lower concentrations, usually acts as a terminal oxidase inhibitor at millimolar concentration. Furthermore, the other substances mentioned can similarly inhibit oxygen uptake at higher concentrations. (Consequently in order to reconcile the work of Ballard and Grant-Lipp with the other work described here, it would be necessary to have information on the effects of these substances on oxygen uptake of subterranean clover seeds—uncouplers would be expected to increase it—and also data on the germination of these seeds when they are subjected to other respiratory inhibitors which have no uncoupling property.)

As a result of her work on barley and rice, particularly with regard to the effects of respiratory inhibitors, Wendy Major (1966) tentatively postulated that the oxidative process involved in loss of dormancy is the oxidative pentose phosphate pathway (PP pathway). There was, in fact, some indirect evidence for the speculation, and a little more indirect evidence was found later. All of this has been summarised in detail (Roberts, 1969) and so here only five major points will be mentioned.

1. Nitrite and methylene blue, effective dormancy-breaking agents, stimulate the operation of the PP pathway in maize roots (Butt and Beevers, 1960; 1961), and methylene blue has the same effect in potato and carrot storage tissue (ap Rees and Beevers, 1960a, 1960b). This stimulation is assumed to be associated with the fact that nitrite and methylene blue preferentially oxidise $NADPH_2$ (reduced form of nicotinamide adenine dinucleotide phosphate (NADP)) (rather than $NADH_2$—reduced form of nicotinamide adenine dinucleotide (NAD)) and that the dehydrogenases of the PP pathway (in contrast with those of conventional respiration—the Embden–Meyerhof–Parnas (EMP) pathway) reduce NADP rather than NAD. Nitrate, which is also a dormancy-breaking agent, is reduced by nitrate reductase which generally operates via $NADH_2$ rather than $NADPH_2$ (Beevers, Flescher and Hageman, 1964); but the nitrate is reduced to nitrite, the further reduction of which will oxidise $NADPH_2$. In this connection it may be relevant that nitrite is often a more stimulatory dormancy-breaking agent than nitrate—e.g. in barley (Major, 1966) and rice (Roberts, 1963a). But in any case, in yeast a change from an amino acid nitrogen source to one containing only KNO_3 results in an increase of glucose-6-phosphate dehydrogenase activity and in the evolution of CO_2 from [14]C-1 labelled glucose (Osmond and ap Rees, 1969).

2. Respiratory inhibitors break dormancy. At the same time there is evidence that PP pathway respiration is not affected by conventional respiratory inhibitors: Gibbs (1954) showed that cyanide and azide do not inhibit the oxidation of glucose-6-phosphate in pea seedling leaves and that cyanide may acceler-

ate it; furthermore, the dehydrogenases of the PP pathway are not sensitive to fluoride or iodoacetate. There is other supporting evidence: in storage root and tuber tissue, respiration via the PP pathway is stimulated by slicing and washing (ap Rees and Beevers, 1960a, 1960b); at the same time the respiration becomes insensitive to cyanide, azide and carbon monoxide (Thimann, Yocum and Hackett, 1954; MacDonald, 1959). Thus in a situation where oxygen is short, as it is in intact seeds, and when at the same time cytochrome oxidase competes for this oxygen for which it has an extremely high affinity, inhibitors of cytochrome oxidase would encourage the operation of the PP pathway. To a lesser extent, this would also apply to inhibitors of the dehydrogenases of the EMP pathway supplying reducing power to cytochrome oxidase.

3. There is some evidence that the early respiration of non-dormant imbibing seeds is carried out mainly via the PP pathway, e.g. in mung beans (*Phaseolus radiatus*) (Chakravorty and Burma, 1959) and in buckwheat (*Fagopyrum esculentum*) (Effer and Ranson, 1967a; 1967b). Furthermore, if one puts together the evidence provided by Poljakoff-Mayber, Mayer and their colleagues, the PP pathway also seems to be important during the early stages of germination in Grand Rapids lettuce seeds (Poljakoff-Mayber and Mayer, 1958), whereas the Krebs cycle and the conventional electron transfer system seems to be of minor importance (Poljakoff-Mayber, 1955; Mayer, Polja-koff-Mayber and Appleman, 1957; Poljakoff-Mayber and Evanari, 1958). It is only after growth has started that meta-bolism shifts from the PP pathway to the system of glycolysis, the Krebs cycle and the cytochrome electron transport chain (Poljakoff-Mayber and Evanari, 1958). During the early stages of germination the oxygen uptake is hardly affected by carbon monoxide (Harel and Mayer, 1963); even a concen-tration of KCN as high as 2×10^{-3} M does not inhibit germination and in fact slightly stimulates it (Mayer, Poljakoff-Mayber and Appleman, 1957). Accordingly, these workers suggested that during the early stages of germination some alternative, unknown electron transport system may be functioning in the seeds. Our evidence strongly suggests that this system is associated with the operation of the PP pathway.

4. Increased oxygen pressure tends to break dormancy. There is some evidence in seedlings of *Fagopyrum esculentum* that increased oxygen pressure may increase the proportion of respiration occurring via the PP pathway (Effer and Ranson, 1967b). This, of course, would be expected if the oxidation associated with

the PP pathway is not operating through cytochrome oxidase, since cytochrome oxidase has a higher affinity for oxygen than any other enzyme and is fully saturated at a very low partial pressure.

5. Stratification (low-temperature treatments of imbibed seeds) breaks dormancy in a wide range of species—even in the tropical species, rice, where it presumably has no ecological function (Roberts, 1962). In stratification treatments which break the dormancy of *Prunus cerasus* seeds, La Croix and Jaswal (1967) reported that after 7 weeks' chilling, when the seeds became non-dormant, there was a change in the C-6/C-1 ratio (the basis of which is explained below) of the embryonic axis from 0·95 to about 0·6, and this suggests an increase in the activity of the PP pathway.

At the same time as La Croix and Jaswàl were working on *Prunus*, the suggestion that the PP pathway may be important in the loss of dormancy in barley seeds was being given its first test in our laboratory by Gaber (1967) using the cultivar Proctor. He used the technique devised by Bloom and Stetton (1953) to obtain evidence as to the importance of the PP pathway relative to conventional respiration (EMP pathway) in dormant and non-dormant seeds during the first 3 h of imbibition. In this technique, glucose-6-^{14}C and glucose-1-^{14}C are fed separately to samples of the same material and the relative amounts of $^{14}CO_2$ evolved are determined. The results are expressed as a C-6/C-1 ratio. The interpretation depends on the fact that in glycolysis the glucose molecule is split into two 3-carbon units and both the carbon-1 and carbon-6 atom of glucose end up in the methyl group of pyruvate and are consequently decarboxylated in an identical fashion. Hence, if all the respiration occurs via the EMP-pathway, the C-6/C-1 ratio would be unity. On the other hand, if glucose is catabolised via the PP pathway, then, during the second dehydrogenation, when 6-phosphogluconate is converted to ribulose-5-phosphate, the carbon-1 atom of the original glucose molecule is decarboxylated. Consequently, any participation of the PP pathway decreases the C-6/C-1 ratio. Gaber's preliminary results showed a C-6/C-1 ratio during the first 3 h of imbibition of 0·17 in non-dormant and 0·44 in dormant seeds, thus confirming that this was a promising line of enquiry.

Smith (1969) has now accumulated a great deal of evidence to show that dormant barley seeds are less capable of operating the PP pathway, even though the rate of gas exchange is greater in dormant than non-dormant seeds during the initial stages of imbibition (Major and Roberts, 1968b). The main early findings (Smith,

1969) which have been previously summarised (Roberts, 1969) showed that in determinations made over 3 h periods from the start of imbibition in Pallas barley, C-6/C-1 ratios of about 0·3–0·4 are typical of dormant seeds, whereas ratios of 0·12–0·15 are typical of non-dormant seeds for the first 24 h. Thereafter, as radicle protrusion takes place, the value in non-dormant seeds rises. Furthermore, in experiments on Proctor barley it was shown that all the dormancy-breaking agents tried—cyanide, nitrite, GA_3 (gibberellic acid), D-threo-chloramphenicol and L-threo-chloramphenicol—decrease the C-6/C-1 ratio of dormant seeds as measured during the second 3 h of imbibition. In general, the agents have little effect on C-6/C-1 ratios in non-dormant seeds since these values are here already extremely low, suggesting that the PP pathway is already operating at maximum capacity.

Any dormancy hypothesis has to account for the fact that in most seeds—including rice (Roberts, 1961) and barley (Harrington, 1923) —dormancy tends to be alleviated by the removal of the covering structures. In general, the covering structures of seeds do not prevent oxygen uptake but they do present some resistance to it. This is indicated by three types of evidence. Firstly, in many species removal or puncture of the covering structures increases the rate of respiration (see reviews by Stiles, 1960; Carr, 1961; Nyman, 1963). Barley, which has been used most in our studies, is no exception (Merry and Goddard, 1941; Urion and Chapon, 1955; Belderok, 1963). Secondly, when the covering structures (lemma and palea) are removed, the respiratory quotient (RQ) decreases: our experiments show a value of about 1·5 for intact dormant, and 2·0 for intact non-dormant barley seeds; in both cases, dehulling (which removes dormancy) reduces the RQ to approximately 1·0 and increases the rate of oxygen uptake (Major and Roberts, 1968b). Thirdly, increasing the partial pressure of oxygen round the seeds to above 21 per cent increases the rate of oxygen uptake (Merry and Goddard, 1941), whereas in most plant tissues in which there is little diffusive resistance, such increases in oxygen pressure usually have little effect on the rate of oxygen uptake—because of the very high affinity of cytochrome oxidase for oxygen which is saturated at low partial pressures.

However, although the covering structures afford a resistance to oxygen uptake, loss of dormancy, at least in barley and rice, cannot be ascribed to an increase in permeability of the seed coats as has sometimes been suggested (e.g. Caldwell, 1959), since it has always been found that during the initial hours of imbibition the gas exchange in dormant seeds is *greater* than in non-dormant seeds (Major and Roberts, 1968b; Smith, 1969).

THE PP PATHWAY DORMANCY HYPOTHESIS:
THE PRESENT POSITION

The hypothesis on which we are continuing to work is as follows (see *Figure 11.1*). In order to germinate, seeds need to operate the PP pathway during the initial stages of germination and this pathway is associated with an oxidase which is relatively insensitive to the conventional terminal-oxidase inhibitors—KCN, NaN_3, CO, H_2S and NH_2OH. Dormant seeds are less capable of operating this pathway than non-dormant seeds, although they show a high activity of conventional respiration (the EMP pathway) involving cytochrome oxidase. Any treatment which increases the activity of the PP pathway tends to alleviate dormancy. Since the condition within the seed is relatively anaerobic, removing the covering is such a treatment, since it reduces the competition between the oxidase of the PP pathway and the highly competitive cytochrome oxidase. Alternatively, the same effect can be obtained by increasing the oxygen pressure, or by reducing the EMP competition by applying respiratory inhibitors. It has been pointed out (Roberts, 1969) that, in addition to the conventional respiratory inhibitors, chloramphenicol could operate in this way rather than in its more usual role as a protein-synthesis inhibitor: both its isomers inhibit conventional mitochondrial oxidation but only the D-threo isomer has a pronounced effect on protein synthesis (MacDonald *et al.,* 1966; Hanson and Krueger, 1966); furthermore, although chloramphenicol breaks dormancy in lettuce seeds, other protein-synthesis inhibitors have no effect (Frankland and Smith, 1967).

In addition to the oxidation of $NADPH_2$ by oxygen through the terminal oxidase involved in the PP pathway, any other agent capable of oxidising $NADPH_2$, and thus stimulating the operation of the PP pathway, will also break dormancy. This, it is suggested, explains the effect of nitrite and methylene blue, and also nitrate since this ion will be converted to nitrite in the seed.

In addition to these dormancy-breaking agents—cyanide, nitrite and chloramphenicol—whose activity could be plausibly accounted for in biochemical terms, the only other dormancy-breaking agent which had been investigated by Smith (1969) at that time, gibberellic acid, was in fact the best agent both in terms of its dormancy-breaking activity and in its ability to decrease the C-6/C-1 ratio.

There is now corroboration of the association of an increase in PP pathway activity with the loss of dormancy in wild oats (*Avena fatua*) (Simmonds and Simpson, 1971; 1972). In this case, because of the very pronounced dormancy in this species, it was possible to work

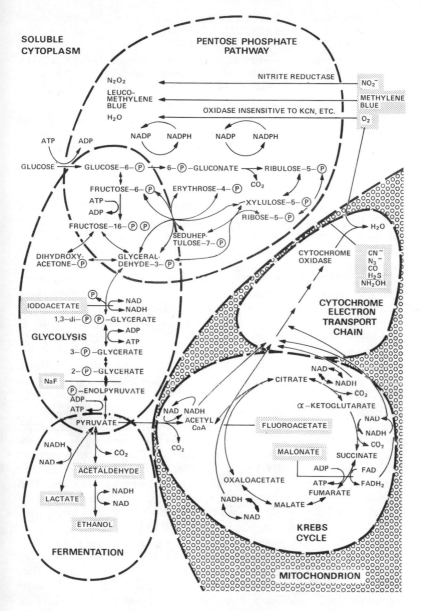

Figure 11.1 Respiratory metabolic pathways. Substances in stippled boxes stimulate germination in dormant seeds (see Table 11.1). The diagram indicates how such substances may directly or indirectly stimulate the activity of the pentose phosphate pathway

H

with isolated embryos rather than whole caryopses. Simmonds and Simpson (1971) found C-6/C-1 values (measured over the first 10 h of imbibition) in the region of 0·76–0·91 for dormant embryos and 0·54–0·59 for embryos after-ripened in dry storage. Similarly, they found that treating the seeds with gibberellic acid reduces the values of dormant embryos to 0·64–0·65 (Simmonds and Simpson, 1971), and that malonate, a respiratory inhibitor which breaks dormancy in this species, also reduces C-6/C-1 ratios from 0·83–0·85 to 0·73–0·83 (Simmonds and Simpson, 1972). It is encouraging to note that on the basis of these experiments they have postulated a dormancy hypothesis (Simmonds and Simpson, 1972) strikingly similar to that outlined in Roberts (1969) which was based primarily on work on barley and rice.

In the meantime, R. D. Smith has been investigating the effects of a number of other dormancy-breaking agents on the C-6/C-1 ratios of dormant barley seeds (cultivar Golden Promise). The detailed results will be published shortly but here the main findings are summarised.

Two sulphydryl compounds have been shown to be particularly effective dormancy-breaking agents, though they have a very narrow range of effective concentrations. 2-mercaptoethanol at 5×10^{-2} M increased germination from 8 to 89 per cent, while at the same time decreasing the C-6/C-1 ratio (during the third to sixth hour after the start of imbibition) from 0·36–0·42 to 0·03–0·04. Similarly dithiothreitol at 10^{-1} M increased germination from 10 to 90 per cent and decreased the C-6/C-1 ratio from 0·2–0·37 to 0·07–0·09.

Stratification treatments also show very clear effects. For example, it was found that in dormant seeds which were allowed to imbibe for 24 h at 25°C and then transferred to 1·5°C, the C-6/C-1 ratio in the 6 h after transfer was reduced to values of 0·29–0·35 as compared with values of 0·59–0·89 in seeds which were maintained at the higher temperature. Furthermore, in dormant seeds which had imbibed at 1·5°C for 24 h, the C-6/C-1 ratio subsequently determined at 25°C over the next 3 h was 0·30–0·37 as compared with 0·55–0·56 in seeds which had imbibed at 25°C.

The effect of dry after-ripening of dormant barley seeds at 43°C is also being investigated, and although more work needs to be done, an extension of the period of the dry heat treatment tends to result in a decrease of the C-6/C-1 ratio (determined 3–6 h after the start of imbibition), and in a concomitant increase in the ability to germinate.

Thus, so far, all the dormancy-breaking treatments which have been applied to barley have resulted in a reduction in the C-6/C-1 ratio long before there are visible signs of germination. These treatments have included nitrite, potassium cyanide, gibberellic acid, two sulphydryl reagents, stratification, high-temperature after-ripening, and normal after-ripening. Despite much evidence in support of the PP pathway hypothesis we have as yet no knowledge of how the PP pathway might be blocked in dormant seeds. S. A. Olusuyi, also working in this laboratory, has examined the quantity and activity of the two dehydrogenases of the PP pathway—glucose-6-phosphate dehydrogenase and 6-phosphogluconate dehydrogenase—in dormant and non-dormant barley seeds. He found that initially there was no significant difference (though there is an increase in the quantity of glucose-6-phosphate dehydrogenase in the non-dormant seeds as germination begins). He has also examined the levels of the oxidised and reduced forms of the co-enzymes involved: NAD and $NADH_2$ which are important in the EMP pathway, and NADP and $NADPH_2$ which are essential to the PP pathway. Again, in the dry seeds and during the first 24 h of imbibition there were no significant differences between dormant and non-dormant seeds. In both cases there were much greater quantities of NAD, typically about 1·7 nM per seed of the oxidised plus the reduced form as compared with a total of about 0·12 nM of both forms of NADP. In both cases the ratio of the oxidised to the reduced form was about 2·5–3·0 for $NAD/NADH_2$ as compared with 0·7–1·0 for $NADP/NADPH_2$. These figures suggest that the relative inactivity of the PP pathway in dormant seeds is not caused by lack of the appropriate dehydrogenases or co-enzyme. But the equilibria between the oxidised and reduced forms of the co-enzymes suggest that although the oxidation of $NADH_2$ is not limiting, the oxidation of $NADPH_2$ may well be. It may be then that the main restraint to germination in dormant seeds resides in the relative inactivity of the postulated, as yet unknown, $NADPH_2$–oxidase system, or in the super-optimal activity of the EMP pathway operating through the cytochrome system so that, although the $NADPH_2$–oxidase system is present and potentially active, it is not sufficiently active for germination to occur, because of competition for oxygen by cytochrome oxidase.

But even if this is where the control might reside, the question as to the significance of the PP pathway during early germination is enigmatic. As already discussed (Roberts, 1969), it does not seem likely that the almost traditional role of the PP pathway—that of producing reducing power in the form of $NADPH_2$ for synthetic reactions—can be invoked, since application of agents which may be expected to oxidise the co-enzyme (NO_2^- and methylene blue) stimulates loss of dormancy. It seems more likely that it is some intermediate, or some substance produced from an intermediate, of

Table 11.1   THE EFFECT

Species	Hydrogen acceptors				reduced N compounds	Respiratory inhibitors							
						of terminal oxidases					of Krebs		
	O_2	NO_3^-	NO_2^-	methylene blue		CN^-	N_3^-	H_2S or NaS	NH_2OH	CO	DIECA	mono fluoro-acetate	mal
Gramineae													
Avena fatua	+(18)	+(18)	+(34)	O(34)	O(34)	+(34)	+(34)						+(
Avena sativa		+(36)			O(36)	+(22)	+(22)	+(22)	+(22)		+(22)	O(22)	+(
Digitaria exelis						+(33)	+(33)		+(33)				
Hordeum distichon	+(22)	+(21)	+(21)	+(21)	O(21)	+(22)	+(22)	+(22)	+(22)	+(22)	+(22)	+(22)	+(
Hordeum spontaneum		+(24)				+(24)							+(
Lolium perenne						+(22)	O(22)	O(22)	O(22)				
Oryza sativa	+(30)	+(31)	+(31)	+(33)	O(31)	+(33)	+(33)	+(33)	+(33)	+(33)	O(33)	O(33)	O(
Zea mays						O(33)	O(33)		+(33)				
Iridaceae													
Neomarica gracilis						O(33)	+(33)		O(33)				
Solanaceae													
Browallia speciosa						+(33)	+(33)		O(33)				
Cruciferae													
Capsella bursa-pastoris	O(29)	+(29)	+(29)		O(29)								
Caryophyllaceae													
Melandrium noctiflorum						+(5)							
Silene noctiflora						+(5)							
Amaranthaceae													
Amaranthus caudatus						+(33)	O(33)		O(33)				
Amaranthus albus		+(17)	+(17)		O(17)	+(17)	+(17)						
Amaranthus hybridus		+(9)											
Compositae													
Lactuca sativa	+(10)	+(17)			O(17)	+(17)	+(17)		+(17)				
Leguminosae													
Trifolium subterraneum					O(2)		+(2)						

+ The agent has been reported to stimulate the germination of dormant seeds (except for light where either a stimulation or inhibition is scored as positive, indicating that the phytochrome system is probably involved; in this table all positive results for light are stimulatory except for *Avena fatua* and *Silene noctiflorum*).
O The agent has been tested but either has no effect or inhibits germination.

Note
The numbers in brackets refer to the following sources of information: (1) Ballard, 1961; (2) Ballard, 1967; (3) Ballard and Grant-Lipp, 1967; (4) Black and Richardson, 1967; (5) Borriss, 1967; (6) Browne *et al.*, 1948; (7) Burger 1965; (8) Cumming and Hay, 1958; (9) Engelhardt *et al.*, 1962; (10) Evenari, 1952; (11) Frankland and Smith 1967; (12) Friesen and Shebeski, 1961; (13) Gaber and Roberts, 1969; (14) Green and Helgeson, 1959; (15) Haber and Tolbert, 1959; (16) Hart and Berrie, 1966; (17) Hendricks and Taylorson, 1972; (18) Johnson, 1931; (19) Kendrick and Frankland,

RIOUS DORMANCY-BREAKING AGENTS

	glycolysis: NaF	uncouplers: DNP	uncouplers: Na_3AsO_4	chloramphenicol	End-products of fermentation: lactate	ethanol	acetaldehyde	CO_2	SH compounds: thiourea, mercaptoethanol, etc.	Hormones: gibberellins	cytokinins	Temperature: low temp. on imbibed seeds	high temp. on 'dry' seeds	Light
	O(34)	O(34)	O(34)					+(16)		+(14)		+(12) +(36)	+(18) +(6)	+(8)
2)	+(22)	+(22)	O(13)	+(21)	+(23)	+(23)	+(23)	+(23)	+(26)	+(25) +(24)	+(25)	+(40) +(24)	+(40) +(24)	+(7) +(7)
3)	O(33)	+(33)	O(33)					+(38)	+(32)	+(32)	+(32)	+(30)	+(30)	+(30)
					+(29)	+(29)	O(29)			+(29)	O(29)	+(29)	+(28)	+(29)
	O(5)	O(5)	O(5)											+(5)
	O(5)	O(5)	O(5)											+(5)
										+(19)		+(19)		+(19)
												+(35)		+(9)
		+(4)		+(11)					+(37)	+(10)	+(15)	+(15)	+(10)	+(10)
		+(2)	+(2)				O(2)	+(1)	+(20)			+(1)		

1969; (20) Grant-Lipp and Ballard, 1969; (21) Major, 1966; (22) Major and Roberts, 1968a; (23) Major and Roberts, 1968b; (24) Ogowara and Hayashi, 1964; (25) Pollock, 1959; (26) Pollock and Kirsop, 1956; (27) Roberts, 1969; (28) Popay, 1968; (29) Popay and Roberts, 1970a; (30) Roberts, 1962; (31) Roberts, 1963a; (32) Roberts, 1963b; (33) Roberts, 1964a; (34) Roberts and Madden (unpublished); (35) Santelman and Evetts, 1971; (36) Schwendiman and Shands, 1943; (37) Simmonds and Simpson, 1972; (38) Thornton, 1936; (39) Tseng, 1964; (40) Wiberg and Kolk, 1959.

the pathway that is likely to be important; but again it is not likely to be one of the obvious precursors, e.g. ribose-5-phosphate for nucleic acid synthesis, since there is evidence that dormant seeds of *Avena fatua* are capable of synthesising nucleic acids (Chen and Varner, 1967) and proteins (Chen and Varner, 1970) during imbibition at least at a rate comparable to that of non-dormant seeds. Furthermore, fractionation of nucleic acids into the components of MAK-column chromatography after incorporation of ^{32}P or glucose-^{14}C-U during the early stages of imbibition, have also indicated that dormant seeds of *Avena fatua* are at least as capable as non-dormant seeds of synthesising all fractions (Madden and Roberts, unpublished).

Even though we cannot be specific about what is required of the PP pathway it may be relevant that in most species, including barley and lettuce, the initial stages of germination involve cell elongation rather than cell division (Haber and Luippold, 1960). Sutcliffe and Sexton (1972) have shown quite clearly, using both enzyme assays on root sections and histochemical techniques, that in elongating cells in pea roots the PP pathway is very active and the EMP pathway is relatively unimportant; whereas in meristematic cells the converse holds. This supports similar conclusions which can be drawn from earlier work by Gibbs and Beevers (1955), who used labelling techniques in castor bean and maize root tips, and by Fowler and ap Rees (1970), who used both labelling techniques and enzyme assays in pea roots; in this earlier work it was suggested that the PP pathway is associated with differentiation, but the results would also fit in with Sutcliffe and Sexton's conclusions that the PP pathway is important in cell ontogeny before this and starts with the onset of cell elongation. Another interesting observation has been made by Kikuta, Akemine and Tagawa (1971) who have shown that the PP pathway is important just before callus from potato explants begins to grow.

THE APPARENT GENERALITY OF THE PENTOSE PHOSPHATE PATHWAY DORMANCY-BREAKING MECHANISM

In the work described so far the emphasis has been on cereal seeds. But there is evidence that a wide range of species show many similar responses both to physical and chemical dormancy-breaking agents. Some indication of this is given in *Table 11.1* where the stimulatory agents may be classified into seven main categories on which brief comment will be made.

HYDROGEN ACCEPTORS

One of the best known chemical agents for breaking seed dormancy in a wide range of species is the nitrate ion (Toole *et al.*, 1956). Steinbauer and Grigsby (1957) studied the germination of 85 weed species in 15 families and half of these showed more germination in the presence of nitrate. Although reduced forms of nitrogen such as urea, ammonium ions and amino acids have been reported to stimulate the breaking of dormancy in some species, e.g. in tobacco (Ogowara and Ono, 1955), in most species for which there is sufficient information only oxidised forms are effective, and this point is brought out in *Table 11.1*. The notion that their effect is due to their hydrogen-accepting ability is emphasised by the response of two species to methylene blue (as well as to nitrate and nitrite); to these a further species, *Chenopodium album*, can be added which was not included in the table (Williams and Harper, 1965). Removal or puncture of the covering structures, which is known to affect many species including most of those listed in *Table 11.1*, may possibly be included in the same category as hydrogen acceptors since it would lead to increased oxygen tensions.

RESPIRATORY INHIBITORS

The possibility has already been discussed here that, in addition to those substances specifically labelled as respiratory inhibitors, chloramphenicol could be included in this group. Since many people find the stimulatory activity of respiratory inhibitors somewhat surprising, it is worth mentioning that S. B. Hendricks and R. B. Taylorson (private communication, 1972) have recently found that (in addition to the species shown to respond in *Table 11.1*) dormancy is broken to various extents by cyanide in the following species: *Lepidium virginicum*, *Rumex crispus*, *Portulaca oleracea*, *Echinochloa crusgalli*, *Chenopodium album* and *Spinacia oleracea*.

END-PRODUCTS OF FERMENTATION

It is possible that these substances could also act as respiratory inhibitors as has been demonstrated for carbon dioxide in seeds (Kidd, 1916; Ranson, Walker and Clark, 1957) and for ethanol in other tissues (Miller, 1934). On the other hand it is possible that CO_2 might act in a different way (and this is why it has not been grouped with ethanol etc. in *Table 11.1*), its fixation leading to the production of oxidised NADP as it would if the fixation occurred as a result of malic enzyme activity. Again, it is known that in *Avena fatua* the CO_2 is fixed into malate (Hart and Berrie, 1967) and the evidence is compatible with this same possibility in subterranean clover (Ballard, 1967). Nevertheless, the notion of the production of oxidised NADP via CO_2 fixation is entirely speculative at present since, as yet, the CO_2 fixation mechanism is not known. It is not always realised how widespread the stimulatory effects might be. In some cases they are slight, e.g. in barley whereas, on the other hand, stimulatory activity of CO_2 has been observed in a number of species in addition to those listed in *Table 11.1*, viz. various *Trifolium* spp., *Medicago* spp. and *Trigonella ornithopoides* (Ballard, 1967), groundnuts (*Arachis hypogea*) (Toole, Bailey and Toole, 1964) *Sorghum halepense* (Harrington, 1917), and *Xanthium pennsylvanicum* (Thornton, 1935). But it should be mentioned that CO_2 can inhibit germination, usually at high concentrations of 20 per cent and above (Thornton, 1944), though in *Capsella bursa-pastoris* and *Senecio vulgaris*, in some circumstances, even at concentrations as low as 2 per cent (Popay and Roberts, 1970a).

SULPHYDRYL COMPOUNDS

Many sulphydryl compounds have a markedly stimulatory effect on dormant seeds. The well-known agent thiourea can be included under this heading because, as Pollock and Kirsop (1956) pointed out, in solution it is in equilibrium with the *iso* form:

$$NH_2CSNH_2 \rightleftharpoons NH_2C(SH):NH$$

But in barley and rice, at least, it is far less stimulatory than many other sulphydryl compounds. One peculiarity of these compounds is that, as R. D. Smith has found, some of them have an extremely narrow concentration range for optimum activity.

HORMONES

At present, four main classes of plant hormones are known—auxins, gibberellins, cytokinins and ethylene. In addition to these there are a number of naturally occurring growth inhibitors of which the best known is abscisic acid. Auxins do not appear to play an important role in seed dormancy. The position of ethylene is not yet clear; it does seem to lead to loss of dormancy in some species, including subterranean clover (Esahi and Leopold, 1969) but we do not know yet how widespread this response is (see Chapter 13 of this volume). Consequently, auxin and ethylene have been omitted from *Table 11.1*.

Khan (1971) has reviewed the effects of gibberellins and cytokinins in seed dormancy and proposes that gibberellins play a major role and are essential, whereas the function of cytokinins is mainly to overcome endogenous growth inhibitors which may or may not be present. In any case there is no doubt that the response to gibberellins is generally more pronounced and widespread. So far we do not know how gibberellic acid (GA) releases dormancy. The possibility that it derepresses the genome to allow RNA synthesis—as has been suggested for potato buds (Tuan and Bonner, 1964)—thus leading to protein synthesis, does not appear to hold in cereals, since samples of chromatin extracted from both dormant and non-dormant barley seeds are equally capable of RNA synthesis (Higham and Roberts, unpublished). Furthermore, as mentioned previously, the evidence suggests that dormant seeds of wild oats are capable of synthesising at least as much RNA as non-dormant seeds.

TEMPERATURE

The effects of temperature on dormancy are quite different, depending on whether the seed is 'dry' or imbibed. In most cases where this has been investigated on dry seeds, it has been found that the higher the temperature the more rapid the loss of dormancy. It may be recalled that the relationship in rice shows a Q_{10} of about 3·1 for rate of loss of dormancy. On the other hand, when the seeds are imbibed, they usually show a pronounced dormancy immediately after harvest, except within a small optimum temperature range. Vegis (1964) has very clearly drawn attention to this phenomenon. In some cases, e.g. *Polygonum persicaria, Datura stramonium* and *Chenopodium rubrum,* the temperature at which most germination takes place is sometimes relatively high (i.e. 35–45°C); but a much more common response is that partially dormant populations show much greater germination at low temperatures (i.e. 3–7°C). An allied aspect of this low temperature stimulation is the well-known stratification treatment in which dormant seeds are held in the imbibed condition at low temperatures (about 3–5°C) in order to break dormancy. In some cases the seed has to be removed to a higher temperature in order to 'bring out' the effect of stratification, since the seeds are incapable of germinating at low temperatures; for example, this is necessary in rice where the minimum temperature that allows germination to take place is about 17°C (Roberts, 1962).

In other cases e.g. in barley (Gaber, 1967), the seed can be left at the stratification temperature for germination to take place. If the seeds are after-ripened in dry storage then there is often no need for them to be stratified or set to germinate at a low temperature. In some cases stratification may produce its effect by activating the mechanism of gibberellin synthesis; this appears to be the case in hazel seeds (*Corylus avellana*) (Bradbeer, 1968; Ross and Bradbeer, 1968).

Finally, there is the important response to fluctuating temperatures. It has long been known that partially dormant populations of seeds show increased germination when set to germinate in fluctuating temperature regimes. In some cases, e.g. in *Capsella bursa-pastoris*, a major part of the effect can be explained by the stratification effect of the low temperature of the regime on dormancy and the effect of the high temperature on the rate of germination (Popay and Roberts, 1970a), but occasionally the actual *change* of temperature seems to be important, e.g. in *Oryzopsis miliacea* (Koller and Negbi, 1959); most commonly, however, there is not sufficient evidence to decide.

LIGHT

In most cases that have been examined in detail, light responses in seeds have been shown to be due to the phytochrome system: with almost no exceptions (but see Singh and Garg, 1971) most light-sensitive seeds appear to conform to the rule that their germination is promoted by red light and inhibited by far red, irrespective of whether it is inhibited or promoted by white light (Evenari, 1965). Consequently, it may be assumed that the same basic mechanism is involved in most of the seeds inhibited or promoted by white light. So far, although a number of theories have been suggested, the biochemical mechanisms involved in the phytochrome response are not known. However, an oxidation reaction appears to be involved, since no response takes place in the absence of oxygen, and on the basis of changes in oxygen uptake and CO_2 output in lettuce seeds, Evenari (1961) suggested that 'it seems obvious that the photomechanism acts at least partly by changing respiration and the respirational pathway.' In addition, active phytochrome may be involved in stimulating the synthesis of GA or its release from a bound form (Loveys and Wareing, 1971), although this would not seem to explain the dormancy of lettuce seeds (Bewley, Negbi and Black, 1968).

We have shown that five of the seven main categories of stimulatory agent described above are involved in stimulating the activity of the

PP pathway in barley seeds: hydrogen acceptors (nitrite), respiratory inhibitors (cyanide and chloramphenicol), hormones (gibberellic acid), sulphydryl compounds (2-mercaptoethanol and dithiothreitol), and temperature (stratification and high temperature 'dry' after-ripening). The effect of stratification had previously been shown on *Prunus cerasus* seeds (La Croix and Jaswal, 1967), and the effects of gibberellic acid, 'dry' after-ripening and of a respiratory inhibitor (malonic acid) have been corroborated in *Avena fatua* (Simmonds and Simpson, 1971; 1972). The main categories of factors left to be investigated for their effects on the PP pathway are light, the end-products of fermentation, and carbon dioxide.

ECOLOGICAL IMPLICATIONS

This final section discusses the ecological implications of the interaction between factors which stimulate the pentose phosphate pathway. In investigating the physiology of dormancy a large number of dormancy-breaking agents have been investigated. They have served not only to shed light on the mechanism involved but also to indicate that the same mechanism may be fundamental to a very wide range of species. But in the field only a few of these stimulatory factors can operate to control germination. The most important of these are likely to be light, temperature (high temperature when the seeds are dry, low temperatures or fluctuating temperatures when the seeds are imbibed), and nitrate (or nitrite) ions in the soil solution (Roberts, 1972b). To these one might add carbon dioxide, for though its possible inhibitory role now seems less important than it once was thought by Crocker (1948) and Bibbey (1948), because CO_2 levels in the soil seldom rise above 1 per cent (Russell, 1961), one should still bear in mind the possibility of CO_2 as a stimulatory factor involved, at relatively low concentrations, in the loss of dormancy of some species. Of these factors, light is probably the most important, as is indicated by the work of Wesson and Wareing (1967; 1969), but none of these factors can properly be considered on its own, since in many cases there is evidence that two or more stimulatory factors can act synergistically.

For this reason, Elizabeth Vincent and S. K. Benjamin, working in this laboratory, have been investigating the effects of the three most important factors—light, temperature and nitrate—and their interactions on the loss of dormancy in some of the more common annual weed species. From the preceding discussion it will be recognised that all three factors may affect an oxidation process and possibly the one concerned with the operation of the PP pathway;

irrespective of the validity of this speculation, however, it is known that these factors can have an important influence on germination in the field. Furthermore it has become obvious that to consider any one of these factors in isolation can give a very misleading impression, for often it is only a combination of factors which produces a dramatic effect and, after all, in the field seeds will seldom be subjected to only one stimulatory factor. The experiments to investigate the possible interactions were based on a simple 2^3 factorial design where the three main stimulatory factors—light, alternating temperature and nitrate—were tested on their own and in all combinations against control treatments which contain no stimulatory factors, i.e. where the seeds are held at constant temperature in the dark without nitrate. Two controls are, in fact, necessary so that both the higher and lower constant temperature can be compared with the four treatments which include alternating temperatures. Details of these investigations will be published in due course, but in the meantime examples of some of the findings are shown in *Figures 11.2* and *11.3*.

In *Figure 11.2* the results for four species are shown and a comparison is made between two different alternating-temperature regimes—one more extreme than the other. Germination tests lasted 28 days, using 200 seeds per treatment. Figures show percentage germination: in circles for single-factor treatments, in squares for 2-factor treatments, and in triangles for 3-factor treatments. Shaded areas indicate significant increases over whichever control shows the greater germination. Percentage germination values for control treatments (constant temperature in the dark without nitrate) are shown beneath each diagram—the value in the upper position was obtained at 25°C and in the lower position at the temperature corresponding to the lower temperature of the alternating-temperature regime. Constant temperature treatments were at 25°C. The symbol ▲ indicates treatment means which are different ($P = 0·05$) and = shows those which are not. Thus if a 2-factor treatment (in a square) has two arrow-heads pointing outwards, there has been a significant positive interaction between the two factors involved; a significant 3-factor positive interaction is shown by three arrow-heads pointing out from the triangle. See key for guide to each treatment. In *Figure 11.3* the effects of different periods of stratification are shown on interactions which become apparent subsequently when the seeds are set to germinate. Alternating-temperature treatments were 8 h at 25°C and 16 h at 15°C; constant temperature treatments were run at 25°C. 400 seeds were used in each treatment. (Further explanation as for *Figure 11.2*.) Summarising briefly, it is evident that strong positive interactions are the rule rather than the exception. Every possible two-factor interaction (light × alternating

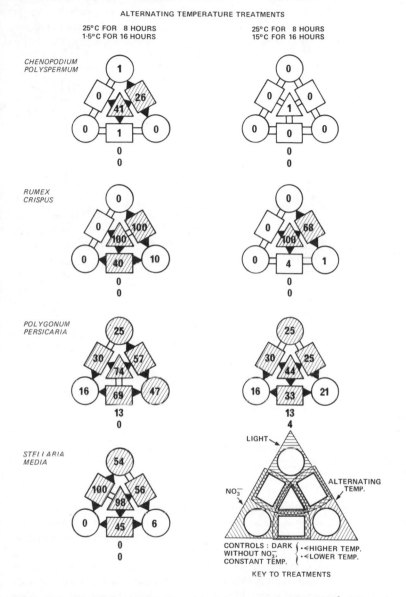

Figure 11.2 The effects of light, alternating temperatures and nitrate (10^{-2} M) on the germination of Chenopodium polyspermum, Rumex crispus, Polygonum persicaria and Stellaria media (previously unpublished work of Elizabeth Vincent)

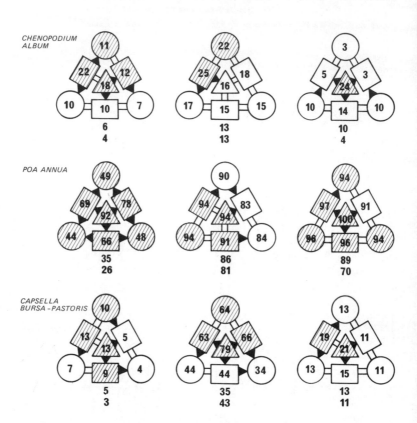

Figure 11.3   The effects of light, alternating temperature and nitrate on seeds of Chenopodium album, Poa annua, *and* Capsella bursa-pastoris *which had been stratified (held imbibed at 1·5°C) for 0, 4 or 21 days (previously unpublished work of S. K. Benjamin)*

temperature; alternating temperature × nitrate; and nitrate × light) is encountered, and a further stimulation of germination often occurs if the three factors are combined. The more extreme temperature alternations are more stimulatory than the less extreme ones. A short period of stratification is more stimulatory than a long period.

From these results it is evident that although light is still probably the most important single factor in controlling the seed germination of annual weeds, very often it has little effect on its own, but needs to be combined with another factor (*Rumex crispus* is an extreme case);

and sometimes light is not essential but can be replaced by a combination of nitrate with alternating temperatures (e.g. in *Polygonum persicaria*). These results emphasise that a laboratory investigation of any stimulatory factor on its own is likely to give an inaccurate impression of its importance in the field.

The most important adaptive physiological response of seeds of annual weeds is probably their tendency to germinate at the soil surface. But the interactions which have been discussed here help to provide some possible explanation (and potentially, at least, some control) of two other characteristic germination responses of these seeds—both of which involve germination beneath the soil surface. Firstly, there are the peculiar germination flushes in undisturbed soil which vary in timing and frequency each year but appear to occur in response to some combination of environmental circumstances—e.g. in *Capsella bursa-pastoris* and *Senecio vulgaris* (Popay and Roberts, 1970b). Secondly, there is the evidence that one of the more, if not most, important factors leading to the depletion of weed seed populations in the soil is the germination of the seeds, and thus the possible death of the seedlings, below the surface (Roberts, 1972b). It seems possible that some of the interactions not involving light could cause some of this subterranean germination in undisturbed soil, whereas soil cultivation and consequent exposure to light provides additional and more powerful interactions which result in the more rapid depletion so clearly demonstrated by H. A. Roberts (1967), which is found under intensive cultivation regimes.

It is hoped that ultimately, the two-pronged approach outlined here the investigation of the physiological mechanisms and the study of interactions between environmental factors—will lead to a greater understanding of field behaviour and possibly to more rational systems of weed control.

ACKNOWLEDGEMENTS

I am very grateful to Dr. J. A. Simmonds and Dr. G. M. Simpson and also to Dr. S. B. Hendricks and Dr. R. B. Taylorson for allowing me to see their recent papers before publication. I should also like to thank R. D. Smith, S. A. Olusuyi, Elizabeth Vincent and S. K. Benjamin for allowing me to quote some of their results before publication elsewhere.

REFERENCES

AP REES, T. and BEEVERS, H. (1960a). 'Pathways of glucose dissimilation in carrot slices', *Pl. Physiol.*, 35, 830–838

AP REES, T. and BEEVERS, H. (1960b). 'Pentose phosphate pathway as a major component of induced respiration of carrot and potato slices', *Pl. Physiol.*, 35, 839–847

BALLARD, L. A. T. (1961). 'Studies of dormancy in the seeds of subterranean clover (*Trifolium subterraneum* L.). II. The interaction of time, temperature, and carbon dioxide during passage out of dormancy., *Aust. J. biol. Sci.*, 14, 173–186

BALLARD, L. A. T. (1967). 'Effects of carbon dioxide on the germination of leguminous seeds', in *Physiology, Ecology, and Biochemistry of Germination*, Vol. 1, (ed. Borriss, H.) 209–219 Ernst-Moritz-Arndt- Universität, Greifswald

BALLARD, L. A. T. and GRANT-LIPP, A. E. (1967). 'Seed dormancy: breaking by uncouplers and inhibitors of oxidative phosphorylation', *Science*, 156, 398–399

BEEVERS, L., FLESCHER, D. and HAGEMAN, R. H. (1964). 'Studies on the pyridine nucleotide specificity of nitrate reductase in higher plants and its relationship to sulfhydryl level', *Biochim. biophys. Acta.* 89, 453–464

BELDEROK, B. (1963). 'Verslag van het in 1962 verrichte onderzoek aangaande kiemrust en schot', *Versl. Tienjarenplan Graanonderz.*, 1962, 9, 63–72

BEWLEY, J. D., NEGBI, M. and BLACK, M. (1968). 'Immediate phytochrome action in lettuce seeds and its interaction with gibberellins and other germination promoters', *Planta*, 78, 351–357

BIBBEY, R. O. (1948). 'Physiological studies of weed seed germination', *Pl. Physiol.*, 23, 467–484

BLACK, M. and RICHARDSON, M. (1967). 'Germination of lettuce induced by inhibitors of protein synthesis', *Planta*, 73, 344–356

BLOOM, G. and STETTON, D. (1953). 'Pathways of glucose metabolism', *J. Am. chem. Soc.*, 75, 5446

BORRISS, H. (1967). 'Experiments on modification of light reactions of seeds by inhibitors and dyes', in *Physiology, Ecology and Biochemistry of Germination*, Vol. 1., (ed. Borriss, H.), 155–170, Ernst-Moritz-Arndt-Universität, Greifswald

BRADBEER, J. W. (1968). 'Studies in seed dormancy. IV. The role of endogenous inhibitors and gibberellin in the dormancy and germination of *Corylus avellana* L., seeds', *Planta*, 78, 266–276

BROWNE, E., STANTON, T. R., WIEBE, G. A. and MARTIN, J. H. (1948). *Dormancy and the Effect of Storage on Oats, Barley, and Sorghum*, U.S. Dept. Agric. Tech. Bull. No. 953, 30

BURGER, W. C. (1965). 'Effect of light on the germination of barley and its relation to dormancy', *J. Inst. Brew.*, 71, 244–250

BUTT, V. S. and BEEVERS, H. (1960). 'Hexose metabolism in maize roots', *Biochem. J.*, 76, 51

BUTT, V. S. and BEEVERS, H. (1961). 'The regulation of glucose catabolism in maize roots', *Biochem. J.*, 80, 21–27

CALDWELL, F. (1959). 'Some notes on dormancy in cereal grains', *Agric. hort. Engng. Abstr.*, 9, 189

CARR, D. J. (1961). 'Chemical influences of the environment', *Handb. PflPhysiol.*, 16, 737–794

CHAKRAVORTY, M. and BURMA, D. P. (1959). 'Enzymes of the pentose phosphate pathway in the mung-bean seedling', *Biochem. J.*, 73, 48–53

CHEN, S. S. C. and VARNER, J. E. (1967). *MSU/AEC Plant Res. Lab. Ann. Rep.*, 1967. Michigan State University, East Lansing

CHEN, S. S. C. and VARNER, J. E. (1970). 'Respiration and protein synthesis in dormant and after-ripened seeds of *Avena fatua*', *Pl. Physiol.*, 46, 108–112

CROCKER, W. (1948). *Growth of Plants: Twenty Years Research at Boyce Thompson Institute*, Reinhold, New York

CUMMING, B. G. and HAY, J. R. (1958). 'Light and dormancy in wild oats (*Avena fatua*, L.). *Nature*, **182**, 609–610

EFFER, W. R. and RANSON, S. L. (1967a). 'Respiratory metabolism in buckwheat seedlings', *Pl. Physiol.*, **42**, 1042–1052

EFFER, W. R. and RANSON, S. L. (1967b). 'Some effects of oxygen concentration on levels of respiratory intermediates in buckwheat seedlings', *Pl. Physiol.*, **42**, 1053–1058

ENGELHARDT, M., VICENTE, M. and SILBERSCHMIDT, K. (1962). 'Acao da luz do nitrato de potassio sobre a germinacao de *Amaranthus hybridus* L.', *Revta. bras. Biol.*, **22**, 1–7

ESAHI, Y. and LEOPOLD, A. C. (1969). 'Dormancy regulation in subterranean clover seeds by ethylene', *Pl. Physiol.*, **44**, 1470–1472

EVENARI, M. (1952). 'The germination of lettuce seeds. I. Light, temperature and coumarin as germination factors'. *Palest J. Bot., Jerusalem Ser.*, **5**, 138–160

EVENARI, M. (1961). 'A survey of the work done in seed physiology by the Department of Botany, Hebrew University, Jerusalem (Israel)), *Proc. int. Seed. Test. Ass.*, **26**, 597–657

EVENARI, M. (1965). 'Light and seed dormancy'. *Handb. Pfl Physiol.*, **15** (2), 804–847

FISCHNICH, O., THIELBEIN, M. and GRAHL, A. (1962). 'Die Beeinflussung des Saatgutweres von Getreide durch äussere Faktoren unter besonderer Berücksichtigung der Keimruhe', *Saatgut-Wirt.*, **14**, 12–14, 39–42

FOWLER, M. W. and AP REES, T. (1970). 'Carbohydrate oxidation during differentiation of roots of *Pisum sativum*', *Biochim. biophys. Acta*, **201**, 33–44

FRANKLAND, B. and SMITH, H. (1967). 'Temperature and other factors affecting chloramphenicol stimulation of the germination of light-sensitive lettuce seeds', *Planta*, **77**, 354–366

FRIESEN, G. and SHEBESKI, L. H. (1961). 'The influence of temperature on the germination of wild oat seeds', *Weeds*, **9**, 634–638

GABER, S. D. (1967). *Investigations into Factors Controlling Water-sensitivity in Barley Seeds*, Ph.D. Thesis, Manchester University

GABER, S. D. and ROBERTS, E. H. (1969). 'Water-sensitivity in barley seeds, II. Association with micro-organism activity', *J. Inst. Brew.*, **75**, 303–314

GIBBS, M. (1954). 'The respiration of the pea plant. Oxidation of hexose phosphate and pentose phosphate by cell-free extracts of pea leaves'. *Pl. Physiol.*, **29**, 34–39

GIBBS, M. and BEEVERS, H. (1955). 'Glucose dissimilation in the higher plant. Effect of age of tissue', *Pl. Physiol.*, **30**, 340–347

GRANT-LIPP, A. E. and BALLARD, L. A. T. (1969). 'Thiourea as a stimulator and inhibitor of germination of seed of subterranean clover (*Trifolium subterraneum* L.)', *Z. Pfl Physiol.*, **62**, 83–88

GREEN, J. G. and HELGESON, E. A. (1959). 'The effect of gibberellic acid on dormant seeds of wild oats', *Proc. 14th N. Cent. Weed Contr. Conf., U.S.A., 1957*, 39

HABER, A. H. and LUIPPOLD, H. J. (1960). 'Separation of mechanisms involving cell division and cell expansion in lettuce seed germination', *Pl. Physiol.*, **35**, 168–173

HABER, A. H. and TOLBERT, N. E. (1959). 'Effects of gibberellic acid, kinetin, and light on the germination of lettuce seed', in *Photoperiodism and Related Phenomena in Plants and Animals* (ed. Withrow, R. B.), 197–206, Amer. Soc. Adv. Sci., Washington

HAGEMAN, R. H., CRESSWELL, C. F. and HEWITT, E. J. (1962). 'Reduction of nitrate, nitrite and hydroxylamine to ammonia by enzymes extracted from higher plants', *Nature*, **193**, 247–250

HANSON, J. B. and KRUEGER, W. A. (1966). 'Impairment of oxidative phosphorylation by D-Threo- and L-Threo-Chloramphenicol', *Nature*, **211**, 1322

HAREL, E. and MAYER, A. M. (1963). 'Studies on the role of cytochrome oxidase and phenolase in germinating lettuce using CO as an inhibitor', *Physiologia. Pl.,* **16**, 804–813

HARRINGTON, G. T. (1917). 'Further studies on the germination of Johnson grass weeds', *Proc. Ass. off. Seed. Analysts. N. Am.,* **9/10**, 71–76

HARRINGTON, G. T. (1923). 'Forcing the germination of wheat and other cereals', *J. agric. Res.,* **23**, 79–100

HART, J. W. and BERRIE, A. M. M. (1966). 'The germination of *Avena fatua* under different gaseous environments', *Physiologia. Pl.,* **19**, 1020–1025

HART, J. W. and BERRIE, A. M. M. (1967). 'Relationship between endogenous levels of malic acid and dormancy in grain of *Avena fatua* L.', *Phytochem.,* **7**, 1257–1260

HENDRICKS, S. B. and TAYLORSON, R. B. (1972). 'Promotion of seed germination by nitrates and cyanides'. *Nature,* **237**, 169–170

JOHNSON, L. V. P. (1931). 'General preliminary studies in the physiology of delayed germination in *Avena fatua*', *Can. J. Res. (C),* **13**, 283–300

KENDRICK, R. E. and FRANKLAND, B. (1969). 'Photocontrol of germination in *Amaranthus caudatus*', *Planta,* **85**, 326–339

KHAN, A. A. (1971). 'Cytokinins: permissive role in seed germination', *Science,* **171**, 853–859

KIDD, F. (1916). 'The controlling influence of carbon dioxide. III. The retarding effect of carbon dioxide on respiration', *Proc. R. Soc.,* **B89**, 136–156

KIKUTA, Y., AKEMINE, T. and TAGAWA, T. (1971). 'Role of the pentose phosphate pathway during callus development in explants from potato tuber', *Pl. Cell. Physiol.,* **12**, 73–79

KOLLER, D. and NEGBI, M. (1959). 'The regulation of germination in *Oryzopsis miliacea*', *Ecology,* **40**, 20–36

LA CROIX, L. J. and JASWAL, A. S. (1967). 'Metabolic changes in after-ripening seed of *Prunus cerasus*', *Pl. Physiol.,* **42**, 479–480

LOVEYS, B. R. and WAREING, P. F. (1971). 'The red light controlled production of gibberellin in etiolated wheat leaves', *Planta,* **98**, 109–116

MACDONALD, I. R. (1959). 'Changes in sensitivity to inhibitors of discs of storage tissue', *Ann. Bot.,* **23**, 241

MACDONALD, I. R., BACON, J. S. D., VAUGHAN, D. and ELLIS, R. J. (1966). 'The relationship between ion absorption and protein synthesis in beet disks', *J. exp. Bot.,* **17**, 822–837

MAJOR, W. (1966). *Investigations into the Physiology of Dormancy in Cereal Seeds,* Ph.D. Thesis, Manchester University

MAJOR, W. and ROBERTS, E. H. (1968a). 'Dormancy in cereal seeds. I. The effects of oxygen and respiratory inhibitors', *J. exp. Bot.,* **19**, 77–89

MAJOR, W. and ROBERTS, E. H. (1968b). 'Dormancy in cereal seeds. II. The nature of the gaseous exchange in imbibed barley and rice seeds', *J. exp. Bot.,* **19**, 90–101

MAYER, A. M., POLJAKOFF-MAYBER, A. and APPLEMAN, W. (1957). 'Studies on the oxidative systems in germinating lettuce seeds', *Physiol. Pl.,* **10**, 1–13

MERRY, J. and GODDARD, D. R. (1941). 'A respiratory study of barley grain and seedlings', *Proc. Rochester Acad. Sci.,* **8**, 28

MILLER, L. P. (1934). 'Time relations in effect of ethylene chlorohydrin in increasing and ethyl alcohol in decreasing the respiration of potato tubers', *Contr. Boyce Thompson Inst. Pl. Res.,* **6**, 123

NYMAN, B. (1963). *Studies on the Germination in Seeds of Scots Pine,* Skogshögskolan, Stockholm

OGOWARA, K. and HAYASHI, J. (1964). 'Dormancy studies in *Hordeum spontaneum* seeds', *Ber. Ohara Inst. Landw, Biol.,* **12**, 159–188

OGOWARA, K. and ONO, K. (1955). 'Effect of various nitrogen compounds and respiratory intermediates on the germination of the light-favoured tobacco seeds', *Bull. School Educ. Okoyama Univ.,* No. 1, 97–104

OSMOND, C. B. and AP REES, T. (1969). 'Control of the pentose-phosphate pathway in yeast', *Biochem. biophys. Acta*, **184**, 35–42

POLJAKOFF-MAYBER, A. (1955). 'Oxidative activity of particles prepared from lettuce seedlings', *J. exp. Bot.*, **6**, 313–320

POLJAKOFF-MAYBER, A. and EVENARI, M. (1958). 'Some further investigations on the oxidative systems of germinating lettuce seeds', *Physiologia. Pl.*, **11**, 84–91

POLJAKOFF-MAYBER, A. and MAYER, A. M. (1958). 'Some further studies on the direct oxidation of glucose in germinating lettuce seeds', *Bull. Res. Coun. Israel*, **6D**, 86–87

POLLOCK, J. R. A. (1959). 'Studies in barley and malt. XV. Growth substances and other compounds in relation to dormancy in barley', *J. Inst. Brew.*, **65**, 334–337

POLLOCK, J. R. A. and KIRSOP, B. H. (1956). 'Studies in barley and malt. VI. Stimulation of the germination of freshly-harvested barley,' *J. Inst. Brew.*, **62**, 323–327

POPAY, A. I. (1968). '*The Ecology of Annual Weeds in Relation to Germination Behaviour*', Ph.D. Thesis, Manchester

POPAY, A. I. and ROBERTS, E. H. (1970a). 'Factors involved in the dormancy and germination of *Capsella bursa-pastoris* (L.) Medik and *Senecio vulgaris* L.', *J. Ecol.*, **58**, 103–122

POPAY, A. I. and ROBERTS, E. H. (1970b). 'Ecology of *Capsella bursa-pastoris* (L.) Medik. and *Senecio vulgaris* L. in relation to germination behaviour', *J. Ecol.*, **58**, 123–139

RANSON, S. L., WALKER, D. A. and CLARK, D. (1957). 'The inhibition of succinic oxidase by high carbon dioxide concentrations', *Biochem. J.*, **66**, 57

ROBERTS, E. H. (1961). 'Dormancy in rice seed. II. The influence of covering structures', *J. exp. Bot.*, **12**, 430–445

ROBERTS, E. H. (1962). 'Dormancy in rice seed. III. The influence of temperature, moisture, and gaseous environment', *J. exp. Bot.*, **13**, 75–94

ROBERTS, E. H. (1963a). 'The effects of inorganic ions on dormancy in rice seed', *Physiologia. Pl.*, **16**, 732–744

ROBERTS, E. H. (1963b). 'The effects of some organic growth substances and organic nutrients on dormancy in rice seed', *Physiologia. Pl.*, **16**, 745–755

ROBERTS, E. H. (1964a). 'The distribution of oxidation-reduction enzymes and the effects of respiratory inhibitors and oxidising agents on dormancy in rice seed', *Physiologia. Pl.*, **17**, 14–29

ROBERTS, E. H. (1964b). 'A survey of the effects of chemical treatments on dormancy in rice seed', *Physiologia. Pl.*, **17**, 30–43

ROBERTS, E. H. (1965). 'Dormancy in rice seed. IV. Varietal responses to storage and germination temperatures', *J. exp. Bot.*, **16**, 341–349

ROBERTS, E. H. (1969). 'Seed dormancy and oxidation processes', *Symp. Soc. exp. Biol.*, **23**, 161–192

ROBERTS, E. H. (1972a). 'Cytological, genetical, and metabolic changes associated with loss of viability', in *Viability of seeds* (ed. Roberts, E. H.), 253–306, Chapman and Hall, London

ROBERTS, E. H. (1972b). 'Dormancy: a factor affecting seed survival in the soil', in *Viability of Seeds* (ed. Roberts, E. H.), 321–59, Chapman and Hall, London

ROBERTS, E. H. (1973). 'Loss of viability: ultrastructural and physiological aspects', *Seed Sci. Tech.*, **1** (in press)

ROBERTS, H. A. (1967). 'Effect of cultivation on the numbers of viable weed seeds in soil', *Weed Res.*, **7**, 290–301

ROSS, J. D. and BRADBEER, J. W. (1968). 'Concentrations of gibberellin in chilled hazel seeds', *Nature*, **220**, 85–86

RUSSELL, E. W. (1961). *Soil Conditions and Plant Growth*, 9th edn., Longmans, London

SANTELMANN, P. W. and EVETTS, L. (1971). 'Germination and herbicide susceptibility of six pigweed species', *Weed. Sci.*, **19**, 51–54

SCHWENDIMAN, A. and SHANDS, H. L. (1943). 'Delayed germination or seed dormancy in Vicland oats', *J. Am. Soc. Agron.*, **35**, 681– 688

SIMMONDS, J. A. and SIMPSON, G. M. (1971). 'Increased participation of pentose phosphate pathway in response to afterripening and gibberellic acid treatment in caryopses of *Avena fatua* L.', *Can. J. Bot.*, **49**, 1833–1840

SIMMONDS, J. A. and SIMPSON, G. M. (1972). 'Regulation of the Krebs cycle and pentose phosphate pathway activities in the control of dormancy of *Avena fatua*, L.', *Can. J. Bot.* (in press)

SINGH, G. and GARG, O. P. (1971). 'Effect of red, far-red radiations on germination of cotton seed', *Pl. Cell Physiol.*, **12**, 411–415

SMITH, R. D. (1969). *Some Aspects of the Respiratory Mechanisms of Dormant and Non-dormant Barley Seeds*, M.Sc.Thesis, Manchester University

STEINBAUER, G. P. and GRIGSBY, B. (1957). 'Interaction of temperature, light and moistening agent in the germination of weed seeds', *Weeds*, **5**, 175–182

STILES, W. (1960). 'Respiration in seed germination and seedling development', *Handb. PflPhysiol.* **12**, (2), 465–492

SUTCLIFFE, J. F. and SEXTON, R. (1972). 'Enzymatic changes during the differentiation of tissues in young pea roots', Tatras Symposium on *Structure and function of primary root tissues* (in press)

THIMANN, K. V., YOCUM, C. S. and HACKETT, D. P. (1954). 'Terminal oxidation and growth in plant tissues. III. Terminal oxidation in potato tuber tissue', *Archs. Biochem. Biophys.*, **53**, 239

THORNTON, N. C. (1935). 'Factors influencing germination and development in cocklebur seeds', *Contr. Boyce Thompson Inst. Pl. Res.*, **7**, 477–496

THORNTON, N. C. (1936). 'Carbon dioxide storage. IX. Germination of lettuce seeds at high temperatures in both light and darkness', *Contr. Boyce Thompson Inst. Pl. Res.*, **8**, 25–40

THORNTON, N. C. (1944). 'Carbon dioxide storage. XII. Germination of seeds in the presence of carbon dioxide', *Contr. Boyce Thompson Inst. Pl. Res.*, **13**, 355–360

TOOLE, E. H., HENDRICKS, S. B., BORTHWICK, H. A. and TOOLE, V. K. (1956). 'Physiology of seed germination', *A. Rev. Pl. Physiol.*, **7**, 299–324

TOOLE, V. K., BAILEY, W. K. and TOOLE, E. H. (1964). 'Factors influencing dormancy of peanut seeds', *Pl. Physiol.*, **39**, 822–831

TSENG, S. (1964). 'Breaking dormancy of rice seed with carbon dioxide', *Proc. int. Seed Test. Ass.*, **29**, 445–450

TUAN, D. Y. H. and BONNER, J. (1964). 'Dormancy associated with repression of genetic activity', *Pl. Physiol.*, **39**, 768–772

URION, E. and CHAPON, L. (1955). 'Releasing barley germinative energy', *Amer. Brewer 88th year, March*, 41

VEGIS, A. (1964). 'Climatic control of germination, bud break, and dormancy', in *Environmental Control of Plant Growth* (ed. Evans, L. T.), 265–287 Academic Press, New York

WESSON, G. and WAREING, P. F. (1967). 'Light requirements of buried seeds', *Nature*, **213**, 600–601

WESSON, G. and WAREING, P. F. (1969). 'The role of light in the germination of naturally occurring populations of buried weed seeds', *J. exp. Bot.*, **20**, 402–413

WIBERG, H. and KOLK, H. (1959). 'Effect of gibberellin on germination of seeds', *Proc. int. Seed Test. Ass.*, **24**, 440–443

WILLIAMS, J. T. and HARPER, J. L. (1965). 'Seed polymorphism and germination. I. Influence of nitrates and low temperatures on the germination of *Chenopodium album*', *Weed Res.*, **5**, 141–150

DISCUSSIONS

Rorison: Is not the nitrate concentration which you use to stimulate germination unduly high, in ecological terms? And why confine yourself to nitrate?

E. H. Roberts: Nitrate concentrations of 10^{-3} M or less which are known to be effective in stimulating germination occur in agricultural soils, in particular at rising temperatures in spring. Virtually no other common ion has the pronounced and widespread effect of nitrate and nitrite. It is true, however, that at present we have no information about the importance of nitrate in the ecological situation.

Wareing: According to Williams and Harper the increase in soil nitrate in spring could stimulate the germination of *Chenopodium* seeds (Williams, J. T. and Harper, J. L., 1965, *Weed Res.*, **5**, 141–150). Admittedly, freshly shed seeds may respond to the interaction of high nitrate and fluctuating temperatures by germinating; but the many seeds which lie buried in soil in a dormant state for many years may well have been selected for light requirement.

E. H. Roberts: Yes, but some seeds may respond to light only when it interacts with nitrate and/or high temperatures (for example, see the results for *Chenopodium album* in *Figure 11.3*).

Harrington: *Striga* (witch weed) seeds are said to germinate only if the host plant exudes cytokinins.

E. H. Roberts: The evidence about root secretions is conflicting. According to R. Brown and colleagues working at Oxford (Brown, R., Johnson, A. W., Robinson, E., and Todd, A. R., 1949, *Proc. Roy. Soc. B.*, **136**, 1–12) sorghum roots secrete xylulose, a simple 5-carbon sugar (an intermediate of the PP pathway) which appears to stimulate *Striga* seed germination in extremely low concentrations down to 10^{-8} M. It also stimulates cell elongation in pea roots; despite its simple nature it seems to act like a gibberellin (Brown, R., Robinson, E., and Johnson, A. W., 1949, *Nature, Lond.*, **163**, 842–843). Interestingly, as shown by Sutcliffe and Sexton (see reference at end of this chapter), those cells which are just beginning to elongate have an extremely high activity of the PP pathway, whereas enzymes of conventional respiration are almost absent or inactive. It is possible that the PP pathway is required for cell elongation which is often also the first step in germination, and poor operation of the PP pathway may be a cause of dormancy.

Smith: Perhaps an intermediate of the PP pathway is important.

E. H. Roberts: The operation of the PP pathway begins within three hours of the application of GA_3, much earlier than any increase in nucleic acids; and nucleic acid synthesis is equally high in dormant and non-dormant wild oat seeds.

Gulliver: Can you break dormancy by supplying intermediates of the PP pathway?

E. H. Roberts: M. Black (Queen Elizabeth College, University of London) (private communication) has shown that hydrolytic enzyme activity can be induced in aleurone and tuber tissue with low concentrations (10^{-5} to 10^{-6} M) of ribose. But we have only in exceptional cases been able to stimulate germination of barley by applications of ribose.

12

LIGHT QUALITY AND GERMINATION: ECOLOGICAL IMPLICATIONS

H. SMITH

Department of Physiology and Environmental Studies, University of Nottingham School of Agriculture

LIGHT AND SEED GERMINATION

The germination of many seeds is sensitive to white light, some seeds requiring light for germination and others being prevented from germinating by light treatment. This phenomenon, which has been known since at least the end of the eighteenth century (see review by Black, 1969), was termed *photoblastism* by Evenari in 1956. Seeds whose germination is stimulated by white light were considered to be positively photoblastic, whereas those whose germination was inhibited by white light were negatively photoblastic. Yet another category, that of non-photoblastic seeds, was recognised whose germination was not affected either way by white light. Since 1956, a great deal of research has centred on the photocontrol of seed germination and it has become clear that the above definitions are no longer meaningful in a mechanistic sense. For example, it is now known that stimulatory and inhibitory effects of white light are simply two opposing facets of a single common mechanism; furthermore, many, and possibly all, non-light-sensitive seeds can be made light sensitive by various stress treatments.

The cause of this unification of ideas was the realisation that different spectral regions of white light are responsible for the stimulatory and inhibitory effects on germination. In the 1950s, action spectra were determined for both responses, clearly showing that red light stimulated germination whilst far red inhibited it. For example, Borthwick and Hendricks and their colleagues, in a now classical paper, showed that the germination of *Lactuca sativa* var. Grand Rapids was stimulated by red light and inhibited by far red (Borthwick *et al.*, 1952). This phenomenon had originally been discovered by Flint and McAllister (1935; 1937) in the 1930s but had remained unnoticed. For seed whose germination is inhibited by white light, Jones and Bailey (1956) showed that far-red light inhibited the germination of *Lamium amplexicaule* seed; subsequent red irradiation resulted in a re-promotion of germination.

Thus, 'positively photoblastic' seeds can be considered as seeds which perceive predominantly the red spectral regions of white light, whereas 'negatively photoblastic' seeds perceive mainly the far-red regions. Since in both cases red/far-red photoreversibility can be demonstrated (Borthwick *et al.*, 1952; Jones and Bailey, 1956) it is reasonable to conclude that phytochrome is the photoreceptor responsible for the perception of the light stimulus. In the simple

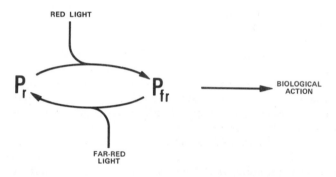

RED LIGHT

P_r P_{fr} BIOLOGICAL ACTION

FAR-RED LIGHT

Figure 12.1   The simple model for phytochrome action

view of phytochrome action shown in *Figure 12.1* red light is considered to convert P_r, the probably inactive form of phytochrome, to P_{fr}, the active form, which then, by some as yet unknown mechanism, sets in train a whole range of developmental phenomena, including seed germination. Far-red light will reverse this process in the early stages by re-converting P_{fr} to P_r. In the original formulation of the hypothesis, only the P_r was considered to be present in dark-imbibed seeds and dark-grown seedlings, and therefore the scheme does not immediately explain how some seeds react to white light as if it were red light, and others as if it were far-red light.

The answer to this intriguing question has come from recent elegant experiments on the behaviour of a range of seeds which normally germinate well in darkness but which are prevented from germination by continuous far-red light. Several seeds, including cucumber (Yaniv, Mancinelli and Smith, 1967), lettuce var. May Queen (Boisard, 1969), *Nemophila insignis* (Rollin, Malcoste and Eude, 1970) and *Amaranthus caudatus* (Kendrick and Frankland, 1969) come into this category and each group has shown that intermittent far red (i.e. repeated short exposures to far red separated by about 20 min darkness) is equally as effective as continuous far red. In order to explain these results, it was suggested that dark-germinating seeds contain significant proportions of P_{fr} which is continually regenerated in darkness after each intermittent far-red light exposure. Thus, P_{fr} in the dark-imbibed seeds allows germination, and only by its continual removal by continuous or intermittent far red can germination be prevented.

This somewhat unorthodox hypothesis was elegantly confirmed by the use of an extremely sensitive *in vivo* spectrophotometer developed at Wageningen (Spruit, 1970), in which it was shown that P_{fr} does indeed exist in dark-imbibed seeds. In lettuce May Queen, 40 per cent of the total phytochrome was present as P_{fr} (Boisard, Spruit and Rollin, 1968); in *Amaranthus*, 25 per cent was P_{fr} (Kendrick, Spruit and Frankland, 1969); and in cucumber, 75 per cent was P_{fr} (Spruit and Mancinelli, 1969). In some cases it was even possible to detect P_{fr} in the dry seeds. Furthermore, it was shown that after reversal of the seed-P_{fr} to P_r, regeneration of P_{fr} occurred in darkness during a period of about 20 min. It should be stated here that the phytochrome detected in the imbibed seed does not appear to be fully representative of that which later becomes detectable in vastly greater quantities in the dark-grown seedling. However, several lines of argument suggest that the seed-phytochrome is likely to be the active phytochrome within the plant under natural conditions (Kendrick, Spruit and Frankland, 1969; Spruit and Mancinelli, 1969).

On the basis of these findings, one can make the generalisation that dark-germinating seeds contain non-limiting proportions of P_{fr}, whilst light-requiring seeds contain limiting quantities of P_{fr}. The existence of a certain amount of P_{fr} in light-requiring seeds of lettuce var. Grand Rapids can be deduced from physiological experiments. If the embryos are excised, germination occurs in darkness; if such excised embryos are exposed to high osmotic pressures, dark germination is reduced and can be re-promoted by red light. Under these conditions, far red reduces germination below that of the dark controls, showing that P_{fr} must be present in the dark-imbibed seeds (Scheibe and Lang, 1965). Thus, whether a seed is, in old terminology, positively or negatively photoblastic is determined by the proportion of P_{fr} present in the imbibed seeds,

and by the activity of the mechanism which generates P_{fr} in darkness. It also seems likely that seeds on which light has no effect
nevertheless have the same hereditary mechanism, since many such
seeds can be rendered light-requiring by various stress treatments
(Evenari, 1965).

THE PHOTORECEPTOR RIDDLE

From the evolutionary and ecological point of view, it is necessary
to ask why seeds should exhibit such striking and varied responses to
light—or to put it less teleologically, what is the adaptive value of
light sensitivity?

One obvious answer to this question is that subsequent seedling
establishment is likely to be enhanced if germination occurs beneath
the soil surface; therefore light-inhibited seeds should have an
adaptive advantage. Paradoxically, the photostimulation of germination, on the other hand, could also confer adaptive advantage
since, as shown by Wesson and Wareing (1967), many buried seeds
will not germinate until the soil has been disturbed. This allows for
the periodic germination of batches of seed at irregular intervals,
enabling the species to take advantage of favourable climatic and
other conditions.

These considerations, however, only show that light sensitivity *per
se* is of value to the species; they do not provide any clues as to the
selection pressures which have led to the evolution of the highly
complex phytochrome system as the effective agency of light

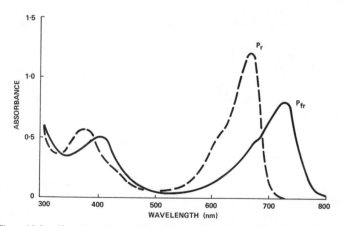

Figure 12.2   Absorption spectra of the two forms of phytochrome (from Hartmann, 1966)

sensitivity. This is, in fact, one of the most important unanswered—
and largely unasked—questions concerning phytochrome. In order
to approach an answer, it is necessary to consider the properties of
phytochrome, and to see if they can be correlated with important
parameters of the natural environment.

SOME PHYSICAL PROPERTIES OF PHYTOCHROME

Phytochrome is a relatively large chromoprotein (molecular weight
approximately 120 000) which in solution exhibits the absorption
spectra shown in *Figure 12.2*. Thus, P_r absorbs maximally at 660 nm,
whereas P_{fr} has maximum absorption at around 730 nm. Both substances, however, absorb significant amounts of light at all wavelengths below 730 nm. The action spectra for the photo-conversion
in both directions have been determined (Butler, Hendricks and
Siegelman, 1964) and have been shown to reflect closely the absorption spectra of the two forms.

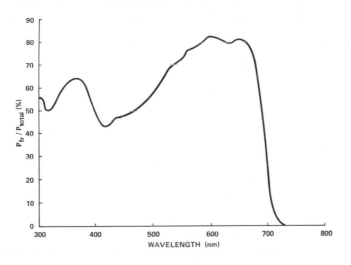

Figure 12.3   Photoequilibria of phytochrome as a function of wavelength (from Hartmann, 1966)

If phytochrome solutions are irradiated with a monochromatic
source below 730 nm, a photoequilibrium is rapidly set up in which
P_r molecules absorb light and are converted to P_{fr}, at the same rate
at which P_{fr} molecules absorb light and are converted to P_r.

At this photoequilibrium, a characteristic ratio of $P_r:P_{fr}$ will be established and maintained, its value being determined by the relative specific absorbance of P_r and P_{fr} at the particular wavelength, and by the relative quantum efficiencies of photoconversion in each direction. Since it is known that the quantum efficiency of the $P_r \rightarrow P_{fr}$ conversion is approximately 1·23 times that of the reverse conversion, and since accurate absorption spectra are available, it is possible to calculate theoretical photoequilibria for each monochromatic wavelength (Hartmann, 1966). The photoequilibria calculated by Hartmann are shown in *Figure 12.3*, expressed in terms of P_{fr} as a proportion of P_{total}, and it should be noted that a striking change is apparent in going from about 680 nm to about 700 nm. Thus, from 600 to 680 nm the proportion P_{fr}/P_{total} is 0·8, whereas at 700 nm it is markedly reduced and at 730 nm is near zero.

Broad spectrum irradiation will obviously act in an exactly analogous manner and thus any particular source of light will establish and maintain a characteristic P_{fr}/P_{total} photoequilibrium. Phytochrome therefore is, *par excellence,* a superb detector of the spectral composition of the incident irradiation; sources containing a preponderance of red light will set up high P_{fr}/P_{total} ratios, whilst sources with a high level of far-red light will set up low P_{fr}/P_{total} ratios. The next question is whether this property is reflected in any important parameter of the natural environment.

SPECTRAL COMPOSITION OF
NATURAL IRRADIATION

It is surprising to discover that the spectral composition of natural daylight has received little attention, either from plant physiologists or from environmental physicists—partly due, no doubt, to the unavailability of satisfactory instrumentation until quite recently. However, the expected spectral distribution of sunlight can be calculated, taking into account the absorptive properties of the atmosphere, light scattering and zenith angle. *Figure 12.4* presents two such computer-generated spectral energy profiles of direct sunlight (ignoring skylight) for summer midday and sunset or sunrise at about the latitude of the southern United Kingdom (calculated by Mr. A. McCartney). It is obvious that very marked daily cycles occur with marked drops in the blue wavelengths at the beginning and end of the day. The relative proportions of red and far-red light energies, however, do not change very markedly during the course of a day, and the observed changes become even less significant when the expected variations due to climatological conditions are taken into account.

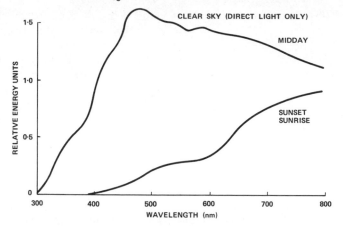

Figure 12.4   Spectral distribution of direct midsummer sunlight at midday and sunset/sunrise (calculated by Mr. A. McCartney for the latitude of Sutton Bonington)

Large changes in red/far red energy levels are therefore not likely to occur in natural daylight. If the transmission of daylight through a forest crop or other vegetation canopy is considered, however, very large changes indeed become apparent. *Figure 12.5* is the transmission spectrum of a single, young leaf of sugar beet showing that most of the red and blue wavelengths do not penetrate the leaf, a small amount of green light is transmitted (although much more is reflected) and, most important, the leaf is largely transparent to far-red light.

In *Figure 12.5* a portion of leaf was placed in the second sample position of a Unicam SP 1800 spectrophotometer as close to the photomultiplier as possible. The spectrophotometer was previously balanced with a piece of thin, white, unglazed paper in both the sample and reference beams and zero and 2·0 absorbance levels calibrated throughout the wavelength range. The leaf was placed adjacent to the incident side of the balancing paper of the sample beam. This arrangement served to equalise the effects of light-scattering in both beams. Results were obtained in terms of absorbances and converted to percentage transmission. (N.B. Older leaves were less transparent at all wavelengths.)

It is clear from *Figure 12.5* that leaves act as excellent far-red filters. Using this transmission spectrum as an example, it is possible to calculate the expected spectral distribution of daylight beneath a canopy consisting of 1 or 2 leaves, either at midday or at sunset/sunrise, by combining the data in *Figures 12.4* and *12.5*.

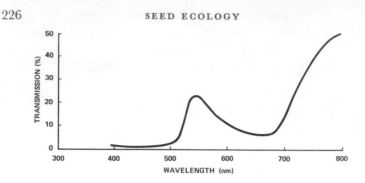

Figure 12.5   Transmission spectrum of a young leaf of sugar beet

These calculated spectral energy distributions are shown in *Figures 12.6* and *12.7* and indicate that canopy light should be largely far red plus a little green at midday, changing to almost pure far red at the beginning and end of the day.

These changes in spectral energy composition in the red/far red region are obviously very large and the possible importance of such canopy effects in relation to the phytochrome control of development have been pointed out in recent years by several other workers (Cumming, 1963; Vézina and Boulter, 1966; Black, 1969; Kasperbauer, 1971).

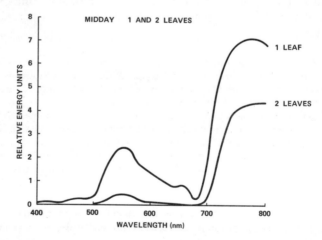

Figure 12.6   Calculated spectral distribution of midday daylight transmitted through 1 or 2 leaves of sugar beet. These curves were obtained by multiplying the curves for midday sunlight spectral distribution and for the transmission spectra of the leaves given in Figures 12.4 and 12.5

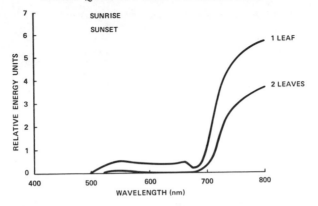

*Figure 12.7   The spectral distribution of sunset/sunrise daylight transmitted through 1 or 2
young sugar beet leaves calculated as for Figure 12.6*

PHYTOCHROME PHOTOEQUILIBRIA UNDER
NATURAL CONDITIONS

In order to test the concept that the function of phytochrome is to
detect the shading effect of neighbouring plants, expected phytochrome photoequilibria have been calculated by integrating the
data of *Figure 12.3* with those in *Figures 12.6* and *12.7* and comparing
the final average values with actual measurements. Since isolated
phytochrome was not freely available, measurements of photoequilibria were carried out using etiolated *Avena* coleoptiles as test
material as follows:

Avena coleoptiles, fully dark-grown and minus the leaves, were placed in the
cuvette of an Asco Ratiospect R2 (a dual-wavelength difference photometer
specifically designed to measure relative phytochrome levels) and exposed
either to direct midday sunlight, or to sunlight filtered through 1 or 2
thicknesses of sugar beet leaves; P_{fr} and total P were then measured in the
Ratiospect in the dark room in the normal manner (see Hopkins and
Hillman, 1966 for methods).

Table 12.1 shows both the calculated values and the range of
measured results, the coincidence between which is acceptable considering the number of assumptions involved in the calculations. It is
quite clear from these data that very large differences in the photoequilibria of phytochrome are to be expected between plants, or
seeds, outside the canopy and those inside. On the question of the
adaptive value of phytochrome, these results are consistent with the

Table 12.1 PHYTOCHROME PHOTOEQUILIBRIA UNDER NATURAL CONDITIONS

Time of day	No. of leaves	P_{fr}/P_{total} calculated	P_{fr}/P_{total} observed
Midday	0	49·6	53–60
	1	20·4	17–26
	2	5·65	5–14
Sunset or sunrise	0	34·7	—
	1	8·7	—
	2	2·1	—

hypothesis that phytochrome serves to detect the shade of other plants and to modulate growth and development accordingly.

PHYTOCHROME AND SEED ECOLOGY

Although the calculations and experimental results are consistent with the view that the phytochrome control of seed germination is a mechanism allowing adaptation to canopy effects, they do not necessarily prove the case. It is important to consider, first of all, whether seed germination, in practice, is affected by the change in P_{fr}/P_{total} predicted for the natural canopy. Hendricks, Borthwick and Downs (1956) before the discovery of phytochrome, showed that 50 per cent pigment conversion is essential in order to obtain full germination of lettuce var. Grand Rapids seeds. Hartmann (1966) arrived at a critical P_{fr}/P_{total} value of 0·1 for the inhibition of Grand Rapids seed germination by continuous far-red light. On a more detailed basis, Kendrick and Frankland (1969) showed that the germination of *Amaranthus caudatus* is sensitive to P_{fr}/P_{total} ratios over a wide range. Thus, it does seem likely that changes of the magnitude predicted here would be sufficient to have profound effects on the germination of at least a range of species. This would operate to inhibit the germination of seeds under canopies and could thus be of distinct survival value in certain cases. It is more likely to be important in later stages of plant development, however, since it is in stem elongation and leaf expansion that the greatest responses to natural shading are observed.

It also seems likely that the irradiation environment experienced by the mother plant at the time of seed maturation may have important effects on the germination behaviour of the mature seed. McCullough and Shropshire (1970) have shown that the germination of *Arabidopsis* seeds can be correlated with the light conditions during seed development. If the seeds are matured under fluorescent light, which contains very little far red, and thus presumably establishes a high P_{fr}/P_{total}, then germination occurs readily in darkness. On the other hand, incandescent light, which contains much far red and sets up a lower P_{fr}/P_{total}, results in seeds with poor dark germination and a high light requirement. Thus, the ecological significance of the spectral distribution of light during the later stages of seed maturation should not be ignored (see also Chapter 4).

Obviously, there is much polymorphism in germination behaviour in natural populations, and it is clear that the spectral distribution of light is only one of many interlocking factors which regulate germination. Nevertheless, consideration of this environmental parameter allows for a significant simplification of the concepts and theories concerning the photocontrol of germination and points the way to future research.

ACKNOWLEDGEMENTS

The author is grateful to Mr. A. McCartney for the data from which *Figure 12.4* was generated, and to Mr. R. Knight for excellent technical assistance in the determination of photostationary states. Professor J. L. Monteith is also thanked for his advice concerning the calculations.

REFERENCES

BLACK, M. (1969). 'Light-controlled germination of seeds', *S.E.B. Symp.*, **23**, 193–217
BOISARD, J. (1969). 'La photosensibilité des akènes de Laitue (variété "Reine de Mai") et son interprétation par spectrophotométrie *in vivo* du photorécepteur', *Physiol. Vég.*, **7**, 119–133
BOISARD, J., SPRUIT, C. J. P. and ROLLIN, P. (1968). 'Phytochrome in seeds and an apparent dark reversion of P_R to P_{FR}', *Meded. :andbHoogesch., Wageningen*, **68**, No. 17
BORTHWICK, H. A., HENDRICKS, S. B., PARKER, M. W., TOOLE, E. H. and TOOLE, V. K. (1952). 'A reversible photoreaction controlling seed germination', *Proc. natn. Acad. Sci. (U.S.A.)*, **38**, 662–666
BUTLER, W. L., HENDRICKS, S. B. and SIEGELMAN, H. W. (1964). 'Action spectra of phytochrome *in vitro*'. *Photochem. Photobiol.*, **3**, 521–528
CUMMING, B. G. (1963). 'Dependence of germination on photoperiod, light quality and temperature in *Chenopodium* spp.', *Can. J. Bot.*, **41**, 1211–1233
EVENARI, M. (1956). 'Seed germination', in *Radiation Biology* (ed. Hollaender, A.), vol. 3, 519–549
EVENARI, M. (1965). 'Light and seed dormancy', in *Encycl. Pl. Physiol.* (ed. Ruhland, W.), vol. 15/2, 804–847, Springer-Verlag, Berlin and Heidelberg
FLINT, L. H. and MCALLISTER, E. D. (1935). 'Wavelength of radiation in the visible

spectrum, inhibiting the germination of light sensitive lettuce seeds', *Smithson. Misc. Coll.*, **94**, No. 5

FLINT, L. H. and MCALLISTER, E. D. (1937). 'Wavelength of radiation in the visible spectrum, promoting the germination of light sensitive lettuce seed', *Smithson. Misc. Coll.*, **96**, No. 2

HARTMANN, K. M. (1966). 'A general hypothesis to interpret high energy phenomena on the basis of phytochrome', *Photochem. Photobiol.*, **5**, 349–366

HENDRICKS, S. B., BORTHWICK, H. A. and DOWNS, R. J. (1956). 'Pigment conversion in the formative responses of plants to radiation', *Proc. natn. Acad. Sci. U.S.A.*, **42**, 19–25

HOPKINS, W. G. and HILLMAN, W. S. (1966). 'Relationships between phytochrome state and photosensitive growth of *Avena* coleoptile segments', *Pl. Physiol.*, **41**, 593–598

JONES, M. B. and BAILEY, L. F. (1956). 'Light effects on the germination of henbit (*Lamium amplexicaule* L.)', *Pl. Physiol.*, **31**, 347–349

KASPERBAUER, M. J. (1971). 'Spectral distribution of light in a tobacco canopy and effects of end-of-day light quality on growth and development', *Pl. Physiol.*, **47**, 775–778

KENDRICK, R. E. and FRANKLAND, B. (1969). 'Photocontrol of germination in *Amaranthus caudatus*', *Planta (Berl.)*, **85**, 326–339

KENDRICK, R. E., SPRUIT, C. J. P. and FRANKLAND, B. (1969). 'Phytochrome in seeds of *Amaranthus caudatus*', *Planta (Berl.)*, **88**, 293–302

MCCULLOUGH, J. M. and SHROPSHIRE, W. jnr. (1970). 'Physiological predetermination of germination responses in *Arabidopsis thaliana* (L.) Heynh', *Pl. Cell. Physiol.*, **11**, 139–148

ROLLIN, P., MALCOSTE, R. and EUDE, D. (1970). 'Le rôle du phytochrome dans la germination des graines de *Nemophila insignis* (L.)', *Planta (Berl.)*, **91**, 227–234

SCHEIBE, J. and LANG, A. (1965). 'Lettuce seed germination: evidence for a reversible light-induced increase in growth potential and for phytochrome mediation of the low temperature effect., *Pl. Physiol.*, **40**, 485–492

SPRUIT, C. J. P. (1970). 'Spectrophotometers for the study of phytochrome *in vivo*', *Meded. LandbHoogesch., Wageningen*, **70**, No. 14

SPRUIT, C. J. P. and MANCINELLI, A. L. (1969). 'Phytochrome in cucumber seeds', *Planta (Berl.)*, **88**, 303–310

VÉZINA, P. E. and BOULTER, D. W. K. (1966). 'The spectral composition of near ultra-violet and visible radiation beneath forest canopies', *Can. J. Bot.*, **44**, 1267–1284

WESSON, G. and WAREING, P. F. (1967). 'Light requirements of buried seeds', *Nature (Lond.)*, **213**, 600–601

YANIV, Z., MANCINELLI, A. L. and SMITH, P. (1967). 'Phytochrome and seed germination. III. Action of prolonged far-red irradiation on the germination of tomato and cucumber seeds', *Pl. Physiol.*, **42**, 1479–1482

DISCUSSIONS

Thompson: Are you suggesting that the response of the seeds to light is not consistent within any given species?

Smith: The old definition of positive and negative photoblasticity no longer holds: one can make seeds light-sensitive by appropriate treatments. The conditions of seed maturation are important in this respect. Some dark-germinating seeds can have up to 75 per cent of their seed phytochrome present as P_{fr}. MacCulloch and Shropshire showed that *Arabidopsis* seeds germinate well in the dark if they have matured in fluorescent light (with a high proportion of red), but poorly if in incandescent light (relatively high in far red) which lowers the proportion of P_{fr} in such seeds.

Gutterman: Considering these effects of the light quality during seed maturation on seed germination, why do Grand Rapids lettuce seeds grown in continuous daylight and *Portulaca oleracea* seeds grown in continuous red light fail to germinate in the dark? What is the relationship between the phytochrome in the mother plant and the phytochrome in the seed?

Smith: We cannot at present answer this complex question.

Karssen: Is there a threshold P_{fr}/P_{total} ratio below which seeds will fail to germinate?

Smith: Borthwick has shown that for maximum germination, lettuce Grand Rapids requires a 50 per cent conversion of P_r to P_{fr} by red light, but other varieties, such as May Queen, need much less, and other species may need much more; conversely there is no one constant percentage of P_{fr} at which germination is inhibited. There is nothing absolute in the P_{fr}/P_{total} proportion which operates the dormancy mechanism in seeds, and other factors are also important. There is much polymorphism between individuals, especially among seeds of weed species.

Côme: Is the seed coat the site of the phytochrome response to light?

Smith: When the coat of Grand Rapids lettuce seeds is removed they germinate in the dark; but if then irradiated with far-red light they become again red light requiring. They therefore basically retain their light sensitivity. The seed coat is equivalent to other stress factors, such as the application of coumarin or a large negative osmotic potential which can make seeds of many varieties light sensitive.

Gray: Piggott was able to produce germination conditions for seeds in petri dishes, analogous to those beneath and those outside woodland canopies, by changing the proportion of red and far red of the incident light through filters. The red–far red balance seems to be the factor controlling germination in woodlands.

E. H. Roberts: Has anyone investigated the role of the seed coat in filtering light?

Smith: Just as the phytochrome ratio is modified when light is filtered through the canopy, so it may also be modified through the seed coat. But at present we have not the ability to determine this.

Harrington: What is the effect of soil as a filter of light?

E. H. Roberts: Where light does penetrate, as in sandy soil, it appears that the red–far red ratio increases greatly and progressively with depth.

13

INTERACTION OF ETHYLENE AND LIGHT ON DORMANT WEED SEEDS

S. T. OLATOYE and M. A. HALL

Department of Botany, University College of Wales, Aberystwyth

Ethylene (C_2H_4) has long been known to elicit a wide range of physiological responses in plants. A promotory effect on seed germination was first shown by Vacha and Harvey (1927) using buckthorn (*Rhamnus catharticus*) and other species. Since that time work has been somewhat sporadic and mainly concentrated on seeds of crop plants, e.g. wheat (Balls and Hale, 1940) and peanut (Toole, Bailey and Toole, 1964). More recently, Esashi and Leopold (1969) have shown that ethylene at concentrations as low as 0·001 ppm can promote germination in seed of *Trifolium subterraneum* L. and also that ethylene production by the seeds themselves might break dormancy under natural conditions.

Smith and Russell (1969) have shown that ethylene levels of up to 10 ppm may occur in the soil, the concentration being correlated with compaction or waterlogging. It seemed possible, therefore, that such levels of ethylene might affect the germination of soil-borne weed seeds; consequently, a range of common weed seeds were screened for their response to ethylene. The results of these preliminary experiments are shown in *Table 13.1*. In the seeds of five of the twelve species examined, germination was promoted, in three it was inhibited and the remainder showed variable and generally insignificant responses. The germination responses shown in *Table 13.1* were obtained under continuous white light at 20°C; however, in parallel experiments in the presence or absence of light, it was observed that with the exception of *T. repens* L. all the ethylene sensitive seeds, whether inhibited or promoted, were light sensitive. This was an absolute requirement in some cases, e.g. *Plantago major* L.,

Table 13.1 EFFECT OF ETHYLENE ON THE GERMINATION OF SEEDS OF SOME WEED SPECIES IN THE LIGHT

Species	Germination (% ± S.E.) at ethylene concentration (ppm)				Time after which germination recorded (days)
	0	10	100	1000	
Germination promoted by ethylene					
*Spergula arvensis L.	12·8 ± 2·4	47·5 ± 3·3	57·5 ± 3·8	57·5 ± 3·6	4
*Hypochaeris radicata L.	4·0 ± 2·2	52·5 ± 2·6	53·5 ± 5·1	16·5 ± 3·3	4
*Chenopodium album L.	0·8 ± 0·3	—	10·5 ± 1·9	13·3 ± 0·9	7
*Rumex crispus L.	9·0 ± 1·7	—	19·0 ± 2·6	22·0 ± 2·6	21
Trifolium repens L.	26·5 ± 4·2	59·0 ± 4·5	59·8 ± 5·6	59·0 ± 6·4	2
Germination inhibited by ethylene					
*Chenopodium rubrum L.	17·5 ± 2·6	—	0·5 ± 0·3	0 ± 0	14
*Plantago major L.	28·3 ± 4·2	8·3 ± 1·9	0·3 ± 0·2	—	7
*Plantago maritima L.	55·0 ± 4·7	38·0 ± 1·2	39·0 ± 3·7	33·0 ± 8·9	3
No effect on germination					
Sonchus oleraceus L.	24·3 ± 1·5	32·8 ± 7·7	21·5 ± 5·8	24·8 ± 3·4	4
Silene dioica L. Clairv.	41·0 ± 3·7	31·5 ± 2·2	34·5 ± 5·7	34·5 ± 4·3	11
Senecio jacobea L.	55·0 ± 3·7	65·0 ± 4·0	52·5 ± 2·9	57·5 ± 2·2	5
Taraxacum officinale Weber	43·0 ± 4·0	39·5 ± 1·7	40·0 ± 1·6	43·5 ± 3·0	4

* Germination light-promoted.

but not in all, e.g. *Spergula arvensis* L. Of the species shown to be ethylene sensitive, two have not previously been reported to require light for germination, viz. *P. maritima* L. and *Hypochaeris radicata* L. *Spergula arvensis* has previously been shown to acquire a light requirement after burial (Wesson and Wareing 1969b) but 'no need for light' has been observed in fresh seeds. In some species the dormancy characteristics change quite rapidly with time. In *H. radicata* the response to ethylene and the light requirement are lost simultaneously some two weeks after harvest (*Table 13.2*). (In this experiment

Table 13.2 EFFECTS OF ETHYLENE AND LIGHT ON THE GERMINATION OF *HYPO-CHAERIS RADICATA*

Ethylene concentration (*ppm*)	Germination (% ± S.E.)			
	Light		Dark	
	(a)*	(b)†	(a)	(b)
0	3·3 ± 0·7	62·7 ± 1·7	3·3 ± 1·2	60·0 ± 2·0
10	50·7 ± 1·3	64·7 ± 1·9	15·0 ± 1·5	63·0 ± 2·1

* 3 days after harvest.
† 14 days after harvest.

germination was carried out at 20°C, with or without continuous white light, after pre-treatment of imbibed seeds in the presence or absence of ethylene. Germination was recorded after four days.) In *P. major*, ethylene ceases to have an inhibitory effect seven days after harvest but the seeds are still completely dependent on light for germination (*Table 13.3*). (Germination tests were carried out at 20°C, with or without continuous red light, after pre-treatment of imbibed seeds with ethylene at the concentrations shown. Germination was recorded after seven days.)

Table 13.3 EFFECTS OF ETHYLENE AND LIGHT ON THE GERMINATION OF *PLANTAGO MAJOR*

Ethylene concentration (*ppm*)	Germination (% ± S.E.) days after harvest		
	Light		Dark
	2	7	2 and 7
0	28·3 ± 4·2	54·3 ± 4·8	0
10	8·3 ± 1·9	43·5 ± 3·9	0
100	0·3 ± 0·2	46·8 ± 2·3	0

Most of the work described hereafter was performed on seeds of *Spergula arvensis*, which retained their response to light and ethylene for long periods—especially if stored at 0°C. *Table 13.4* shows the effect of ethylene upon germination of such seeds in the light and in

Table 13.4 EFFECT OF ETHYLENE AND LIGHT ON GERMINATION OF *SPERGULA ARVENSIS* SEEDS

Ethylene concentration (*ppm*)	Light	Dark
0	8·3 ± 1·5	7·8 ± 0·9
10	40·8 ± 3·4	14·5 ± 2·4
100	42·0 ± 4·8	20·5 ± 2·5

the dark. Seeds were treated with ethylene immediately after moistening with water and incubated at 20°C either in darkness or under continuous white light: germination was recorded after five days. In the absence of ethylene, germination is very low in both the light and the dark; ethylene stimulates germination in the dark but a much more marked effect is observed in the light. The level of dark

Table 13.5 EFFECT OF ETHYLENE ON GERMINATION OF *SPERGULA ARVENSIS* SEEDS

Ethylene concentration (*ppm*)	Germination (% ± S.E.)
0	12·8 ± 2·4
0·1	12·0 ± 1·2
0·2	14·5 ± 0·9
0·5	24·3 ± 2·7
1·0	34·8 ± 2·6
10	47·5 ± 3·3
100	57·5 ± 3·8
1 000	57·5 ± 3·6

germination of seeds varies from batch to batch, but in freshly harvested seeds it is always substantially less than in the light.

Germination may be stimulated by very low concentrations of ethylene, the threshold of response being of the order of 0·2 ppm, although maximum effect is achieved between 10 and 100 ppm (*Table 13.5*). Seeds were imbibed, treated with ethylene at the concentrations shown and incubated at 20°C under continuous

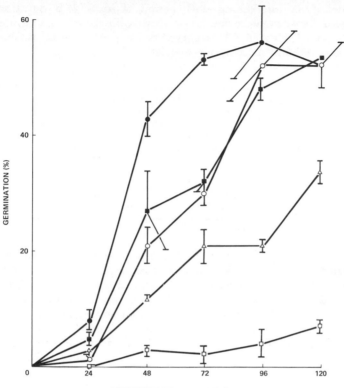

Figure 13.1   The effect of ethylene concentration on the germination rate of Spergula arvensis *seeds. Seeds were imbibed and treated with ethylene under continuous white light. Percentage germination was recorded at daily intervals (□—no C_2H_4, △—1 ppm, ○—10 ppm, ●—100 ppm, ■—1000 ppm)*

white light: germination was recorded after five days. The rate of germination is also markedly affected by ethylene concentration (*Figure 13.1*), about 100 ppm ethylene giving a maximal response. At 10 and 1000 ppm, the rate of germination is lower but the total germination after four days is the same as at 100 ppm.

Brief irradiation with red light elicits a similar response to that of continuous illumination with white light. In very dormant seeds red/far-red reversibility occurs, indicating that phytochrome is involved in the light response. As the seeds age, however, not only does the level of dark germination increase but, in addition, irradiation with red or white light does not fully reverse the effects of previous exposure to far red. Reversibility can, however, be demonstrated if an intervening dark period of some 24 h duration is allowed between successive irradiations with far-red and white or red light (*Table 13.6*). This effect has been observed by Kadman-Zahavi (1960) in *Amaranthus retroflexus*.

Table 13.6 EFFECT OF AN INTERVENING DARK PERIOD ON THE GERMINATION OF *SPERGULA ARVENSIS* SEEDS TREATED WITH FAR-RED LIGHT

Light treatments	Germination (% ± S.E.)
(a) W continuous	66 ± 8
(b) FR (24 h)/W continuous	45 ± 5
(c) FR (24 h)/dark (24 h)/W continuous	72 ± 6
(d) R (1 h)/dark continuous	69 ± 1
(e) FR (24 h)/R (1 h)/dark continuous	37 ± 5
(f) FR (24 h)/dark (24 h)/R (1 h)/dark continuous	71 ± 7

FR = far red, R = red, W = white.

It appears that the effect of ethylene is to remove some inhibition to the expression of the phytochrome response. In an effort to identify the process affected, one batch of *Spergula arvensis* seeds was allowed to imbibe in diffuse light for two hours, exposed to red light for 30 min and then treated with 100 ppm ethylene for five days in the dark. A further batch, dark-imbibed with ethylene for 24 h, was exposed to red light for 30 min and then incubated in the dark for five days. Germination in both cases was similar (*Table 13.7*).

Table 13.7 RELATIONSHIP BETWEEN ETHYLENE AND PHYTOCHROME IN THE GERMINATION OF *SPERGULA ARVENSIS*

(a) Red-irradiated before treatment	(b) Red-irradiated after treatment	(c) Dark-imbibed for 24 h, then exposed to red light; no ethylene
55·0 ± 3·4	61·0 ± 4·4	16·5 ± 5·5

This suggests that the inhibition removed by ethylene acts on processes occurring subsequent to phytochrome conversion; were it an inhibition of phytochrome conversion then ethylene treatment following red irradiation (treatment (a) in *Table 13.7*) would not result in promotion of germination.

Since part of the ethylene-promoted germination is light dependent, what is responsible for the relatively high level of germination in the dark? To clarify this point seeds were incubated in the dark at 20°C or 30°C for 24 h followed by incubation in the dark for five days at 20°C. One of the batches heat treated at 30°C was subsequently given a 30 min exposure to red light. Each of the batches was divided into two, one half being incubated in 100 ppm ethylene, the other in air. *Table 13.8* shows that heat treatment inhibits ethylene-promoted germination in the dark but that this inhibition is more than reversed by exposure to red light. This suggests that

Table 13.8 EFFECT OF HEAT TREATMENT UPON GERMINATION OF NON-IRRADIATED SEEDS OF *SPERGULA ARVENSIS*

	Germination (% ± S.E.)	
	Ethylene (100 ppm)	Control
(a) Not heat-treated; dark germinated	38·5 ± 5·1	12·5 ± 0·5
(b) Heat-treated; dark germinated	12·0 ± 3·6	9·0 ± 2·4
(c) Heat-treated and irradiated	72·5 ± 2·5	13·5 ± 1·7

these seeds contain some pre-existent P_{fr}. Treatment at 30°C for 24 h appears to cause substantial decay of P_{fr} in this seed (see *Figure 13.2*). Thermal reversion of P_{fr} was used rather than far-red irradiation because of the difficulty experienced in reversing exposure to far red in older seed, mentioned above in connection with *Table 13.6*.

Where a process is controlled by two separate environmental factors the situation may arise where one is limiting, and it is thus important to determine the effect of a temporal separation of the supply of the two promoting factors. An effort was thus made to determine the effect of light and ethylene treatments separated, at different temperatures, by various times. Seeds were allowed to imbibe, exposed to red light for 30 min and maintained at 5, 10, 20 or 30°C for varying periods prior to incubation with 100 ppm ethylene at 20°C for five days (*Figure 13.2*). At higher temperatures there is a relatively sharp decrease in response to ethylene over the first 24 h, followed by a levelling off. This appears to be due to thermal reversion of P_{fr} to P_r, because even for seeds held at 30°C for four days the effect may be almost completely reversed by a 30 min irradiation with red light and incubation with ethylene for five days. The converse situation, where ethylene is applied at different times before exposure to light, is shown in *Table 13.9*. Dark-imbibed seeds were treated with ethylene and incubated in the dark for various periods, followed by exposure to white light at 20°C. The control was also treated with ethylene but kept in the dark for the duration of the experiment. Germination was recorded five days after the conclusion of the dark treatment in each case. In these

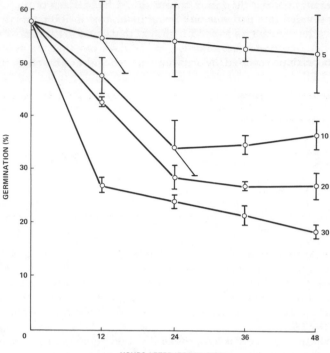

Figure 13.2   Effect of temperature (°C shown on the right) and of time elapsing between irradiation and treatment with ethylene on subsequent germination at 20°C of Spergula arvensis *seeds pre-irradiated with red light*

circumstances there is little change in total germination over a period of four days. After eight days there is a smaller response but this is still substantially greater than when ethylene-induced germination takes place in the dark.

In the experiment just described seeds were left imbibed in the

Table 13.9   EFFECTS OF PROLONGED DARK-INCUBATION FOLLOWING ETHYLENE TREATMENT ON THE GERMINATION OF *SPERGULA ARVENSIS* SEEDS

Duration of dark incubation (h)	Germination (% ± S.E.)
24	80 ± 4
48	77 ± 7
96	73 ± 5
144	58 ± 2
∞ (Control: no light)	14 ± 4

interval between the ethylene and the light treatments. In a subsequent experiment the seeds were allowed to imbibe, treated with ethylene for 24 h and then air-dried for various periods. They were then remoistened and set to germinate under white light for five days. The germination response to ethylene was lowered by drying the seeds after treatment, but the level of germination was always higher than that obtained with untreated seeds in the absence of ethylene—even after six days' drying (*Table 13.10*). The ethylene effect is therefore not readily reversible by the drying treatment.

Table 13.10   EFFECTS OF DRYING AND WETTING ON THE GERMINATION OF SEEDS OF *SPERGULA ARVENSIS* UNDER WHITE LIGHT

Duration of drying (h)	Germination (% ± S.E.)
0 (no ethylene)	12 ± 2
0 (with ethylene)	54 ± 2
24 (with ethylene)	29 ± 1
48 (with ethylene)	26 ± 4
96 (with ethylene)	28 ± 6
144 (with ethylene)	35 ± 9

Table 13.11   REQUIREMENT OF *SPERGULA ARVENSIS* SEEDS FOR EXPOSURE TO ETHYLENE (100 PPM) UNDER WHITE LIGHT

Duration of exposure (h)	Germination in white light (% ± S.E.)
0 (Control)	29 ± 7
1	28 ± 10
2	37 ± 1
4	25 ± 1
6	44 ± 0
12	61 ± 3
24	82 ± 4
48	75 ± 1
72	83 ± 3

Ethylene levels in some soils may fluctuate widely over fairly short time intervals, (Dowdell *et al.*, 1972) and it was important therefore to determine how long seeds need to be exposed to elicit an effect. Thus, *Spergula arvensis* seeds were exposed to ethylene for varying times under white light and subsequent germination measured under white light in the absence of ethylene (*Table 13.11*).

242

Table 13.12 EFFECT OF POTASSIUM NITRATE
(KNO_3) UPON GERMINATION OF
SPERGULA ARVENSIS SEEDS

Treatment	Germination (% ± S.E.)	
	Dark	5 min red light
No additions	1 ± 1	6 ± 0
KNO_3	4 ± 4	17 ± 5
C_2H_4	2 ± 2	7 ± 1
$KNO_3 + C_2H_4$	19 ± 3	78 ± 4

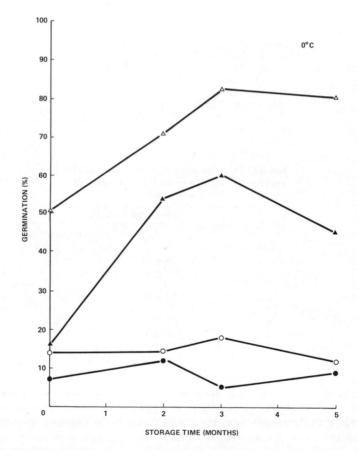

Figure 13.3 Effect of storage temperature (0°C) on germination characteristics of Spergula
arvensis *(●—Dark, ○—Light, ▲—Dark +C_2H_4, △—Light +C_2H_4)*

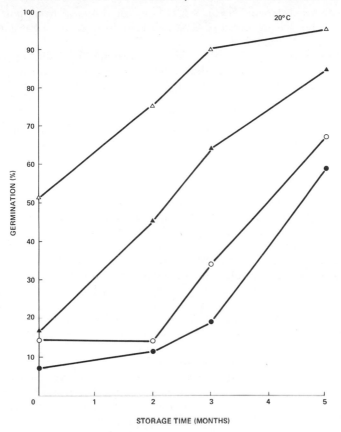

Figure 13.4   Effect of storage temperature (20°C) on germination characteristics of Spergula arvensis *(●—Dark, ○—Light, ▲—Dark +C_2H_4, △—Light +C_2H_4)*

It can be seen that at an ethylene concentration of 100 ppm, significant promotion over controls is observed after 6 h.

It has long been known that potassium nitrate (KNO_3) may promote the germination of light-sensitive seeds (Toole, 1941) and this effect may have an important bearing on the germination of weed seeds in agricultural soils. In *S. arvensis* KNO_3 has no significant effect on germination in the dark. A brief (5 min) exposure to red light alone or in the presence of 10 ppm ethylene (itself not promoting) has likewise quite a small effect. By combining red light and ethylene with KNO_3 (0·2 per cent), however, a marked synergistic response is obtained (*Table 13.12*). This would appear to support the contention that the effect of KNO_3 is to reduce the red light energy requirement of the seed.

Experiments were next carried out to determine what effect, if any, prolonged storage at different temperatures has on the response of the seeds to light and ethylene, alone and in combination. The results of such experiments are shown in *Figures 13.3* and *13.4*.

For seed stored at 0°C the germination in the light or dark in the absence of ethylene remains essentially constant even after five months. In the presence of ethylene, however, there is, up to three months, a progressive increase in total germination. Germination is always better in the light but much of the improvement occurs irrespective of whether light is present. For seeds stored at 20°C a different pattern emerges. In the absence of ethylene no effect is observable up to two months; thereafter, however, there is a sharp increase in germination in both the light and the dark, with little difference between treatments. Seeds treated with ethylene show an increase in germination in the light similar to that observed at 0°C; in this case, however, dark germination continues to improve and after five months there is very little difference between light and dark treatments. Furthermore, total germination is much higher than after storage at 0°C. It appears, therefore, that for seeds stored at low temperature both light and ethylene remain important in the control of germination at least for several months; at higher temperatures, however, the ethylene effect is brought out further and is the one which is ultimately of importance. The observed increases in dark germination suggest two possible mechanisms. Either the processes normally dependent on P_{fr} are 'escaping' from this requirement, or the seeds acquire a greater capacity for *de novo* synthesis of P_{fr} as they age; alternatively, there may be an increased capacity for pre-existing P_{fr} to become rehydrated; both the latter suggestions have been made by Mancinelli, Yanir and Smith (1967). In *S. arvensis*, as in the tomato seed used by Mancinelli, exposure to far-red irradiation of very old seeds substantially inhibits germination, even though they may have lost their light requirement; this indicates that they are still under phytochrome control. As in seeds which have not lost their light requirement (*Table 13.6*), the effect is reversed by exposure to light, provided a dark period of 24 h duration is allowed between irradiations.

Seeds which have been stored for a year at room temperature showed very different characteristics to those described previously (*Table 13.13*). In the absence of ethylene, germination in the light was substantially less than in the dark; inclusion of ethylene, while having no effect on dark germination, overcame the observed light inhibition. It seemed possible that the capacity of the seeds themselves to produce ethylene might increase with time but that this capacity might be adversely affected by light—an effect well documented in a number of other biological systems. Preliminary experiments have indeed indicated that old seeds produce more ethylene

Table 13.13 EFFECT OF LIGHT AND ETHYLENE UPON GERMINATION OF ONE YEAR OLD *SPERGULA ARVENSIS* SEEDS

Ethylene concentration (*ppm*)	Germination (%)	
	Light	Dark
0	35	75
10	60	80

on imbibition than fresh seeds and that this process is inhibited by exposure to continuous white light.

The nature of the inhibition that is removed by ethylene is still not clear. Jones (1968) showed that, in barley, ethylene enhances the release of gibberellin-induced α-amylase, presumably by an effect on membrane permeability. Ethylene and light do induce changes in levels of endogenous cytokinins (van Staden, Olatoye and Hall, unpublished) and gibberellin-like substances (Olatoye and Hall, unpublished) in imbibed seeds of *Spergula arvensis* prior to germination. The effects of ethylene and light on the cytokinins are partially separable. No significant difference in inhibitor levels between treatments has so far been observed.

Neither cytokinins nor gibberellins break dormancy if applied to intact *S. arvensis* seeds. Acid scarification, however, leads to an improvement of germination in the absence of ethylene, and gibberellic acid will promote the germination of acid-scarified seeds. The gibberellin-induced effect appears to be partly under the control of pre-existent P_{fr}, since pretreatments expected to cause reversion of P_{fr} to P_r, such as holding the seeds at 30°C for 24 h or irradiation with far-red light, reduce the response even though, in the absence of such treatments, maximal germination will occur in the dark. Furthermore, the effects of high temperature and far-red light are partly reversed by white light treatment (*Table 13.14*). It seems possible, therefore, that the effect of ethylene in this system is to stimulate the synthesis or release of growth promoters necessary for germination to proceed.

What significance do these results have in relation to dormancy of *S. arvensis* and other seeds responding similarly? It has still to be proved that ethylene evolution is of universal occurrence in all or

Table 13.14   EFFECT OF GIBBERELLIC ACID ON GERMINATION OF SCARIFIED *SPERGULA ARVENSIS* SEED IN RESPONSE TO VARIOUS TREATMENT

Treatment	Germination (% ± S.E.)	
	without gibberellic acid	with gibberellic acid
(a) 20°C dark 5 days	19·5 ± 2·6	85·0 ± 2·1
(b) 30°C 24 h, then 20°C dark 5 days	13·0 ± 3·0	26·5 ± 2·2
(c) 30°C 24 h, then 20°C light 5 days	16·0 ± 0·0	50·0 ± 6·5
(d) Far red 24 h, then dark 5 days[*]	4·0 ± 0·0	14·5 ± 1·5
(e) Far red 24 h, then light 5 days[*]	14·0 ± 2·0	35·0 ± 7·0

[*] 20°C

most soils; there is, however, little doubt that it could be an important factor controlling germination in a number of seeds under natural conditions. If ethylene production is indeed associated with high moisture levels then the advantage for the seed would seem to be related to the presence of sufficient water to support seedling established after germination. That the situation is, however, more complex than this, is shown by the results concerning ethylene evolution by seeds of different ages. Thus, while fresh seeds will be dependent on exogenous ethylene, older seeds appear to be capable of producing sufficient endogenous ethylene to trigger germination.

The relative effect of this changing capacity will perhaps be mediated to some extent by the condition of the soil around the seed. Thus, the very small amounts of ethylene produced by fresh seeds may be sufficient to induce germination if the soil is so packed that free gaseous diffusion is inhibited. If ethylene formation proves to be autocatalytic in this system, as it is in many others, the effect would be reinforced. These observations that light inhibits ethylene production by *S. arvensis* seed, which in turn inhibits germination, may provide a partial explanation for light inhibition in other species also.

Yet another facet to the problem is introduced by the work of Wesson and Wareing (1969a; 1969b) who showed that burial of non-dormant seeds of a number of species—including *S. arvensis*—leads to the induction of a light requirement. Some of their results do show a light inhibition in *S. arvensis*, although much less than in our year-old seeds. Another species—*Stellaria media* L.—which shows a similar induction of a light requirement, is also light inhibited prior to burial. The induction of the light requirement appeared to be due to the production of some volatile factor by the seeds themselves, the nature of which has not yet been elucidated. The evidence presented here on the loss of light requirement in seeds stored at room temperature, probably as a result of increased availability of P_{fr}, suggests that the effect observed by Wesson and Wareing may represent a reversion or destruction of this P_{fr}.

These considerations do not represent the whole picture, however. There has always been observed some variation in the degree to which ethylene and light affect seeds collected at different sites at different times of the year. Whether this is associated with environmental conditions or represents natural variation between populations remains to be elucidated and this aspect is currently being investigated. The significance of ethylene in the inhibition of seed germination must await further work. However, such data as are available on one of these species indicate that the effect is transitory.

The ethylene effect on *S. arvensis* seeds is shown to be affected to a greater or lesser extent by a number of treatments which might be experienced under natural conditions, viz. duration of exposure to ethylene, temporal separation of exposure to light and application of ethylene, presence or absence of KNO_3 and desiccation. Nevertheless, for the first few months after maturation the seeds are almost absolutely dependent on an exogenous supply of ethylene for germination. Seed storage experiments indicate that the period of dependence may be more prolonged in seeds set late in the season which will be subjected to several months of low temperature. In old seeds, although dark germination is unaffected by exogenously supplied ethylene—apparently due to their capacity to produce endogenous ethylene levels sufficient to induce germination— germination in the light is still partially dependent on an external ethylene supply.

Since there is no reason to suspect that the species examined here and shown to be ethylene sensitive are in any way unique, it seems probable that a large number of weed species are affected in the same way—at least for some part of their lifetime. For this reason, ethylene, whether supplied exogenously by the soil or synthesised endogenously by the seed, should always be considered as a possible factor when seed germination is to be investigated in the laboratory or the field. In addition, the possible adaptive significance and ecological implications of the mechanism should be examined more closely.

ACKNOWLEDGEMENTS

We are grateful to Professor P. F. Wareing, F.R.S. for much helpful advice and discussion and to the British Council for providing a research studentship for one of us (S.T.O.).

REFERENCES

BALLS, A. K. and HALE, W. S. (1940). 'The effect of ethylene on freshly harvested wheat', *Cereal Chem.*, **17**, 490–494

DOWDELL, R. J., SMITH, K. A., RESTALL, S. W. F. and CREEF, R. (1972). 'Field studies of ethylene in the soil atmosphere—equipment and preliminary results', *Soil Biol. & Biochem.*, **4**, 325–332

ESASHI, Y. and LEOPOLD, A. C. (1969). 'Dormancy regulation in subterranean clover seeds by ethylene', *Pl. Physiol.*, **44**, 1470–1472

JONES, R. L. (1968). 'Ethylene enhanced release of α-amylase from barley aleurone cells', *Pl. Physiol.*, **43**, 442–444

KADMAN-ZAHAVI, A. (1960). 'Effects of short and continuous illumination on the germination of *Amaranthus retroflexus* seeds', *Bull. Res. Coun. Israel*, **9D**, 1–20

KETRING, D. L. and MORGAN, P. W. (1969). 'Ethylene as a component of the emanations from germinating peanut seeds and its effect on dormant Virginia-type seeds', *Pl. Physiol.*, **44**, 326–330

MANCINELLI, A. L., YANIR, Z. and SMITH, P. (1967). 'Phytochrome and seed germination. I. Temperature dependence and relative P_{fr} levels in the germination of dark-germinating tomato seeds', *Pl. Physiol.*, **42**, 333–337

SMITH, K. A. and RUSSELL, R. S. (1969). 'Occurrence of ethylene and its significance in anaerobic soil', *Nature*, **222**, 769–771

TOOLE, V. K. (1941). 'Factors affecting the germination of various dropseed grasses (*Sporobolus* spp.)', *J. agric. Res.*, **62**, 691–715

TOOLE, V. K., BAILEY, W. K. and TOOLE, E. H. (1964). 'Factors influencing dormancy of peanut seeds', *Pl. Physiol.*, **39**, 822–832

VACHA, G. A. and HARVEY, R. B. (1927). 'The use of ethylene, propylene and similar compounds in breaking rest period of tubers, bulbs, cuttings and seeds', *Pl. Physiol.*, **2**, 187–192

WESSON, G. and WAREING, P. F. (1969a). The role of light in the germination of naturally occurring populations of buried weed seeds', *J. exp. Bot.*, **20**, 402–413

WESSON, G. and WAREING, P. F. (1969b). 'The induction of light sensitivity in weed seeds by burial', *J. exp. Bot.*, **20**, 414–425

DISCUSSIONS

A Questioner: Since ethylene, which stimulates germination, is produced under anaerobic soil conditions which also inhibit root elongation, does this amount to ecological suicide?

Hall: Ethylene levels in the soil fluctuate naturally. They may be sufficient to stimulate germination without affecting subsequent root growth adversely; also, certain plants, such as rice, can live under water-logged conditions and their root growth is not retarded by ethylene.

Rorison: What is the ecological significance of ethylene as an environmental factor in determining the distribution of individual species?

Hall: We are finding a wide range of responses; it is hard to determine whether these are genetically inherent or related to environmental conditions.

Storey: Some soil fungi, e.g. *Fusarium* spp., produce ethylene; hence the question, is the ethylene produced by the seed itself or by the microflora on the seed coat?

Hall: All we know is that there is a change in the capacity to produce ethylene, either of the seeds or of something attached to them, which might well be their microflora.

A Questioner: In view of the considerable biological degradation of ethylene in soil, do the concentrations found at depths close to the soil surface really play a part in germination?

Hall: We do not yet know how ubiquitous and how stable the concentrations of ethylene in soil are; without ethylene, seeds of *Spergula* will remain dormant for long periods. But, as pointed out, it appears that the seed itself changes in its capacity to produce ethylene and has therefore a built-in germination-delaying mechanism.

Thurston: Nevertheless *Spergula* seeds, which are small and often buried deeply by ploughing, usually germinate only within the top two inches of soil.

Gutterman: Is anything known about any relationship between ethylene and phytochrome?

Hall: Nothing at present.

Moore: Freshly harvested peanuts are dormant; when placed in a plastic bag together with mature apples (which produce ethylene) they become non-dormant.

14

PROBLEMS OF SEED STORAGE

J. F. HARRINGTON

Department of Vegetable Crops,
University of California, Davis, U.S.A.

INTRODUCTION

It must be clearly understood that seed storage is not just the time the seed remains in the warehouse. Seed is being stored from the moment it matures on the plant until it is sown. Also, it must be clearly understood that all seeds are deteriorating. We now know ways to minimise the rate of seed deterioration but so far we have not learnt how to stop it; therefore, every lot of seed in commerce has aged to a greater or lesser extent and if kept long enough will finally contain no viable seed.

Although this paper will deal mainly with seed stores, the other periods of storage must also be recognised. Even while on the plant, seeds deteriorate; a process known as weathering. High moisture, high temperatures, sunlight, insects and diseases can all adversely affect the seed before harvest. Harvesting and cleaning, if improperly done, further injure the seed. If the seed is held at a high moisture content awaiting cleaning, decline in viability will occur. Even the period between storage and sowing can be deleterious to the seed. The time during transit in the retail store and in the farmer's shed is also critical. Proper drying with packaging in moisture-proof containers can minimise seed deterioration during this time (Harrington, 1963; 1972).

IMPORTANCE OF MOISTURE AND TEMPERATURE FOR SEED LONGEVITY

The two most important factors which affect seed longevity are the seed moisture content and the seed temperature. Below are suggested two rules-of-thumb which express in an easily understood manner the influence of the moisture and temperature of the seeds on the rapidity of their deterioration.

1. Each 1 per cent reduction in seed moisture doubles the life of the seed.
2. Each 5°C reduction in seed temperature doubles the life of the seed.

These rules-of-thumb are not as exact as the formulae of Roberts and Abdalla (1968) but they should give a quick grasp of the importance of low seed moisture and low seed temperature in preserving high seed germination.

MOISTURE CONTENT

It is necessary to discuss some qualifications to these rules-of-thumb. It is obvious that if the seed moisture content is high enough (something over 30 per cent), non-dormant seeds will germinate. In the range from about 18 per cent to about 30 per cent, heating due to micro-organisms will occur if oxygen is present, resulting in rapid death of the seeds. In the range from about 10 per cent seed moisture for oily seeds or about 13 per cent for starchy seeds to about 18 per cent, storage fungi grow actively and destroy the seed embryo. Christensen and Kaufmann (1969) cover this problem thoroughly. Thus, above 10–13 per cent seed moisture, sprouting, heating and fungal invasion can quickly destroy seed viability, probably more quickly than the first rule-of-thumb would indicate. Therefore, seeds should be dried as quickly as possible to below 14 per cent seed moisture and should be kept below this moisture content at all times.

On the other hand, if seeds are dried below 4–5 per cent seed moisture, it appears that deterioration is somewhat faster than with seeds of 5–6 per cent seed moisture. This is probably due to damage from lipid autoxidation (Koostra and Harrington, 1969). Unsaturated lipids in seed cells may break, producing two free radicals at the double bonds. These free radicals can react with other lipids, destroying the structure of cell membranes; with proteins, inactivating enzymes; and with nucleic acids, causing chromosomal abnormalities and even mutations. In imbibed cells tocopherols made by enzymes combine with the free radicals rendering them harmless. All seeds contain tocopherols, but these are used up during storage. Since enzymes are inactive at these low moisture contents no more tocopherols are produced during storage. The free radicals, which are produced non-enzymically, then become destructive. It is believed (Schultz, Day and Sinnhuber, 1962) that at about 5 per cent moisture the monomolecular layer of water which surrounds macromolecules in seeds ceases to be continuous and this facilitates the destruction of macromolecules, such as enzyme and membrane proteins, by the free radicals. Therefore, it appears that about 5–6 per cent seed moisture is the ideal moisture content for maximum storage life.

TEMPERATURE

The rule-of-thumb for temperature is applicable down to at least 0°C. If the seed moisture is below 14 per cent, no ice crystals form below the temperature at which a seed could freeze; so storage of dry seed at sub-zero temperatures should improve longevity. Unfortunately, most sub-zero stores have a high relative humidity (RH) and after a period of storage the seeds gain moisture and ice crystals form. These damage cells, causing loss in viability. Even if crystallisation did not occur during storage it may occur during thawing, destroying viability. If the seeds are first dried and then placed in moisture-proof containers then they will not regain moisture and should survive for a long period in sub-freezing storage.

Even though cold storage (0–5°C) of seeds is desirable, unless the seeds are sealed in moisture-proof containers or the store is de-humidified, the storage RH at low temperatures will be high. The seeds will gain moisture and when brought out to higher temperatures for transport they will deteriorate quickly because of their high moisture content. Lowering the storage temperature reduces or stops the activity of fungi and insects, but not activity of rodents.

There is a small group of species whose seeds cannot be dried. Many nuts, such as walnuts and acorns, seeds of citrus species, of the palms such as coconuts and of several aquatic species such as wild rice cannot be dried and are therefore very short lived. The cacao seed can neither be dried not stored in cold storage (10°C or colder) without irreversible injury. Its maximum life span is only a few weeks.

MOISTURE EQUILIBRIUM

The moisture content of a seed normally comes to equilibrium with the ambient RH. Seeds which are hard coated, including seeds of many legume species, are exceptions. For most species each seed has its own moisture equilibrium value for a given RH. This is because the molecules in seeds vary in the amount of water they adsorb. Proteins adsorb most water, starch and cellulose less but still a considerable quantity, while lipids adsorb no water. Thus, at the same RH oily seeds will contain less water than starchy seeds. Even within one species the moisture content of individual seeds in equilibrium with a given RH varies because seeds vary in size (and therefore surface to volume ratio) and seed coat thickness. These and other factors influence the chemical composition of each seed and, in turn, its equilibrium moisture content at a given RH.

Further, like soils, seeds drying to equilibrium with a given RH

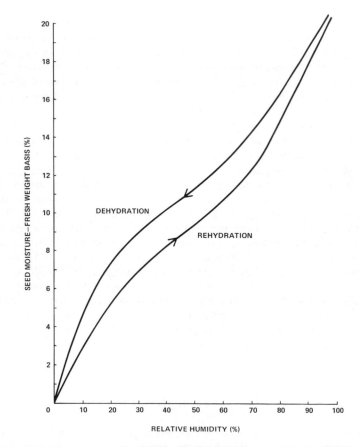

Figure 14.1   Hysteresis or the effect of dehydration and rehydration on the seed moisture equilibrium at different relative humidities

will not end up at the same moisture content as seeds gaining moisture to equilibrium with the same RH. This phenomenon is called hysteresis. It is common to many plant products, not only to seeds. The hysteresis effect is believed to occur here because, on rehydration, dried macromolecules do not completely unfold at intermediate moisture levels and thus do not have as many sites available for adsorption of water molecules. Hysteresis is illustrated in *Figure 14.1*. The moisture content of a seed in equilibrium with a given RH is slightly higher at cooler storage temperatures. This is because the energy of the water molecules is less at lower temperatures and fewer escape the attractive forces of the macromolecules.

For all these reasons, the seed moisture in equilibrium with ambient RH is different with each seed lot and may vary even for the same seed lot. The differences among lots of the same species may be as great as 2 per cent seed moisture. Thus, tables giving equilibrium values for various kinds of seed to more than one decimal place may be true for the particular sample tested but are only approximate indications of the equilibrium moisture contents of other samples of the same kind of seed.

RELATION BETWEEN RELATIVE HUMIDITY AND STORAGE TEMPERATURE

The RH of the air is a measure of the actual weight of moisture in a given weight of air in relation to the total weight of moisture the air will hold when saturated. As the temperature of the air increases, the weight of moisture a given weight of air can hold also increases. If the absolute weight of moisture remains constant, RH decreases on heating.

Table 14.1 MOISTURE IN AIR AT SATURATION (100% RH)

Temperature °C	g water/kg air
0	3·9
10	7·6
20	14·8
30	26·4
40	41·4

Using the figures in *Table 14.1*, if a kg of air at 20°C contains 7·4 g of moisture, its RH is 50 per cent (7·4/14·8). Heating this air to 30°C reduces its RH to 28 per cent (7·4/26·4). Cooling this air to 10°C increases its RH to 97 per cent (7·4/7·6). Cooling it a little lower will increase the RH above 100 per cent and moisture will condense on a surface such as a seed coat. Therefore, as the storage temperature rises, the RH decreases. Since the moisture content of the seed changes to remain in equilibrium with the ambient RH, the seed moisture content decreases also. The converse is also true. If the storage temperature is lowered the RH and the seed moisture both rise.

A question often raised is: 'Does this not also happen to seeds in moisture-proof containers when placed in cold storage?' The answer is that there is almost no increase in seed moisture content in this case. The reason is that while in a refrigerated seed store the air that flows through with its moisture content is essentially infinite so there can be a gradient of moisture from the air to the seed until the seed moisture reaches equilibrium with the air RH. In a sealed container there is a finite amount of moisture in the air which can move into the seed. For example, in a moisture-proof can containing 1 kg of seed of 6 per cent moisture content the seed contains 60 g of water. Even if the can was full of saturated air at 20°C it would contain less than 0·06 g of water. So if the water were completely absorbed by the seed it could only gain less than 0·06 per cent moisture, an insignificant amount. Therefore, when seeds in moisture-proof containers are placed in cold or sub-freezing storage, they do not gain moisture, while those in porous containers do.

DRYING SEEDS

To dry seeds the RH of the air must be below equilibrium with the seed moisture so there will be a moisture gradient from the seed to the air. If the RH of the air is higher than this equilibrium then the

Table 14.2 MOISTURE CONTENT OF WHEAT AND CABBAGE IN EQUILIBRIUM WITH CERTAIN RH

Air RH %	Seed moisture, fresh weight basis	
	Wheat—starchy	Cabbage—oily
15	6·5	3·5
30	8·5	4·5
45	10·0	6·0
60	11·5	7·0
75	14·5	9·0

moisture gradient will be to the seed and the seed will gain moisture instead of drying. Examples of seed moistures in equilibrium with a given RH are given for wheat (a starchy seed) and cabbage (an oily seed) in *Table 14.2.*

LOWERING RELATIVE HUMIDITY

Seed can be dried by unheated air only as long as the moisture gradient is from the seed to the air. If the seed moisture is 20 per cent or higher, as is often the case at harvest, then blowing unheated air of 75 per cent RH will dry the seed considerably, as can be seen from *Table 14.2*, but not to a safe moisture content. However, if the air with 75 per cent RH has a temperature of 20°C then heating it to 30°C will reduce the RH to about 40 per cent and the seed can be dried to a safe moisture content for storage. However, using air with too high a temperature to dry seed can kill the seed, especially if the seed moisture is high. A little-understood danger is that even at air temperatures not high enough to kill the seed immediately, the seed can be injured with loss of vigour and shortening of storage life. A temperature of 35°C is maximal for many seeds, especially vegetable and flower seeds. Drying, if high vigour and longevity is desired, should not be done at temperatures above 45°C for any crop seed. Drying can be hastened as follows:

1. By using higher temperatures (within the maxima stated above) because the vapour pressure deficit (vpd) is increased and the moisture gradient is made steeper.
2. By increasing the air flow because replacing moistened air with dry air keeps the gradient steep. About 5 cm^3 of air per minute per cm^3 of seed is the maximum economical air flow, however.
3. By allowing all sides of each seed to be exposed to the air flow as in a tumble drier or a baffled continuous flow drier.

However, there is danger in too rapid drying. If the moisture gradient from the seed surface is steeper than the moisture gradient from the interior of the seed to the seed surface there will be more rapid drying of the surface than of the interior. Cracking of the surface tissues may then occur in seeds of some species such as rice, causing loss of viability or vigour. 'Case hardening' may occur in seeds of other species such as cabbage seeds. In case hardening the outer cells shrink and become almost impervious to moisture, thus temporarily preventing movement of moisture out of the still moist centre of the seeds. Case hardening should not be confused with hardseededness because case hardening is only a temporary situation that disappears as the seed structures gradually regain equilibrium with each other after the drying operation.

The alternative to reducing the RH in the air by heating is actually to remove the moisture from the air, thereby reducing the RH. This can be done by refrigeration or by desiccation.

Refrigeration

Using refrigeration to dehumidify the air is economical at warm ambient temperatures down to about 20°C. At 20°C a RH of 45 per cent is achievable. The air is passed over the refrigeration coils, where the moisture condenses as ice. Periodically, a heating element around the coils is turned on, melting the ice, and the water is drained out of the room. At lower temperatures than 20°C dehumidification by refrigeration becomes uneconomical.

Desiccation

Dehumidification by desiccation is possible by two means. A solid inert desiccant such as silica gel or activated alumina can be used or a saturated solution of a desiccating salt can be used. If a solid desiccant is used, the air is blown from the storage or drying room through the desiccant, which removes moisture, and then is re-circulated through the seed in the room, absorbing moisture from the seed. Periodically, the desiccant must be recharged by heating it to a high temperature and driving the moisture into the outside air. If a desiccating solution is used, such as a saturated solution of lithium chloride, the air from the room is blown through a spray of the solution which absorbs moisture. For best efficiency the LiCl solution is cooled to 10°C when it is in equilibrium with 14 per cent RH. This, of course, at the same time cools the air which is also desirable for storing the seed. The drawbacks with using LiCl solution are that the solution is corrosive, particularly to the pump parts, and is toxic. A problem with the solid desiccants is that the absorption of water and the regeneration cycle cause the circulating air to rise in temperature. Frequently, an air-cooling device is placed in the air flow between the desiccator and the storage room.

DRY STORAGE

It must be firmly stated that any drying or storage room which depends on dehumidification must be completely moisture-proof or the dehumidifier will be drying the whole vicinity instead of the room and seed. For proper seed storage the seed is dried first to the low moisture content desired. For one season's storage the seed moisture content must be dried down to at least equilibrium with 65 per cent RH. For 2–3 years' storage it should be dried to equilibrium with 45 per cent RH. For long-term storage, or for packaging in moisture-proof containers, the seed should be dried to equilibrium with 25 per cent RH or to 5–6 per cent moisture content.

MAINTAINING SEED IN DRY STORAGE

After the seed is dried to the desired moisture content it must be kept at this level or the cost and benefit of the drying is lost. Maintaining the seed in a dry condition can be done in three different ways, although the principles are the same in each method.

1. The store itself may be made moisture-proof and have dehumidification to maintain the desired RH.
2. The seeds may be packaged in moisture-proof containers.
3. The seeds may be placed in gasketed containers enclosing also an indicator desiccant.

Each method will now be more fully described.

Moisture-proof storage rooms

The construction of a moisture-proof storage room should include the following points. There should be no windows or other openings and only one door, which should be of the gasketed refrigerator type. It is best if the entrance has an anteroom between two doors, with the inside door not opened until after the outside door is closed. All walls, the ceiling, the floor and the doors must be completely sealed against moisture penetration. This should preferably be on the inside of the structural frame. Moisture-proof seals include an asphalt coating at least 3 mm thick, 1000 gauge polyethylene and aluminium foil bonded to paper or plastic. All seams must be sealed. On the inside or dry side of the moisture-proof barrier insulation can be added. A layer to protect the moisture-proof barrier against punctures or tears is also desirable.

Moisture-proof containers

Moisture-proof containers range in capacity from aluminium foil laminated packets, through tin cans, polyethylene (700 gauge) bags, aluminium foil laminated drums, to steel bins with gasketed lids. None of these containers is moisture-proof until properly sealed (Harrington and Douglas, 1970).

Gasketed containers

For many small seed lots which must be opened from time to time, such as those stored for plant breeders and those serving as stock seed or official test samples, a desirable moisture-proof container is a steel box with a gasketed lid. For every 5 kg of seed 1 kg of dry indicator silica gel is added in a cloth bag. If, on frequent opening, the indicator colour changes from blue to pink, the silica gel is replaced by a dry silica gel and the replaced silica gel is reactivated in an oven. In a cool room (20 °C or lower) this method of storage will keep seeds of most species without loss of germination capacity for several years.

PROBLEMS OF UNINSULATED STORAGE ROOMS

It should be mentioned here that common storage rooms (un-insulated) have obvious and not so obvious disadvantages. Obviously the seed temperature will quickly change to the ambient temperature and, more slowly, the seed moisture will change to equilibrium with the ambient RH. In many places the ambient temperature and/or RH are undesirable for seed storage though not in all places and not at all times of year. There are disadvantages that are not so obvious. Even though the ambient RH may be satisfactory, if the soil under the floor is wet, diffusion of moisture through the floor into the seeds will take place. Also, the average moisture content may be satisfactory, but if it turns cold outside there will be a temperature gradient from the centre of the store to the walls and air convection currents will be created. The warm air of the centre will flow upward and then to the walls where it will be cooled and increase its RH. The moisture will diffuse into the outer seeds, the cool air will flow downward, across the floor and up through the centre where it will warm, and moisture will move from the centre seeds to the now drier air. If either the moisture movement from the damp soil or from the centre seeds to the outside seeds continues for long there will be spots of high moisture at the outside. Mould growth, insect build-up, heating and even sprouting may occur in the seeds situated there, causing deterioration of the seed with eventual loss of viability.

PROBLEMS ARISING OUT OF DRY STORAGE OF SEEDS

Although dry seeds maintain viability better in storage, there are two cautionary points to remember. First, dry seeds are more easily damaged by the handling that will occur in transport and planting. Second, from initially dry seeds seedlings do not emerge as quickly as from seeds of higher moisture content. This is not because they are less vigorous but because they require more water and a longer time to imbibe and germinate. Consideration might be given to conditioning dry seeds back to a higher moisture content before transport and planting. Conditioning to equilibrium with 65 per cent RH would minimise transport damage and increase the rate of emergence without endangering the seed by fungal growth or heating.

MEASUREMENT OF SEED MOISTURE

Since seed moisture is such an important aspect of seed longevity, the accurate measurement of the moisture of a seed lot becomes important. The simplest moisture test, and one used for centuries, is the bite test. Although not too precise it is useful to determine if the seed is too high in moisture for even short time storage. If the seed feels rubbery when bitten, the moisture content is dangerously high. On the other hand the most exact moisture test employs the Karl Fischer reagent. The reagent combines stoichiometrically with water and the amount of reagent combined is measured colorimetrically. All other methods of measuring moisture use the Karl Fischer method as the standard for comparison. However, the Karl Fischer method requires a chemist and is tedious and expensive.

The International Seed Testing Association approves oven methods of moisture determination (105°C for 24 h or 130°C for 1 h) or the oil distillation method, depending on the kind of seed. These methods can be accurate to ± 0.1 per cent seed moisture. Infra-red balances and various electrical methods are quick but are subject to considerable error. They should only be used as a guide and should be checked by oven moisture determinations.

K

An indirect method is to measure the RH of the store with a psychrometer. If the seeds have reached equilibrium with the RH of the store, this should be a better measure of the storability of the seeds than the seed moisture content, since so many factors influence the seed moisture content, and ageing of seed, fungal growth and insect activity are more directly correlated with RH than with seed moisture.

CONCLUSIONS

Seed moisture and seed temperature are the most important factors affecting viability of sound seed. Within limits, the lower the seed moisture and seed temperature, the longer viability is maintained. Storage of seed is not only the time spent in a seed store but includes the time from seed maturity to planting; so proper storage conditions must be considered at all times in the life of the seed. In order to preserve viability of seeds, as soon as possible seed moisture and seed temperature should be reduced to a desirable level, and maintained at this level. Proper use should be made of controls to ensure that the seed moisture and seed temperature are kept at the levels desired.

REFERENCES

CHRISTENSEN, C. M. and KAUFMANN, H. H. (1969). *Grain storage, the Role of Fungi in Quality Loss*, Univ. of Minnesota Press, Minneapolis, Minn.

HARRINGTON, J. F. (1963). 'The value of moisture-resistant containers in vegetable seed packaging', *Bull. Calif. agric. Exp. Stn., 792*, J–23

HARRINGTON, J. F. (1972). 'Seed storage and longevity', in *Seed Biology*. Vol. 3 (ed. Kozlowski, T. T.) 145–245. Academic Press, N.Y.

HARRINGTON, J. F. and DOUGLAS, J. E. (1970). *Seed Storage and Packaging—Applications for India*. National Seeds Corp. and Rockefeller Foundation, New Delhi, India

KOOSTRA, P. T. and HARRINGTON, J. F. (1969). 'Biochemical effects of age on membrane lipids of *Cucumis sativus* L. seed', *Proc. int. Seed Test. Ass., 341*, 329–340

ROBERTS, E. H. and ABDALLA, F. H. (1968). 'The influence of temperature, moisture and oxygen on period of seed viability in barley, broad beans and peas', *Ann. Bot., 32*, 97–117

SCHULTZ, H. W., DAY, E. A. and SINNHUBER, R. O. (eds.) (1962). *Lipids and their Oxidation*, Avi, Westport, Connecticut

DISCUSSIONS

Gordon: What temperature and what flow rate of air do you recommend for a seed store?

Harrington: Seeds intended for sowing (including malting barley)

should be dried rapidly (within 6 to 24 h), but preferably with cool air (maximum 35°C), with a maximum flow of 5 m³ of air per minute per ton of seed. Afterwards only a slow flow of air is required to prevent convection currents from hot to cold spots in the store.

MacKay: Christensen (Minnesota) has found that the detrimental microflora needs a relative humidity higher than 65 per cent to become active. The critical drying temperature tends to increase as the seed moisture level drops. 60°C is not too high provided seed moisture is less than 15 per cent.

Harrington: But such high drying temperatures can reduce storage life even though this may not be noticeable within the first 6 months.

Bowman: *Phaseolus* beans crack badly if dried too rapidly. We increase the temperature in steps of 5°C, but not above 35°C, and do not permit the RH to drop below 65 per cent.

Harrington: Similarly, rice seeds have to be dried in steps of 2 per cent H_2O and then reconditioned, otherwise they crack. Seed producers in Idaho, where the relative humidity is very low (less than 30 per cent), raise the moisture content of pea, bean and soya bean seeds before depatch in order to avoid injury during transit owing to their brittleness.

Gutterman: The rate of germination may be influenced by storage conditions. *Lactuca scariola* seeds stored in high humidity and *Portulaca oleracea* seeds stored at high temperatures (40°C) germinate faster than after drier or cooler storage, respectively.

Harrington: Lipid autoxidation is likely to be more damaging at extremely low seed water contents because the water layer surrounding macromolecules is then broken and they are more easily attacked by free radicals. Also, at this low water content the antioxidant tocopherols (e.g. Vitamin E) cannot be synthesised because there is no enzyme activity, and once the tocopherol reserve in the cell is used up the damage by the free radicals becomes much greater.

E. H. Roberts: One report (from the U.S.A.) exists on the use of an antioxidant tocopherol on okra and some other seeds.

Stein: How do you envisage that antioxidants can be introduced into seeds so as to reach the appropriate sites of action?

Harrington: It is admittedly a difficult problem.

Heydecker: It may be possible to incorporate antioxidants into dry seeds by means of organic solvents. Alternatively, deterioration due to oxidation may be prevented by storing seeds in an oxygen-free atmosphere, preferably pure CO_2, as was shown by McLeish and Harrison at the John Innes Institute.

A Questioner: Certain seeds are injured by cold storage. Is such injury reversible?

Harrington: Chilling injury, to cacao beans (which do not survive when dried) and to partially germinated cucumber and *Capsicum* seeds at least, is not reversible.

15

AGEING AND THE LONGEVITY OF SEEDS IN FIELD CONDITIONS

T. A. VILLIERS

Department of Plant Biology, University of Natal, Durban, South Africa

INTRODUCTION

It is well known that, in general, low moisture content, cool temperatures and low oxygen tension increase the longevity of seeds in
storage (Barton, 1961). Simpson (1942) showed that the longevity
of cotton seeds decreased with rising temperature and rising moisture
content from 7 to 16 per cent, such that seeds with 14 per cent
moisture content did not survive one year at room temperature or
above. The viability of seeds of *Primula,* onion, lettuce, tomato and
parsnip was increased by storage in carbon dioxide or nitrogen, and
by reducing their moisture content (Harrison, 1966).

At first sight this appears to be entirely contradictory to ecological
observations, which have shown that many seeds can remain buried
but viable for very many years in the soil, where they are presumably
imbibed with water, but held in a state of induced or secondary
dormancy. Thus, Turrill (1957) has reported that buried arable
weed seeds existed in grassland soil which had not been cultivated
for periods of between 30 and 40 years, and even up to 300 years in
one instance. In an experimental study, Kjaer (1948) found that 9
species of seeds which still germinated after 10 years in the soil failed
to germinate after dry storage for a similar period.

Barton (1961) stated that whilst it may not be surprising that seeds
which are dry, and therefore have their metabolic processes greatly
reduced, are still capable of germination after years of storage, the
extended life-span of some species in the soil, where they are presumably fully imbibed, is more difficult to explain. Wareing (1966)

commented that, in view of the generally-accepted requirements of seeds in dry storage conditions, it is surprising that many seeds apparently retain their viability longer when buried in moist soil than when kept in dry storage.

Field observations on the extended longevity of seeds in the soil have been repeated in the laboratory, where the seeds have not been buried. Toole and Toole (1953) maintained seven varieties of lettuce seeds at a range of moisture contents and also fully imbibed. The seed was kept in the dark at 30°C in order to prevent the germination of the imbibed seeds. The results of viability tests clearly supported the ecological observations in that a high germination capacity is retained by fully imbibed seeds, whilst over similar periods of time, seeds stored air-dry had shorter life-spans and when stored air-dry at higher relative humidities were all dead within a few weeks.

It would thus be more satisfactory if explanations of the causes of ageing in seeds could also include the apparently anomalous behaviour of seeds in field conditions. This paper attempts to reconcile these two types of seed behaviour and in doing so provides evidence for a general theory of seed ageing.

GENERAL THEORIES OF AGEING

These have been grouped under three headings by Curtis (1963) as follows:

1. The accumulation of deleterious products of metabolism. In this group are included the various theories involving cross-linkage of macromolecules which would render enzymes and nucleic acids inactive, membranes non-functional, and cause the accumulation of metabolically inert material and mutagenic substances.
2. Wear-and-tear theories, where it is assumed that with increasing use, organelles, cells and organs become inefficient. This could also be due in part to cross-linkage.
3. Somatic mutation theories have attracted most attention and have the most convincing experimental support. The work of Curtis (1967) has shown that the number of mutations increases with age, and that species and inbred strains of animals with shorter life-spans have higher mutation rates. An organism therefore becomes gradually less efficient as it carries an increasingly large number of mutations, many of which may cause the production of faulty proteins, and post-mitotic cells

will eventually die. In addition to strand breakages, deletions and point mutations, damage can again be caused by cross-linkage between DNA molecules and also between DNA and its associated protein.

RELEVANCE TO AGEING OF SEEDS

Cell division cannot take place during dry storage, nor presumably in seeds lying dormant in the soil for long periods. Therefore, in this respect at least, the cells in non-germinating seeds are similar to post-mitotic cells. Ageing damage to macromolecules and structures must therefore accumulate, since there is no possibility of selection of undamaged cell lines which can occur in a population of dividing cells.

It is well known that chromosome damage increases in incidence during the ageing of seeds in dry storage. Roberts, Abdalla and Owen (1967) review this subject and come to the conclusion that ageing in seeds can best be explained by the accumulation of genetic damage which manifests itself as chromosomal aberrations seen in the first mitotic divisions in early germination. They point out that this visible chromosome damage is only a small aspect of the total damage which has probably occurred. Gross damage would never be seen because mitosis would not take place at all, and many mutations would presumably occur which would not result in visible structural alterations to the chromosomes.

In addition, Abdalla and Roberts (1968) found that under conditions of stress during storage of seeds there was evidence for the operation of an ageing factor additional to that of chromosome damage. Reviewing the evidence that ageing is caused by somatic mutations, Curtis (1966) suggested that while considerable support for the theory exists, a number of lines of evidence point to the fact that it cannot be the only cause of ageing.

It is known that cell division is not essential to the process of germination as the emergence of the embryo takes place by elongation of the cells already present. In certain cases, a seedling can even be formed without cell division (Haber and Luippold, 1960). It has also been shown that certain quiescent seeds probably contain the complete protein-synthetic machinery necessary for germination (Marcus and Feeley, 1964; Dure and Waters, 1965).

It is therefore questionable whether the genetic material is involved in the actual germination-initiating process in quiescent seeds, and therefore mutations and chromosomal aberrations should not prevent the activation and emergence of the embryo. Damage to the chromatin undoubtedly occurs, but is more likely to affect cellular differentiation and seedling morphogenesis rather than germination. It is concluded, therefore, that in addition to the ageing changes which take place in DNA, another system, more closely concerned with the cytoplasm and the cell extension process, also becomes damaged during seed ageing.

In seeds stored air-dry, metabolism is reduced to a very low level in the desiccated tissues, and longevity is directly related to the moisture content of the seed and to environmental factors such as relative humidity, temperature and oxygen tension. Within the range of moisture contents achieved by seeds in equilibrium with atmospheric water, increases in seed moisture content, which would be expected to increase the rate of metabolic processes, drastically reduce the life expectancy of the embryo. Together with the effects of warm temperature and availability of oxygen, this tends to support the prevalent idea that seeds are able to survive for long periods as a result of their low metabolic rate.

We may therefore be forced to consider that the behaviour of seeds in moist soil is anomalous. If we extrapolate the results of storage experiments, the high water content of seeds in the soil should allow survival for only short periods of time. The fact that these fully imbibed seeds have greatly extended life-spans seems to indicate that their ageing is controlled by a mechanism which is different from that operating in stored seeds. The only explanation so far reported has been that seeds in the soil are held dormant and so are enabled to survive (Toole and Toole, 1953; Roberts, 1972).

However, perhaps contrary to general expectation, dormancy does not necessarily cause a cessation of metabolism. It has been shown that in seeds stored fully imbibed and deeply dormant in the true sense of the term, a whole range of metabolic activities, including food interconversions, membrane formation and organelle production, can take place (Villiers, 1971a).

REQUIREMENTS FOR A SEED-AGEING THEORY

It would be highly satisfactory if a general theory of seed ageing could explain the following points:

1. The deleterious effects of (a) increase in moisture content, (b) increase in temperature, and (c) the presence of oxygen.
2. The extended life-span of fully-imbibed seeds, and seeds in the soil.

3. The extra-chromosomal ageing factor postulated above.
4. The appearance of chromosomal aberrations, and also mutations.
5. The reduction in vigour of plants grown from old seeds.
6. The delay in germination of old samples of seeds.

It becomes possible to reconcile the deleterious effects of increasing
moisture content in the low range (5–20 per cent) with the increased
longevity of fully-imbibed seeds, if it is accepted that (a) in addition
to chromatin, the cytoplasmic organelles and their macromolecular
components are also subject to deterioration, and (b) in normal
tissues, most cytoplasmic and nuclear components are maintained
structurally and functionally in good order by repair mechanisms,
or are replaced as whole units by newly-formed organelles. Evidence
for the existence and operation of such maintenance systems will be
discussed below.

Systems for the repair and replacement of organelles would
necessitate a complex pattern of enzyme activities which would be
unlikely to operate in air-dry seeds, but could take place in fully-
imbibed seed tissues. Thus, in air-dry tissues the damage to macromolecules, membranes and whole organelles must accumulate
without the possibility of repair, until the amount of damage
sustained becomes too great for recovery when the tissue eventually
imbibes water, and disorganisation therefore takes place.

In seeds fully imbibed with water throughout the storage period,
however, the damage may be corrected as and when it occurs. It is
also possible for the repair enzymes themselves to be damaged
during ageing in dry storage, and consequently disorganisation of
the cytoplasm would take place during early imbibition before the
repair systems could be replaced and come into operation.

MECHANISM OF DETERIORATION

In considering the possible mechanisms whereby membranes and
enzymes could deteriorate, it would appear most likely that they
become damaged by peroxidation reactions (Barber and Bernheim,
1967). Briefly, cell membranes contain a relatively high proportion
of polyunsaturated lipids and in the presence of oxygen these may
react to form free-radical intermediates and unstable peroxides. This
can result in the destruction of the lipid itself, the formation of
insoluble lipid-protein complexes by cross-linkage, and damage to
amino acids (Desai and Tappel, 1963).

In addition to the free-radical intermediates of lipid peroxidation, free radicals may also be produced by the ionisation of water to form proteins which react with reduced molecular oxygen. In this way ·OOH and ·OH radicals are formed, which are highly reactive and may also damage macromolecules.

The fact that stored seed tissues are air-dry would not prevent such peroxidation reactions. In normal tissues, water has been suggested to be in a semi-crystalline state, especially in the vicinity of macromolecules (Verzar, 1968). As tissues become air-dry, the remaining water becomes tenaciously bound to the cell colloids (Askenov, Askochenskaya and Petinov, 1969). The majority of enzyme-controlled reactions, especially those involving large structural units, would be unlikely to occur under these conditions. Being bound to the seed colloids, however, the water would still be able to permit peroxidation reactions requiring only the proximity of water and oxygen to double bonds in the unsaturated lipids of the cytoplasmic membranes.

An important function of cytoplasmic membrane systems is to serve as selectively permeable barriers (O'Brien, 1967). Damage occurring during storage would be enormously amplified during the imbibition process if organelles containing hydrolytic enzymes (lysosomes) were damaged so that they released their contents into the ground cytoplasm. It has been shown that mitochondria and endoplasmic reticula (ER) peroxidise more rapidly than lysosomes, due to their higher content of polyunsaturated lipid and the presence of haemoprotein which acts as a catalyst of peroxidation reactions. However, disintegration takes place more rapidly in lysosomes than in mitochondria, presumably because they possess only a single unit membrane (Tappel, 1965). Peroxidation of isolated lysosomes was shown to release hydrolytic enzymes in the order aryl sulphatase, glucuronidase, acid phosphatase, and ribonuclease, which is similar to the effects of breaking the lysosomal membrane by osmotic shock or detergents.

The implications of this for animal cells have been discussed by Hochschild (1971) and postulated to amplify the original peroxidation damage by the release of hydrolases damaging other membrane systems, and also enzymes, RNA and DNA. It should be noted here that some of the chromosomal damage seen in germinating aged seeds could have been caused in this way during imbibition, and the loss of RNA, including the long-lived m-RNA considered by Osborne, Roberts and Hallam (1971), could also take place during imbibition.

In the normal course of events, lysosomal membranes must be kept in good repair for the protection of the cells which contain them, but in air-dry seeds this would not be possible. Upon full imbibition of the seed with water, the damaged membranes would leak hydro-lytic enzymes into the cytoplasm, which would rapidly undergo lysis.

It is an important point for the present argument that the full consequences of damage accumulated during the storage period are not felt until imbibition, when the damage might be too extensive for repairs to be effected in time to prevent autolysis of the cells. In this way, as well as in the production of free radicals, the effect is similar to that of irradiation, where a large amount of damage may be inflicted within a short space of time.

This is not the place to enter into a discussion of the existence and nature of the plant lysosome, but much evidence has accumulated which equates the plant cell vacuole with the lysosome (Matile, 1968). Elsewhere it has been concluded that the many small vacuoles in undifferentiated plant cells have a normal autophagic activity and in the mature cell they become confluent and take on the secondary function of providing turgor support, and the large volume typical of plant cells, with great economy of materials (Villiers, 1971b).

In a study of ageing in the embryo and endosperm of wheat, Aspinall and Paleg (1971) found that these age independently of each other, but at identical rates. They concluded that ageing is due to some basic aspect of metabolism or to intracellular integrity. The loss of membrane integrity has previously been proposed as an effect of ageing in seeds following observations on the leakage of organic solutes from aged seeds during imbibition. Ching and Schoolcraft (1968) suggested that this leakage reflects proteolytic activity during dry storage. However, by the mechanism proposed here, this would easily be explained by non-enzymic peroxidation damage during storage, resulting in further damage to the cells by the activity of proteases released *during imbibition*. Koostra and Harrington (1969) have suggested lipid oxidation during storage as a cause of solute leakage during imbibition.

An objection to the possibility of free-radical damage in stored seeds may be found in the work of Conger and Randolph, who were able to demonstrate free radicals in irradiated seeds (1959), but not in aged seeds (1968). However, free radicals are very shortlived because of their high reactivity. While an excess of free radicals might be produced in a short time by irradiation, the ageing process is fortunately slow, and the free radicals formed would be removed by reaction and therefore would not build up to a level sufficient for detection. In addition, irradiation produces hydroxy-radicals in quantity, in addition to those formed as intermediates of lipid peroxidation.

REPAIR MECHANISMS

The existence of repair and maintenance systems is fairly generally accepted. Damage to DNA by u.v. irradiation can be reversed by enzyme action in bacteria (Setlow and Carrier, 1964), and in fungi (Rupert, 1960; Gampel and Toha, 1969). In higher organisms the increased incorporation of DNA precursor into liver cells during ageing of rats was suggested to be evidence for DNA repair systems operating in post-mitotic cells (Samis, Falzone and Wulff, 1966). Pelc (1970) has suggested that self-replication of metabolic DNA (multiple copies of the genes active in a particular cell) might occur, replacing molecules damaged during cellular activity.

Omura, Siekevitz and Palade (1968) have postulated that continuing turnover of membrane proteins and lipids occurs, and Luzzati *et al.* (1969) have suggested that any given portion of a membrane may be transient. Components of the ER have been shown to be removed and replaced independently of each other (Tarlov, 1968). Mitochondria do not remain functional throughout the life of a cell, but have been shown to be replaced, possibly as entire units, with a turnover time of approximately 10 days in rat liver (Fletcher and Sanadi, 1961). It has subsequently been suggested that at least the soluble enzyme components may be renewed independently of the structural components. Gross, Getz and Rabinowitz (1969) suggest that the soluble fraction and the inner and outer membranes turn over at different rates, and that some components of the inner membrane and matrix may be deleted and inserted independently of the synthesis or destruction of the mitochondrion as a unit.

There is, in addition, a lot of other evidence which shows that continuous exchange of membrane components occurs, and that much of this may involve synthesis of additional or replacement components *in situ*. For the purposes of the present argument, it is necessary to show that such synthesis can take place in imbibed, but non-germinating, seeds. As stated earlier, it has been shown by the present writer in a cytological study of dormant embryos that membrane synthesis and organelle formation can take place during deep dormancy, when no germination is possible (Villiers, 1971a).

It is postulated above, that although chromosomal aberrations increase during storage, this may have little direct effect upon the germination process itself. However, following germination, aberrant cell divisions, abnormalities of the plant, pollen abortion and the loss of vigour are probably expressions of damage accumulated in the genome during storage. The delay frequently encountered before germination in old seeds is possibly due to a period during which repairs and replacements are effected to organelles and enzymes (Berjak and Villiers, 1972), while complete failure to germinate is connected with membrane damage as described.

Evidence is offered below that while chromosomal aberrations and propensity towards seedling mutations increase during dry storage of lettuce seeds, they do not increase in seeds stored fully imbibed for the same time at similar temperatures. These results were obtained following an accelerated ageing treatment, and the author is therefore grateful for the findings of Cartledge and Blakeslee (1935) that *Datura* plants grown from seeds aged 22 years in the soil had only slightly higher mutation rates than control plants grown from normal seeds; whereas plants grown from seeds aged in dry storage had high mutation rates which increased with rise in seed moisture content, temperature and time of storage (Cartledge, Barton and Blakeslee, 1936).

SUPPORTING EVIDENCE

If the above proposal is accepted as a working hypothesis, it is possible to provide some supporting evidence.

SEED LONGEVITY

Previous experiments on moist storage of seeds have been repeated in the author's laboratory, with exactly similar results. When lettuce seeds, variety Arctic King, were stored at 30°C, their longevity decreased with increase in moisture content from 3 to 14 per cent and seeds with 14 per cent moisture were all dead after 4 weeks. However, when the seeds were fully imbibed in the dark at 30°C (an environment in which they were unable to germinate), a very high germination capacity was retained, although a short delay occurred before germination when the seeds were brought out into the light at 22°C (*Figure 15.1*).

Similarly, seeds of *Fraxinus americana* were stored air-dry with moisture contents of 6·0, 9·5 and 18·6 per cent, and also fully imbibed in the dark (to prevent germination) at 22°C. Here again, viability decreased in the air-dry seeds as the moisture content increased, such that seeds containing 18·6 per cent water were all dead by the third month, whereas those stored fully imbibed gave full germination when brought out into the light after the same period of time (*Table 15.1*).

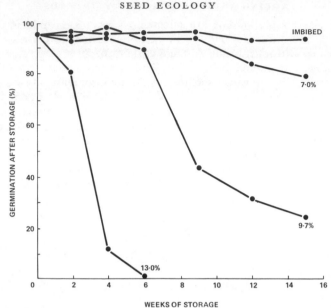

Figure 15.1　Germination of batches of seeds of lettuce variety Arctic King withdrawn at intervals from large samples stored at 30°C as follows: 7·0 per cent seed moisture content, 9·7 per cent moisture content, 13·0 per cent moisture content, and fully imbibed (in the dark to prevent germination). Germination tests were carried out at 22°C under fluorescent lighting

NUCLEAR DAMAGE

Chromosomal aberrations

These increase in number in seeds stored air-dry (Roberts, Abdalla and Owen, 1967). If this is due to the inability of DNA repair mechanisms to operate in air-dry tissues, then seeds stored fully imbibed should possess few such aberrations.

Root tip squashes were examined from germinated lettuce seedlings at about the time when the first cell divisions were taking place. It was found that the incidence of chromosomal damage increased with time of storage and especially with rise in moisture content of the seeds during storage. However, there was little increase in the number of chromosomal aberrations in seeds from the same original harvest, but stored fully imbibed over a period of 4 months, by which time seeds with moisture contents of 9·7 and 13·0 per cent were all dead (*Table 15.2*).

Table 15.1 GERMINATION OF SEEDS OF *FRAXINUS AMERICANA* AFTER STORAGE AT 6·0, 9·5 AND 18·6 PER CENT SEED MOISTURE CONTENT AND ALSO FULLY IMBIBED. SEEDS WERE STORED IN THE DARK AT $22°$C TO PREVENT GERMINATION OF THOSE FULLY IMBIBED

Moisture content	Percentage germination			
	1 month	2 months	3 months	4 months
6·0%	98	92	96	94
9·5%	94	88	76	4
18·6%	81	22	0	0
Fully imbibed	96	95	98	96

Table 15.2 ABERRANT CELL DIVISIONS, EXPRESSED AS A PERCENTAGE OF ALL LATE ANAPHASE FIGURES OBSERVED, IN RADICLE TIPS OF *LACTUCA SATIVA* VAR. ARCTIC KING, AFTER STORAGE AT 7·0, 9·7 AND 13·0 PER CENT SEED MOISTURE CONTENT, AND ALSO FULLY IMBIBED. SEEDS WERE STORED IN THE DARK AT $30°$C TO PREVENT GERMINATION OF THOSE FULLY IMBIBED

Moisture content	Percentage aberrant anaphases			
	3 weeks	6 weeks	9 weeks	12 weeks
7·0%	3	3	8	13
9·7%	12	19	17	22
13·0%	19		no germination	
Fully imbibed	2	2	3	2

Mutations

Whilst the experiments have not been under way for a sufficient time to observe the characteristics of mature plants, it is considered significant that a high percentage of deformities was observed in the young seedlings produced from seeds stored air-dry at higher relative humidities. These included stunted growth, sub-divided cotyledons and first leaves, aberrant pigmentation, swollen roots and necrosis of the root meristems. It is presumed that the majority of these abnormalities were caused by damage to the genome. Seeds stored fully imbibed for the same length of time showed none of these deformities during the four months of the storage experiment.

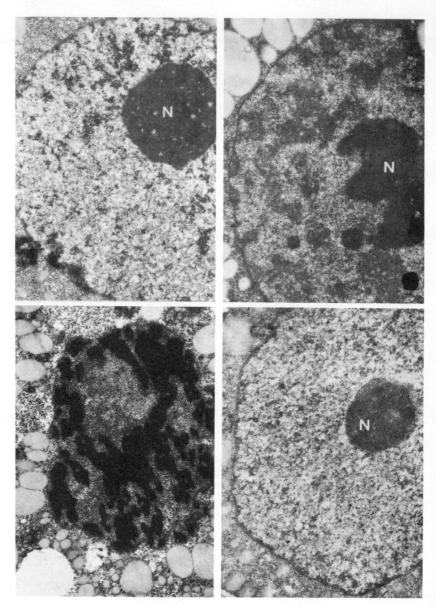

Nuclear staining reaction

In electron microscope studies of aged embryos, changes in the staining reaction of the nucleus to electron stains such as uranyl acetate and lead citrate are one of the most noticeable features. In embryo root tips from normal, unaged embryos, the nuclei stained fairly homogeneously except for a fine network of darker material identified as the heterochromatin. As ageing proceeded, the hetero-chromatin stained more deeply, until in embryos from batches of seeds with low viability, dense patches of chromatin occurred and eventually the nuclear membrane became difficult to distinguish as a continuous membrane system. However, the staining reaction of the nuclei in seeds stored fully imbibed for similar periods appeared to be quite normal (*Figure 15.2*).

MEMBRANE DAMAGE

Cytology of imbibition in aged seeds

For the reasons outlined above, the immediate damage which renders the aged seed incapable of germination is considered to be extra-nuclear. Free-radical damage to membranes and enzyme systems could cause autolysis and failure of essential metabolic processes when the seeds become imbibed. Examination of aged embryos during the imbibition and germination processes indicates that a large amount of membrane damage does occur during storage. There appear to be two fairly distinct stages in the ageing of mem-brane systems in lettuce embryos as they are later affected by water imbibition.

Where seeds are examined from samples in which the germination percentage is low, and the majority of the newly-emerged embryos die without further growth, the cells of the embryo radicle appear to be well organised in the very early stages. Often the only sign that the seeds are aged is the dense staining and clumping of the chromatin

Figure 15.2   Nuclei from radicle meristem cells of lettuce (cultivar Arctic King), 24 h after placing in germination test conditions. Glutaraldehyde/osmium fixation, followed by lead citrate and uranyl acetate staining. N, nucleolus. Magnification: × 16 000. Seeds stored as follows before germination test: Top left, shortly after received from supplier—chromatin stains as fine network; Top right, after two weeks at 30°C with 13 per cent seed moisture content—chromatin more heavily stained; Bottom left, after four weeks at 30°C with 13 per cent moisture content— chromatin clumped together and deeply stained, embryo probably non-viable; Bottom right, after four weeks' storage at 30°C but fully imbibed—chromatin gives normal staining reaction; embryo viable

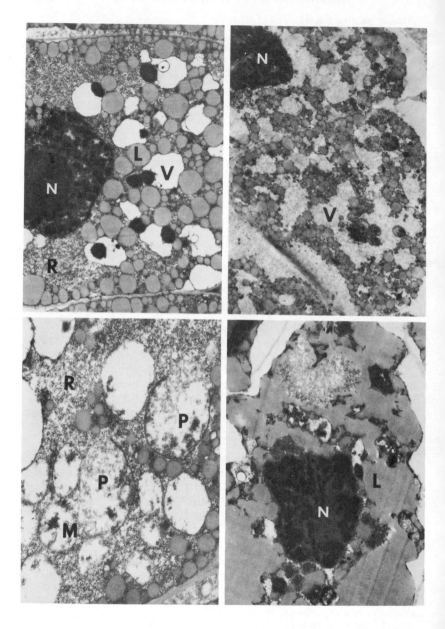

already described. However, as imbibition proceeds during the first 24 hours the vacuolar membranes appear to burst and disperse. This is followed by the disintegration of the mitochondria and plastids, and then the nuclear envelope. The plasma membrane becomes dissociated from the cell wall and disperses, the lipid body membranes disintegrate and the lipid coalesces into large, confluent masses. Microbodies are often the last organelles distinguishable in the disorganised cytoplasmic debris (*Figure 15.3*).

Seeds kept in storage for longer periods behave differently during the imbibition period. In these, the protoplast is not in contact with the cell wall, but appears to be dispersed at the earliest time of observation as the tissues take in water. Similarly, vacuoles never swell but appear as though ruptured from the very beginning of imbibition. Other organelles are hardly recognisable and the stored lipid appears to be in confluent masses even in the air-dry seeds (*Figure 15.4, top left and top right*).

A progressive change in the sequence of events can be followed from the first to the second stage. The first stage is thought to be due to the membranes of the organelles being only slightly damaged and the vacuoles (lysosomes) leaking hydrolytic enzymes into the cytoplasm. The initial stages of imbibition, therefore, have a normal appearance, and disorganisation takes place subsequently. However, in later stages of ageing, the membranes have become so damaged that they are no longer functional as differentially-permeable membranes and water is not taken up by osmosis, but only by imbibition of the colloids. The coalescence of the lipid into large masses must also be the result of damage to the lipid body membranes allowing the oils to become confluent even in dry seeds.

Germination behaviour and membrane damage

The germination behaviour of seeds sampled during ageing of a batch can also be explained by the above argument. First, there is

---

Figure 15.3   Cells from radicle meristems of Arctic King lettuce, at increasing times after beginning of imbibition. Seeds stored at 30°C with 13 per cent moisture content for four weeks before test. L, lipid; M, mitochondrion; N, nucleus; P, plastid; R, ribosomes; V, vacuole. Top left, 8 h after beginning of imbibition; cytoplasm still organised, but nucleus dense and abnormal. × 8 000. Top right, 16 h later; the vacuoles appear to have disintegrated, the cytoplasm appears disorganised. × 5 700. Bottom left (same time as top right), mitochondria and plastids appear 'empty' and plastid membranes have dispersed. × 15 500. Bottom right, 48 h after beginning of imbibition; cytoplasm totally disorganised, lipid body membranes burst and lipid has become confluent. × 8 000

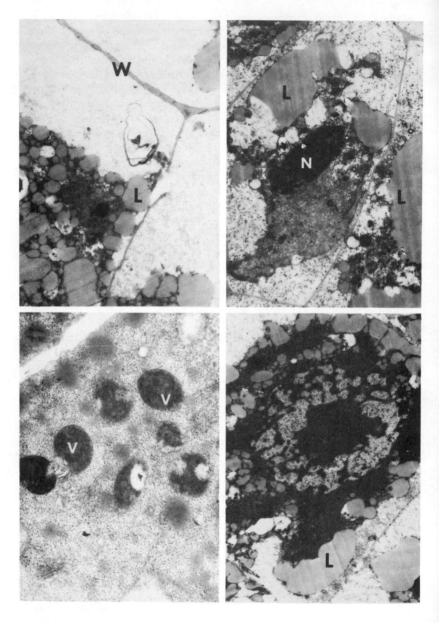

noticed an increase in the time needed for germination to begin. Later, as viability declines, many of the seeds commence germination by emergence of the radicle, do not develop further, and later die. Lastly, the seeds do not enter the germination process at all, have no firm structure and exude water when compressed slightly.

The decreased germination rate might be due to the need for repair and replacement processes to occur. The second stage, where late emergence occurs without further development, is due to the germination mechanisms already present and operative within the stored seed, but damage to the genome does not allow cell division following this initial extension nor the provision of further information for the continuance of the growth and differentiation of the seedling. Internal cellular disorganisation in this stage is due to the membrane damage described above. The final stages, where no germination changes are detectable and rapid internal disorganisation takes place, are due to severe membrane damage preventing the process of cell elongation necessary to germination by not allowing the osmotic uptake of water into the cells.

Release of hydrolytic enzymes

If embryos of lettuce from the ageing experiments previously described are subjected to the Gomori test for acid phosphatase, the reaction shows that in the early stages of imbibition the enzyme is confined to the vacuoles, which are small in the radicle root tip cells (*Figure 15.4, bottom left*). As imbibition proceeds, and the vacuoles disperse, the reaction for acid phosphatase becomes spread throughout the ground cytoplasms of the cells, showing that dispersal of the membranes has released acid phosphatase into the cytoplasm and therefore presumably other dangerous hydrolytic enzymes also (*Figure 15.4, bottom right*).

More comprehensive results, similar to these described for lettuce embryos, have been obtained by Berjak and Villiers (1972) using *Zea mays* seeds from an ageing sequence. Here also, membrane aberrations increased with increasing age of the seed. In those

Figure 15.4   Cells from meristems of Arctic King lettuce. L, lipid; N, nucleus; V, vacuole; W, cell wall. Top left and top right, seeds stored at 30°C with 13 per cent moisture content for 8 weeks before test. Plasma membrane appears dispersed from earliest time of imbibition and lipid bodies confluent. 2 h after beginning of imbibition. × 8 000. Bottom left, seeds stored 4 weeks at 13 per cent moisture content. Appearance of small vacuoles in Gomori test for acid phosphatase modified for electron microscopy. Deposit is confined to vacuoles 8 h after beginning of imbibition. × 18 000. Bottom right (same conditions as bottom left), 24 h after beginning of imbibition, reaction for acid phosphatase has spread throughout the cytoplasm. × 9 000

embryos less severely damaged, evidence was obtained for the operation of compensatory mechanisms and organelle repair.

Fully imbibed embryos

The cytology of the meristems of lettuce embryos maintained fully imbibed with water for similar periods appeared to be normal in all respects even when they were kept without germinating until seeds from the same harvest, but stored air-dry at 9·7 and 13·0 per cent moisture content, were all dead. When embryos from these long-imbibed seeds were subjected to the Gomori test, modified for electron microscopy, activity did not appear in the ground cytoplasm at any stage of imbibition. In this respect they were similar to normal, unaged embryos and presumably the functional integrity of the membrane systems had been retained during the period of wet storage.

ENVIRONMENTAL FACTORS

In seeds which are not fully imbibed with water the relative humidity of the storage atmosphere is very important. This is because the seed colloids come into equilibrium with the moisture content of the atmosphere, and the greater amounts of water adsorbed to the colloids permit more rapid rates of reaction. Rise in temperature will also increase the rate of reaction and enable free-radical peroxidation damage and polymerising reactions to proceed more readily. The viability tests performed during the present investigation fully confirmed that high humidity and warm temperatures had deleterious effects upon stored seeds. However, fully imbibed seeds remained viable over a range of temperatures and presumably this was because full imbibition permitted enzyme action, including repair and replacement activities. These would also increase in rate with rise in temperature unless lethal temperatures were encountered.

Oxygen tension was not varied during these investigations, but it is well known that the availability of oxygen causes an increase in the rate of seed deterioration (Harrison, 1966). If the mechanism of damage to the macromolecules is by free-radical peroxidation, then vacuum storage or replacement of the air by an inert gas such as nitrogen would greatly reduce such damage and should lead to a greater life-expectancy of the seeds. Tappel (1965) states that an enhancing effect of damage by oxygen is itself an indication of the involvement of lipid peroxidation reactions.

The low oxygen content of the soil atmosphere has been suggested to act as a factor maintaining seeds in a state of dormancy, thereby increasing their longevity. However, while this may well be one of the factors preventing germination of buried seeds, low oxygen tensions would greatly retard free-radical damage and in this way be responsible also for the great longevity of seeds in the soil.

CONCLUSIONS

It appears that no attempt has yet been made to reconcile the apparently contradictory findings that whilst increase in moisture content reduces the longevity of seeds stored 'air-dry', full imbibition of seeds with liquid water either in the laboratory or in the field allows a greatly extended life-span. However, acceptance of the premise that macromolecule and organelle repair or replacement occur as a normal cellular activity gives a plausible reason for this difference in longevity, as such maintenance activities would probably not occur in dry seeds.

In addition, it has been argued that the genome does not directly affect early germination, but rather seedling establishment and stages of differentiation. Considered together with the suggestion that cellular maintenance mechanisms exist but are not operative in air-dry seeds, this would strongly suggest that the primary cause of ageing in air-dry seeds is the deterioration of membrane systems.

The accumulation of damage to membranes and macromolecules can be compared to irradiation damage because in both cases the cellular maintenance mechanisms are required to deal suddenly with extensive damage rather than being able to replace damaged components, as occasion arises, over an extended period of time. In addition, in both cases the repair mechanisms themselves may have been damaged and have become incapable of proper functioning.

It is proposed that the most likely mechanism of membrane damage is the production of free radicals in the presence of oxygen. Free radicals are very reactive and can initiate hydrogen abstraction reactions and polymerisation, cause cross-linkage of enzymes, structural proteins and lipids, and cause polypeptide molecule scission and deleterious changes in amino acids. Even in 'dry' seeds the polyunsaturated lipids of the membranes and the water bound to these cell colloids are situated where the greatest damage can be caused by free radicals.

The quantity of substances lost by leaching from seeds during soaking increases with ageing. This has provided the basis for tests in which the quality of seeds can be assessed by the increase in conductivity of the water used for soaking. While this provides evidence that damage occurs to cell membranes during dry storage, the fact that seeds in the soil have great longevity suggests that their cell membranes retain their structural integrity, and do not leak solutes. If substances were lost from such seeds at similar rates, severe shortages of substrates would occur and make later germination and growth impossible.

SUMMARY OF PROPOSALS

1. In seeds stored air-dry:
   (a) failure of the germination mechanism is caused by membrane and enzyme damage, whereas chromatin damage affects seedling morphogenesis and later development;
   (b) damage to membranes, enzymes and chromatin is caused by free-radical peroxidation reaction, and accumulates in air-dry seeds;
   (c) membrane and DNA repair mechanisms cannot operate, and enzyme and organelle replacement cannot occur;
   (d) chromatin damage cannot be eliminated by cell-line selection;
   (e) the operation of repair mechanisms might be responsible for the delay in germination of ageing seeds after imbibition;
   (f) extensive membrane deterioration allows hydrolytic enzymes to be released during imbibition, amplifying the damage;
   (g) the repair mechanisms themselves might suffer damage.
2. In seeds fully imbibed in the soil:
   (a) enzyme, membrane and DNA repair mechanisms are able to eliminate damage as it occurs;
   (b) membranes are retained intact, ensuring the compartmentation of enzymes and metabolic processes;
   (c) the low soil oxygen content is beneficial in retarding peroxidation reactions;
   (d) ultimate death might occur either upon exhaustion of one or more essential substrates, or upon mutations rendering the essential repair mechanisms inaccurate.

One of the obvious deficiencies of the membrane deterioration theory as it is applied to metabolising cells and tissues is that any system capable of repair, or of being replaced, is not likely to be important in ageing. However, in the case of air-dry tissues such as seeds, spores, cysts and pollen grains this cannot be an objection. In dry seeds, repair systems cannot function and this seems to offer the most reasonable explanation for the observed differences in longevity between seeds stored dry and those in the soil.

ACKNOWLEDGEMENT

This work was supported by funds from the Council for Scientific and Industrial Research.

REFERENCES

ABDALLA, F. H. and ROBERTS, E. H. (1968). 'Effects of temperature, moisture and oxygen on the induction of chromosome damage in seeds of barley, broad beans, and peas during storage', *Ann. Bot.*, **32**, 119

ASKENOV, S. I., ASKOCHENSKAYA, N. A. and PETINOV, N. S. (1969). 'The fractions of water in wheat seeds', *Fiziologiya Rast.*, **16**, 58

ASPINALL, D. and PALEG, L. G. (1971). 'The deterioration of wheat embryo and endosperm function with age', *J. exp. Bot.*, **22**, 925

BARBER, A. A. and BERNHEIM, F. (1967). 'Lipid peroxidation: its measurement, occurrence, and significance in animal tissues', in *Advances in Gerontological Research*, Vol. 2, (ed. Strehler, B. L.) 355, Academic Press, New York

BARTON, L. V. (1961). *Seed Preservation and Longevity.* Leonard Hill, London

BERJAK, P. and VILLIERS, T. A. (1972). 'Ageing in plant embryos, II. Age-induced damage and its repair during early germination', *New Phytol.*, **71**, 135

CARTLEDGE, J. L., BARTON, L. V. and BLAKESLEE, A. F. (1936). 'Heat and moisture as factors in the increased mutation rate from *Datura* seeds', *Proc. Am. phil. Soc.*, **76**, 663

CARTLEDGE, J. L. and BLAKESLEE, A. F. (1935). 'Mutation rate from old *Datura* seeds', *Science*, **81**, 492

CHING, T. M. and SCHOOLCRAFT, I. (1968). 'Physiological and chemical differences in aged seeds', *Crop Sci.*, **8**, 407

CONGER, A. D. and RANDOLPH, M. L. (1959). 'Magnetic centers (free radicals) produced in cereal embryos by ionizing radiation', *Radiat. Res.*, **11**, 54

CONGER, A. D. and RANDOLPH, M. L. (1968). 'Is age-dependent genetic damage in seeds caused by free radicals?' *Radiat. Bot.*, **8**, 193

CURTIS, H. J. (1963). 'Biological mechanisms underlying the aging process', *Science*, **141**, 686

CURTIS, H. J. (1966). 'A composite theory of aging', *Gerontologist*, **6**, 143

CURTIS, H. J. (1967). 'Radiation and ageing', in *Aspects of the Biology of Ageing*, S.E.B. Symposium XXI (ed. Woolhouse, H. W.) 51, Cambridge University Press

DESAI, I. D. and TAPPEL, A. L. (1963). 'Damage to proteins by peroxidized lipids', *J. Lipid Res.*, **4**, 204

DURE, L. S. and WATERS, L. C. (1965). 'Long-lived messenger RNA: evidence from cotton seed germination', *Science*, **147**, 410

FLETCHER, M. J. and SANADI, D. R. (1961). 'Turnover of rat-liver mitochondria', *Biochim. biophys. Acta*, **51**, 356

GAMPEL, Z. and TOHA, J. C. (1969). 'Aging and repair in *Neurospora crassa* studied by ultraviolet irradiation', *Radiat. Res.*, **40**, 525

GROSS, N. J., GETZ, G. S. and RABINOWITZ, M. (1969) 'Apparent turnover of mitochondrial DNA and mitochondrial phospholipids in the tissues of the rat', *J. biol. Chem.*, **244**, 1552

HABER, A. H. and LUIPPOLD, H. J. (1960) 'Separation of mechanisms initiating cell division and cell expansion in lettuce seed germination', *Pl. Physiol.*, Lancaster, **35**, 168

HARRISON, B. J. (1966) 'Seed deterioration in relation to storage conditions and its influence upon germination, chromosomal damage and plant performance', *J. natn. Inst. agric. Bot.*, **10**, 644

HOCHSCHILD, R. (1971) 'Lysosomes, membranes and ageing', *Exp. Gerontol.*, **6**, 153

KJAER, A. (1948). 'Germination of buried and dry stored seed, II. 1934–44', *Proc. int. Seed Test. Ass.*, **14**, 19

KOOSTRA, P. T. and HARRINGTON, J. F. (1969). 'Biochemical effects of age on membranal lipids of *Cucumis sativus* L. seed', *Proc. int. Seed Test. Ass.*, **34**, 329

LUZZATI, V., GULIK-KRZYWICKI, T., TARDIEU, A., RIVAS, E. and REISS-HUSSON, F. (1969). 'Lipids and membranes', in *The Molecular Basis of Membrane Function* (ed. Tosteson, D. C.) 79, Prentice-Hall Inc., Englewood Cliffs, N.J.

MARCUS, A. and FEELEY, J. (1964). 'Activation of protein synthesis in the imbibition phase of seed germination', *Proc. natn. Acad. Sci. USA*, **51**, 1075

MATILE, P. H., (1968), 'Vacuoles as lysosomes of plant cells', *Biochem. J.*, **111**, 26P

O'BRIEN, J. S. (1967). 'Cell membranes—composition: structure: function', *J. theoret. Biol.*, **15**, 307

OMURA, I., SIEKEVITZ, P. and PALADE, G. E. (1968). 'Turnover of constituents of the endoplasmic reticulum membranes of rat hepatocytes', *J. biol. Chem.*, **242**, 2389

OSBORNE, D. J., ROBERTS, B. E. and HALLAM, N. D. (1971). 'Protein synthesis and viability in rye', *Biochem. J.*, **124**, 7P

PELC, S. R. (1970). 'Metabolic DNA and the problem of ageing', *Exp. Gerontol.*, **5**, 217

ROBERTS, E. H. (ed) (1972). *Viability of Seeds*, Chapman and Hall, London

ROBERTS, E. H., ABDALLA, F. H. and OWEN, R. J. (1967). 'Nuclear damage and the ageing of seeds', in *Aspects of the Biology of Ageing*, S.E.B. Symposium XXI. (ed. Woolhouse, H. W.) 65, Cambridge University Press

RUPERT, C. S. (1960). 'Photoreactivation of transforming DNA by an enzyme from baker's yeast', *J. gen. Physiol.*, **43**, 573

SAMIS, H. V., FALZONE, J. A. and WULFF, V. J. (1966). '³H-thymidine incorporation and mitotic activity in liver of rats of various ages', *Gerontologia*, **12**, 79

SETLOW, R. B. and CARRIER, W. L. (1964). 'The disappearance of thymine dimers from DNA: an error-correcting mechanism', *Proc. natn. Acad. Sci.*, *USA*, **51**, 226

SIMPSON, D. M. (1942). 'Factors affecting the longevity of cottonseed', *J. agric. Res.*, **64**, 407

TAPPEL, A. L. (1965). 'Free-radical peroxidation damage and its inhibition by vitamin E and selenium', *Fed. Proc.*, **24**, 73

TARLOV, A. (1968). 'Turnover of mitochondrial phospholipids by exchange with soluble lipoproteins *in vitro*', *Fed. Proc.*, **27**, 458

TOOLE, V. K. and TOOLE, E. H. (1953). 'Seed dormancy in relation to seed longevity', *Proc. int. Seed Test. Ass.*, **18**, 325

TURRILL, W. B. (1957). 'Germination of seeds: 5. The vitality and longevity of seeds', *Gdnrs', Chron.*, **142**, 37

VERZAR, F. (1968). 'Intrinsic and extrinsic factors of molecular ageing', *Exp. Gerontol.*, **3**, 69

VILLIERS, T. A. (1971a). 'Cytological studies in dormancy, I. Embryo maturation during dormancy', *New Phytol.*, **70**, 751

VILLIERS, T. A. (1971b). 'Lysosomal activities of the vacuole in damaged and recovering plant cells', *Nature, London*, **233**, 57

WAREING, P. F. (1966). 'Ecological aspects of seed dormancy and germination', in *Reproductive Biology and Taxonomy of Vascular Plants*, BSBI Conference Reports, **9**, 103

DISCUSSIONS

Wareing: The high rate of biosynthesis of non-dormant seeds would hardly be compatible with longevity; the metabolic rate of dormant seeds must be much lower. The two kinds of metabolism may therefore be very different.

Villiers: The metabolic rate is not high but imbibed 'dormant' seeds are capable of undergoing enzymic reactions and of manipulating large structural units more readily than if they were air dry. For instance, when lettuce seeds imbibe at high temperatures in the dark their metabolic rate first rises but then settles down to a low steady state until they are given light.

Villiers: There is a large jump in seed water content between the air dry state, even in equilibrium with a saturated atmosphere, and the state in which the colloids of a seed are fully imbibed. When dormant seeds lie buried in the soil, slight amounts of damage which accumulate during a drought period can very likely be repaired after the next rain because the damage occurs in small doses. The effect of a long period of dry storage is analogous to the effect of cytological collapse due to atomic radiation, because here a large amount of damage is done and the repair mechanism cannot operate quickly enough either to retain lysosomal enzymes or to keep the compartmentation of the cytoplasm intact. In both cases, a large amount of accumulated damage must be made good over a very short time, otherwise hydrolytic activity ensues during imbibition.

B. Roberts: In the system you visualise, what are the relative rates of the breakdown and repair processes?

Villiers: The chromosomal aberrations in the radicle tips in germinating seeds after dry storage may even be caused *during germination* by the release of DNAse, and the accumulated damage is amplified during imbibition because of the loss of cytoplasmic compartmentation. mRNA can be destroyed very rapidly by the release of ribonuclease.

Harrington: Some enzyme activity occurs even at the low water content of air-dry seeds, but Stumpf (Davis, California) has shown that enzymes involved in lipid production are only activated at water contents of 30 per cent or above.

Villiers: Yes, and membrane repair (possibly involving the excision and replacement of sheets of cell membrane) would require this.

Thompson: Might the damage sometimes associated with re-hydration after dry storage be prevented by re-hydrating seeds under conditions unfavourable to germination and only afterwards placing them in favourable ones?

Villiers: Yes, in fact delayed germination may occur in part *because* the repair mechanisms are at work.

Edwards: Do dry seeds really contain so little water that no repair processes can take place?

Villiers: The water which remains in the vicinity of macromolecules after drying is tenaciously bound to the colloids and has, according to Aspinall, a lattice structure rather like ice. This would tend to impede the manipulation of structural units in the seed, although the water could still take part in oxidative reactions.

Villiers: Many air dry seeds fall prey to micro-organisms, and there is much less microbial damage to seeds which are fully imbibed. Unfortunately, long term storage in the imbibed state is probably not a practical proposition except when the seeds are dormant (e.g. lettuce seeds at high temperatures).

Heydecker: It might well become a practical way of storing at least germ plasm for a long time, by incorporating germination inhibitors into seeds and then incorporating germination stimulants at the end of the imbibed-storage period.

Thurston: But wild oat seeds survive about twice as long in dry storage as in the field (where the extreme is about 7 years).

Cocking: This discussion has shown that we have much to learn about plant membranes, especially membranes in seeds.

16

PHYSIOLOGICAL DISORDERS IN GERMINATING SEEDS INDUCED BY THE ENVIRONMENT

J. D. MAGUIRE
Department of Agronomy and Soils,
Washington State University, U.S.A.

INTRODUCTION

The ecological situations that seeds may encounter have definite effects upon the expression of their viability potential and the resultant seedling growth and development. An interrelationship exists between the various environmental factors and the germination of seeds. Physiological disorders result from the intolerance of the seed's metabolic system to certain environmental changes and are characterised by a narrowing amplitude of response or by the inability to function normally. The manifestations of these disorders are the development of abnormal morphological structures and the internal disruption of biochemical mechanisms associated with seed germination and development of the seedling prior to the initiation of photosynthesis. For example, induced dormancy may be thought of as a temporary physiological disruption caused by an intolerable environment. But environmental fluctuation can be promotive as well as disruptive. The promotion of seed germination by chilling moist seeds for a period of time is a procedure often used for some species and yet the embryonic plants of other species may suffer chilling injuries from low temperatures.

The effects of some internal and external conditions on seed germination were reviewed by Lang (1965), and Austin (1972) indicated that a seed inherits the effects of certain environmental influences

290 SEED ECOLOGY

as well as its ability to respond to them. Unless these conditions are
extreme, there is little effect on viability, provided that the ripening
processes have not been interrupted by premature harvesting.
Pollock (1972) summarised the effects of the major environmental
factors that modify seed viability. Seed potential may be impaired
while the seeds are still developing on the mother plant and further
modification may be imposed during harvesting, processing and
storage. Temperature regulates the rate of seed germination within
the limits tolerated; these limits vary among species. Temperature
sensitivity of seeds occurs during imbibition and some work has
indicated that adverse conditions at that time may also influence
subsequent plant development (Pollock and Toole, 1966; Lang, 1965).
Although water is essential for hydration and activation of the
germination system, an excessive supply of water may limit the
availability of oxygen for seed respiration. The changes due to
oxygen sensitivity that occur in germinating seeds may induce
enzymes that result in the production of compounds unsuitable for
radicle development and further germination processes. Other
enzymes may be inactivated or disrupted. Light sensitivity during
germination also occurs in many seed species. Mediated by the
phytochrome system, the quality and intensity of light encountered
in the environment may have profound effects upon the seed's
expression of viability. Endogenous substances may inhibit germination until modified by environmental factors. Air pollutants and
certain natural substances, such as ozone, detrimentally influence the
metabolic processes in growing plants and, *perhaps*, also in germinable
seeds.

Ageing of seeds is a natural sequence; senescence begins at, or
shortly after, harvest and the seeds undergo irreversible changes that
lead to loss of vigour and, ultimately, viability (Anderson, 1971).
Harrington (1971) has studied the biochemical bases for seed longevity
and Ching (1971) has pointed out the changes that occur in seeds
with ageing. Enzymes, proteins, mitochondria and membranes are
adversely affected by certain environmental conditions encountered
by the seeds, particularly high temperatures. Substrates for early
enzyme activities may be reduced by high humidity and temperature.
Seed maturity seems to affect the functional status of messenger
ribonucleic acid which encodes germination events. Overdrying of
seeds can denature proteins and inactivate enzymes. Low germination temperature may fail to reactivate the pre-existing ribosomes
and the messenger RNA, preventing polysome formation and protein
synthesis. The weakest link in the chain of metabolic processes or
the lowest concentration of a substrate, effector, co-enzyme or
co-factor can disrupt normal germination.

PHYSIOLOGY OF GERMINATION

The physiology of seed germination has been delineated by Toole *et al.* (1956), Mayer and Poljakoff-Mayber (1963) and, more recently, Street and Öpik (1970) have outlined the seed structures, chemical composition and metabolism associated with germination. In the hydration process the seed absorbs water, and this may trigger into activity endogenous substances that initiate germination processes. This hydration may result in specific enzymic activities that hydrolyse storage materials and synthesise new substances for growth and development. These reactions are programmed by RNA and are controlled by various enzymes which may respond to environmental conditions. Water sensitivity, oxygen availability and light quality influence the rate and direction of the reactions. Energy for these reactions is supplied through respiration. Normal progress of seed metabolism during germination results in growth of the embryo into a vigorous seedling, capable of establishing itself in the given environment and of developing further into a normal plant. Natural deterioration of the seeds or injury through external factors may cause physiological disorders that disrupt germination and seedling growth. These biochemical changes associated with weakened germinability and loss of vigour are characterised by a decline in metabolic activities (lower respiration, reduced germination and slower growth rate). There is greater total activity of hydrolytic enzymes—phytase, protease and phosphatases, and reduced activity of respiratory enzymes—catalase, peroxidase, dehydrogenase, cytochrome oxidase and glutamic acid decarboxylase. Cell membrane permeability is increased, and leakage of sugars, amino acids, inorganic solutes and fatty acids results (Anderson, 1970; 1971; Aspinall and Paleg, 1971; Abdul-Baki, 1969; Heydecker, 1972). Moore (1966; 1968) and MacKay (1972) described the visible symptoms of seed deterioration and injuries from environmental factors. The intensity of tetrazolium staining within the viable area of the seed, along with malformation of seedling structures, can be analysed to diagnose the causes of seed weakness.

Temperature plays an important role in the initiation of germination and the development of seedlings. Natural after-ripening of seeds results in the removal of physiological 'blocks' of the germination processes. Chilling moist seeds of many plant species at 5–10° C may allow germination to proceed. Bradbeer and Coleman (1967), who studied the metabolic changes in hazel (*Corylus avellana* L.) seeds, found that the ability to synthesise nucleotides was enhanced by chilling the seeds through activation of the tricarboxylic acid (TCA) cycle.

The metabolic processes occurring in peas (*Pisum sativum* L.) have been studied by various research workers. Pritchard, Quartley and Turner (1967) have shown that high oxygen pressure inhibits the respiratory mechanism of the seeds, disrupting the TCA cycle. The oxygen supply to the embryonic axis may be influenced by the seed coat. Bain and Mercer (1966) and Nawa and Asaki (1971) have found a rapid increase of mitochondrial activity in pea seed cotyledons as the key enzymes, namely those involved in glycolysis and pentose phosphate reactions, are activated with the imbibition of water. With the energy derived from respiration, nucleic acid and protein synthesis takes place. McDaniel (1969) has related respiratory metabolism to the vigour associated with heterosis and has found that the oxidative and phosphorylative activities of mitochondria are most important in the seedling vigour of barley.

GERMINATION DISORDERS

Heydecker and Feast (1969) investigated 'hollow heart' of pea seeds, a physiological deterioration of the storage tissue occurring near the centre of the cotyledons. Apparently this condition is related to the effects of water uptake by seeds that were subjected to adverse environmental conditions during maturation. Perry and Howell (1965) attributed the inherent cause of this condition to rapid drying of pea seeds prior to maturity.

Water sensitivity of seeds during imbibition may be temperature related and may also interact with oxygen supply. Pollock and Toole (1966) showed that the imbibition period was the critical temperature sensitive stage in the germination of lima bean seeds during which low temperatures resulted in reduced germination and in seedling injuries. Orphanos and Heydecker (1968) have attributed the actual injury that occurs in imbibing bean (*Phaseolus vulgaris* L.) seeds during soaking in water to a deficient oxygen supply to the interior of the seeds.

Roos and Pollock (1971) reported that the soaking injury encountered in lima bean seed germination is also related to low temperatures and that this anaerobic condition allows lactic acid to accumulate and reduces RNAase activity. Obendorf and Hobbs (1970) found that low-temperature treatments during the imbibition of soya bean seeds that are initially very low in moisture reduces seedling survival and growth. Lyons, Wheaton and Pratt (1964) considered the physical condition of the mitochondrial membranes is responsible for the adverse reactions to chilling observed in imbibing seeds. Fatty acid saturation decreases the inherent capacity of the membranes to swell, thus increasing the relative sensitivity of the seeds to cold temperatures. Janssen (1962) postulated that the water sensitivity (i.e. failure to germinate in super-optimal amounts of water) in barley seeds is caused by a toxic build-up of indole-acetic acid (IAA) with a resultant inability of the seed to uncouple oxidative phosphorylation. This water sensitivity can be overcome by the application of hydrogen peroxide or Fe^{++} ions; pure oxygen or scarification of the seed coat also promotes germination. Gaber and Roberts (1969), in more recent work, found no such IAA effects, but they were able to stimulate germination of such seeds with the use of fungal and bacterial antibiotics. They concluded that oxygen is limited for seed germination and the micro-organisms can utilise the available oxygen to the detriment of germination. On the other hand Paleg (1961) showed that gibberellic acid induces amylase synthesis in germinating barley seeds and osmotic regulation of amylase production is regarded by Jones and Armstrong (1971) as the mechanism of water sensitivity involved in germination of barley seeds.

Thomas (1972) stated that the control mechanism in resting pea seeds is related to their lipid content. Changes in the state of cell membranes affect the activities of organelles such as mitochondria and glycosomes, and the water content of the seed regulates the activation and rates of these metabolic processes. Ovcharov (1969) discussed the physiology of seeds of different quality. According to him the environmental conditions encountered during seed maturation result in differences in the content of vitamins, e.g. nicotinic acid and ascorbic acid, and in other substances that can alter the nitrogen metabolism and oxidation–reduction reactions in germinating seeds. With seed deterioration, damaged mitochondria may lose their ability to esterify phosphorus. Also, disruption of the membranes may allow leakage of soluble sugars, amino acids and other metabolites before the embryonic axis can utilise them (Roberts, 1972). This leakage may be related to relative concentrations of glucose in the seeds as well as to membrane integrity. Such losses of metabolites have been measured by electro-conductivity tests on the steeping water of soaked pea seeds (Matthews and Bradnock, 1967; Perry, 1969) and though these tests only measure the concentration of leached electrolytes, their results are reported to correlate well with seedling emergence in the field.

Larson (1968) studied water imbibition by pea seeds with and without seed coats. When the seed coats were removed the seeds imbibed water more rapidly and lost more soluble carbohydrates and total nitrogen than those with intact seed coats. He concluded that the rapid uptake of water disrupted membrane organisation and destroyed tissue integrity. Flentje (1964), in Australia, found that

L

pea seeds subjected to rain two days before harvest showed much reduced germination and emergence compared to seeds harvested before the rain. Similar results were reported with lima beans by Pollock and Toole (1964), who noted that the less vigorous seeds were often bleached.

Simon (1972), in studies with round-seeded peas (cultivar Meteor), showed that removal of the testa results in greater solute leakage from within the seed. Rapid leaching occurs immediately after immersion of seeds in water but then, after 15 min, leaching declines rapidly. Subsequent drying and re-wetting of the seeds results in similar patterns of solute losses. Simon theorised that the liquid moves through the dry cell wall matrix until the single lipid layer in the membranes becomes water filled. Manohar and Heydecker (1964) separated the pea testa into two parts and showed it to be differentially permeable. The entry of solutes of higher molecular weight may occur exclusively through the micropyle. The seed coat may provide entry for smaller molecules and may act as a wick conducting water (and solutes) around the periphery of the inner seed.

Moore (1963) advanced the theory of natural crushing of tissues, that occurs when mature seeds are subjected to alternate wet and dry conditions, as causing the resultant leakage through membranes.

BLEACHING OF PEAS—A PROBLEM

The problem of bleaching of pea seeds occurs in the eastern part of the state of Washington and in adjacent Idaho, USA, where large acreages of peas are grown for seed production. Bleaching—the loss of chlorophyll (and other constituents)—usually occurs when the maturing seeds are subjected to rains followed by hot sunny weather. This bleaching should not be confused with blonding which is the lack of chlorophyll from the start, associated with rainy cloudy weather during development of the seed (Duncan *et al.*, 1965). Blonding does not reduce the vigour of pea seeds although it is objectionable in marketing quality. Bleaching, by contrast, is due to adverse environmental conditions at harvest time and may result in reduction of germination and loss of vigour. Hendricks (1971) stated that photo-oxidation of lipids in the pea seed membranes could be brought about through exposure of the seeds to intense sunlight. A low molybdenum content of pea seeds might also be a contributing factor to such bleaching. Some workers have noted that wrinkled pea seeds bleach but do not lose vigour like the round seeds. Attempts have been made in our studies to relate the occurrence of bleaching in round pea seeds with the various physiological processes related to the observed loss of germination and vigour.

ELECTRO-CONDUCTIVITY

Samples of pea seeds (*Pisum sativum* L.) of the cultivars Alaska (round seed type) and Perfection (wrinkled seed type), grown in 1970 and 1971, were separated into green and bleached portions. These seeds were soaked in distilled water for 24 h and then germinated in rolled paper towels at 20°C for 6 days to determine their viability. The steeping water from these imbibed seeds was checked for electro-conductivity as a measure of the quantity of seed leachate and this was compared with the corresponding seed viability, using the methods outlined by Matthews and Bradnock (1967). These results are shown in *Table 16.1*. The electro-conductivity readings for green

Table 16.1 ELECTRO-CONDUCTIVITY, SUGAR ANALYSIS AND GERMINATION OF PEAS *PISUM SATIVUM* L.) : THE ROUND SEEDED CULTIVAR ALASKA AND THE WRINKLED SEEDED CULTIVAR PERFECTION

Seeds		Conductivity of leachate $(micromho/cm^2/g)$	Total sugars in 50 seeds (g)	Sugars in leachate* $(mg/50\ ml)$	Sugar loss $(\%)$	Germination per 50 seeds
Alaska (round)	Green	18	0·46	19	0·41	47
	bleached	47	0·42	69	1·62	3
Perfection (wrinkled)	Green	6	2·60	2	0·08	47
	bleached	11	2·80	10	0·32	48

* After steeping in water for 24 h

wrinkled, bleached wrinkled and green round seeds were less than 24 micromhos/cm²/g of seed, values indicative of high vigour (Biddle, 1970; Perry, 1969). The high germination values obtained for these groups of seed correlated well with their electro-conductivity. By contrast, bleached round pea seeds gave high conductivity readings and germinated poorly. Initially, green pea seeds that had been bleached artificially by placing them on moist blotters in petri dishes under fluorescent light for two weeks gave similar results. Germination was reduced markedly and electro-conductivity readings were similar to those of naturally bleached peas, indicating leaching of metabolites and loss of vigour. When dishes containing green seeds were covered with green plastic under the same conditions, bleaching occurred in only three days with similar reduction in germination and vigour. When bleached seeds were checked again several months after treatment a further reduction in germination was found (*Table 16.2*) and conductivity readings had risen accordingly.

Table 16.2 EFFECT OF ARTIFICIAL BLEACHING FOR 3 DAYS ON ELECTRO-CONDUCTIVITY OF LEACHATE AND GERMINATION OF PEAS (*PISUM SATIVUM* L.) CULTIVAR ALASKA (ROUND TYPE)

Seeds	3 days after bleaching		6 months after bleaching	
	Conductivity (micromho/cm²/g)	Germination per 50 seeds	Conductivity (micromho/cm²/g)	Germination per 50 seeds
Bleached	26	37	64	7
Partially bleached	19	45	—*	—
Control, Not bleached	16	48	17	46

* Insufficient seeds to continue evaluation

SUGAR ANALYSIS

Total sugar analyses conducted on ground unsoaked samples of pea seeds showed that green wrinkled seeds contained five times as much sugars as green round seeds. The bleached round seeds had a slightly smaller total sugar content than the green round seeds, but the total sugar content of the bleached wrinkled seeds was actually somewhat higher than that of the green wrinkled ones. In order to determine the quantities of soluble sugars that had been leached from the pea seeds, the steeping water in which the seeds had been soaked was analysed by the anthrone test (Dent, 1963; see *Table 16.1*). The results showed that, similar to electro-conductivity, sugar leakage from the imbibed seeds also correlates with loss of vigour. The bleached round pea seeds exuded approximately three and one-half times as much sugar as the green round seeds, whereas bleached wrinkled seeds, despite their initial six-fold sugar content, lost only about half the amount that was lost by green round seeds. Very small amounts of sugars were leached from the green wrinkled seeds.

WATER UPTAKE

A study was conducted on the various types of pea seeds to evaluate their imbibition rate and the influence of the seed coats upon the rate. The seeds were placed in rolled paper towels, moistened with distilled water and allowed to imbibe over a 24 h period. At designated intervals some seeds were removed, blotted dry on the surface and weighed. *Figure 16.1* shows the water uptake of these seeds. The bleached round seeds initially imbibed more rapidly than the green round seeds although their total uptake was somewhat less over the 24 h period (9·8 g v 10·2 g/100 seeds, respectively). This rapid initial

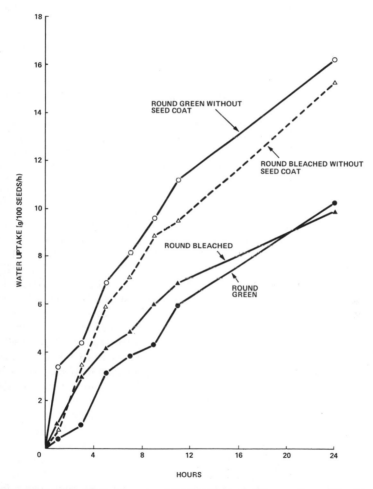

Figure 16.1   Water uptake of green and bleached pea seeds (Pisum sativum *L.*) *cultivar Alaska (round type)*

uptake of water could indicate some disruption within the seed tissue of the bleached peas. Both bleached wrinkled and green wrinkled seeds absorbed water at steady rates, similar to the green round seeds, and there was virtually no difference in water uptake between the bleached and green seeds. Removal of seed coats from the wrinkled seeds resulted in much more rapid water uptake with essentially the same increase in rates for both types of wrinkled seeds.

In fact, when the seed coats were removed, or the seeds were broken in half, water uptake became two to three times as fast for all types of seeds, but the wrinkled seeds absorbed water more rapidly than the round seeds. Final water uptake was similar for all comparable seeds, irrespective of bleaching.

RESPIRATION

Respiration patterns of germinating seeds can reflect the degree of viability and the progress of the metabolic processes involved. Woodstock and co-workers (Woodstock and Feeley, 1965; Wood-stock and Pollock, 1965; Woodstock and Combs, 1967; Woodstock, 1969) utilised such techniques to evaluate the seedling vigour of maize, lima beans and pea seeds and found a high correlation between respiration levels of the seeds and seedling growth.

Maguire and Kropf also, in the studies of bleached Alaska peas, used respiration rates of imbibed seeds as an indication of their viability and vigour (see *Figure 16.2*). The bleached round seeds absorbed 30 μl O_2/10 seeds/h after 24 h of imbibition; the rate dropped slightly (to 25 μl O_2) at 48 h but then increased slowly to 48 μl O_2 after 72 h, although (at 66 h) there was another slight intermittent reduction in respiration. The green round seeds respired at an increasing rate up to 48 h (96 μl O_2), then decreased to 75 μl O_2 at 66 h. 72 h after the imbibition of water began, the green seeds absorbed 102 μl O_2/10 seeds. The bleached pea seeds here responded similarly to the 'Newport' Kentucky bluegrass seeds (as will be shown later), in that the respiratory activity declined after an initial period, and then continued at a reduced rate. Such respiration patterns may well reflect the leaching of sugars and other metabolites, reflecting a disruption of cell membranes brought about by the bleaching of the seed coat, and a general slowing down of the physiological processes involved in germination. The temporary decline of the respiration rate of both types of pea seed may indicate a depletion of storage materials or alternatively a change in substrate could have been responsible for changes in the metabolic activities within the seeds. This possibility was further indicated in that the radicles emerged from bleached pea seeds but seedling growth ceased after 2–3 days.

Natural crushing of tissues within the seeds could also contribute to the physiological disorders observed in the bleached peas. When these bleached pea seeds were placed in tetrazolium chloride solution,

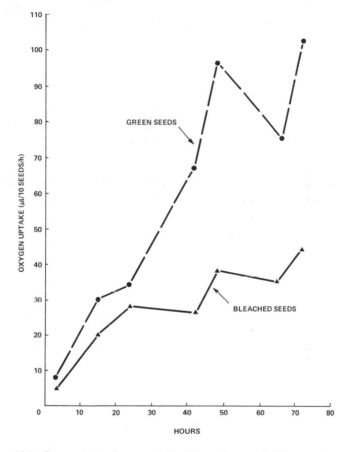

Figure 16.2   Oxygen uptake of green and bleached round pea seeds (Pisum sativum *L.*) *cultivar Alaska*

necrotic areas appeared along the embryonic radicle and within the bleached cotyledons. Deep transverse cracks appeared across the face of the cotyledons. The green round pea seeds stained normally throughout the viable tissue. The intensity of tetrazolium staining appeared to be greater with the green round, compared to the bleached round pea seeds. When the seed coats were removed from bleached seeds, the cotyledons often fell apart, indicating again a deterioration of the seed induced by the environment. When the bleached seeds were moistened, the inrush of water into the cotyledons could cause crushing of the tissue—a condition like the 'hollow heart' described by Heydecker (1972) and Perry (1969). In addition, the bleached seeds may be more susceptible to mechanical injury occurring during harvesting.

RESPONSE OF KENTUCKY BLUEGRASS CULTIVARS

Seeds are often pretreated with chemicals, low temperatures and other means to obtain prompt and complete germination. Dilute solutions of potassium nitrate have been used with grass seeds for this purpose. The physiological role of KNO_3 is not clearly under-stood although various workers (e.g. Taylorson, 1969; Roberts, see Chapter 11) have related its effect to enhancement of the light action, presumably through the phytochrome system.

Table 16.3 GERMINATION OF 1969 COUGAR AND NEWPORT KENTUCKY BLUEGRASS SEED TESTED AT 1, 4, AND 9 MONTHS AFTER HARVEST

Seed age	Days after imbibition	% Germination			
		Cougar		Newport	
		KNO_3	H_2O	KNO_3	H_2O
1 month	14	17	16	0	0
	28	62	57	53*	32
4 months	14	56	56	0	0
	28	90	89	74*	63
9 months	14	69	62	18	19
	28	87	84	84*	75

* $P = 0.05$

Maguire and Steen (1971) measured respiration, germination percentage, rate of germination and the seedling growth of the Ken-tucky bluegrass (*Poa pratensis* L.) cultivars Cougar and Newport in order to determine their respective viability and vigour. These grasses vary inherently in the degree of their post-harvest dormancy and apparent seedling vigour (Maguire, 1969). Studies of the effects of KNO_3 on the physiological response of these seeds were also included.

Table 16.3 shows that in these experiments, carried out in autumn, 1969, the 1969 Newport seeds did not germinate at all for 14 days after imbibition while a portion of the Cougar seeds had germinated by then. After 28 days these Cougar seeds showed no apparent germination differences between the KNO_3 and the water treatment (62 per cent as against 57 per cent). However, KNO_3-treated Newport seeds germinated significantly better (53 per cent) than those treated with water (32 per cent). Similar tests 4 months after harvest showed an increased early germination of Cougar seeds, but still no early germination of Newport. After 28 days' imbibition, however, the Newport seeds treated with KNO_3 again germinated significantly higher than the water-imbibed seeds (74 per cent compared to 63 per cent). There was no difference in the germination of Cougar seeds, both treatments resulting in high germination (KNO_3—90 per cent *v* water—89 per cent).

Even 9 months after harvest, some dormancy was still evident in the 1969 Newport seeds. The germination of Newport seeds after 14 days was still low for both treatments, and early germination of Cougar seeds was much higher, both with and without KNO_3. But here 28 days after imbibition a higher proportion of the seeds from both cultivars had germinated in KNO_3 than in water alone, marginally so for Cougar (87 per cent and 84 per cent, respectively, with KNO_3 and water) but significantly so for Newport (84 per cent and 75 per cent).

Respiration of all Cougar seeds increased rapidly after imbibition, then at a slower rate for the remainder of the test (*Figure 16.3*). Respiration rates of all seeds treated with KNO_3 were higher than in water; however, 1965 and 1968 Cougar seeds respired at appreciably higher rates throughout the test period than the freshly harvested (month old) 1969 Cougar seeds which also showed a characteristic lag period for the first 3 days.

Newport seeds showed little respiration response to KNO_3, compared to water, except the 1968 seeds (*Figure 16.4*) which absorbed more O_2 than seeds in water. Again, as with Cougar, the 1965 and 1968 seeds respired at higher rates than the 1969 seeds. The respiration rates of the 1965 and 1969 Newport seeds decreased after the initial 4-day rise. A comparison of 1965 Cougar and Newport seeds indicates much higher respiration rates for Cougar. Since germination tests of the 1965 seeds showed no apparent dormancy in either cultivar at this time, it would seem that Cougar seeds have greater inherent potential levels of respiration. Newport seeds reached their peak respiration rate at 3–4 days while Cougar seeds, in addition to having reached a higher level of respiration by then, continued to respire at an increasing rate. This fact could explain the observed superiority in seedling vigour of Cougar over Newport (Maguire, 1969).

The 1965 seeds of Cougar, but not of Newport, treated with KNO_3 gave higher respiration rates compared to seeds imbibed in water. However, respiration of seeds from the 1968 crop of both cultivars

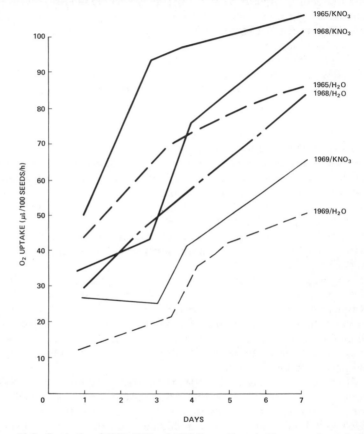

Figure 16.3   Respiration of 1965, 1968 and 1969 Cougar Kentucky bluegrass seed treated with KNO_3 and water

was stimulated by KNO_3 (*Figures 16.3* and *16.4*). Freshly harvested (1969) seeds of both cultivars showed lower respiration rates than 1965 and 1968 seeds. With the Newport seeds respiration peaked at 4 days, then declined to earlier levels, observed 2 days after imbibition.

The respiration of the 1969 seeds during the 14-day period after

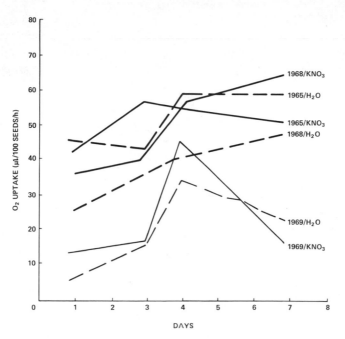

Figure 16.4   Respiration of 1965, 1968 and 1969 Newport Kentucky bluegrass seed treated with KNO_3 and water

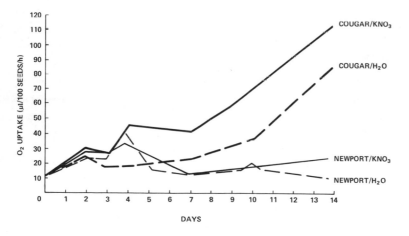

Figure 16.5   Respiration of 1-month-old 1969 Cougar and Newport Kentucky bluegrass seed treated with KNO_3 and water

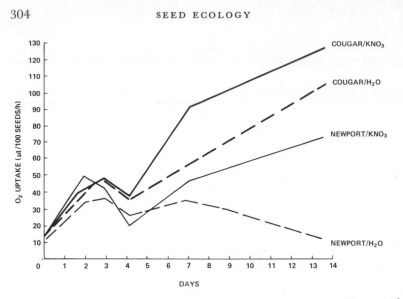

Figure 16.6   Respiration of 4-month-old 1969 Cougar and Newport Kentucky bluegrass seed treated with KNO₃ and water

the seeds were soaked (*Figures 16.5, 16.6, 16.7*) was measured at intervals of 1, 4 and 9 months after seed harvest. Variations in respiration at 1 and 4 months were similar to those observed in previous 7-day studies. Respiration of one-month-old Newport seeds remained very low. Except for a slight peak at 4 days they showed

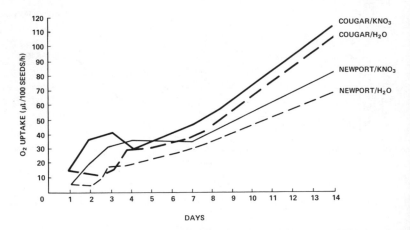

Figure 16.7   Respiration of 9-month-old 1969 Cougar and Newport Kentucky bluegrass seed treated with KNO₃ and water

virtually no difference between KNO_3 and water media, and the seeds absorbed only 18 and 12 μl of O_2 per 100 seeds per hour. On day 14 Cougar seed respiration increased, however, and the KNO_3-imbibed seeds absorbed 115 μl O_2, compared to 89 μl O_2 for water-imbibed seeds. Rate of moisture uptake was not considered to be a factor since previous studies (Maguire, 1969) had shown that seeds of the two cultivars have similar rates of water absorption. The initial uptake of water is primarily a physical process, occurring similarly with live and dead seeds. The observed difference in respiration was probably due both to inherent vigour differences and to the prolonged post-harvest maturation of the Newport seeds, before the end of which period these seeds were not fully capable of metabolising endogenous substances and achieving steady growth. The effect of KNO_3 was clearest 4 months after harvest, especially with Newport seeds which then respired at the dormant (1 month) rate in water but at the non-dormant (9 month) rate in KNO_3.

Nine months after harvest, the Newport seeds had largely lost their dormancy as shown in the higher germination of KNO_3 and water-treated seed (*Table 16.3*). The seed respiration rates (*Figure 16.7*) revealed similar results. Respiration of both the Cougar and the Newport seeds now increased steadily over 14 days. Cougar seed respiration showed no difference due to KNO_3 or water, absorbing 107 μl and 104 μl of O_2, respectively. The respiration level of Newport seeds was slightly raised by KNO_3 but the differences between respiration rates of the two cultivars were still evident, Cougar seed being higher than Newport seed for both the KNO_3 and water treatments.

The germination of 1965 and 1968 seeds from both cultivars was high at the time of this test and no differences were noted between the KNO_3- and water-treated seeds. KNO_3 increased the respiration of Cougar bluegrass seeds at an earlier date than those of Newport seeds, apparently because a certain degree of post-harvest maturation must occur before KNO_3 can become effective. However, the wide differences between seeds of the two cultivars indicate that there are other inherent, metabolic differences in dormancy levels that influence germination. KNO_3 does not overcome dormancy *per se* but may act in conjunction with dormancy-breaking treatments, such as light and alternating temperatures, to increase the proportion of seeds capable of prompt germination.

Further studies are being conducted to determine the effects of KNO_3 upon the metabolic processes of seeds of these cultivars during germination and seedling growth. Such information is needed to develop further knowledge of attributes connected with seedling vigour and of the essential metabolic processes in seeds.

CONCLUSIONS

Adverse environmental factors can bring about the disruption of metabolic processes within germinating seeds, producing physiological disorders with resultant abnormal growth and development of the seedlings. In some crops man has increased such susceptibility through the breeding of varieties with higher levels of soluble carbohydrates and labile proteins. Within the interacting stages of seed germination, certain environmental factors are most critical in regulating or curtailing the germination processes. Initially, germination involves reactivation of systems conserved from the seed maturation period; the rate of water imbibition is critical for hydration of reserve tissues and activation of enzyme systems. Investigators have demonstrated the osmotic regulation of the hydrolytic enzymes involved in the conversion of these stored substrates (Jones and Armstrong, 1971).

Temperature enters into the scheme as well; low temperatures may inhibit the catabolic activities in seeds that require warm temperatures to germinate or may fail to activate pre-existing enzymes. When seeds have not been fully after-ripened, low temperatures apparently may alter the metabolism of these seeds, when imbibed, towards greater synthesis of nucleotides and more effective growth and development. Excessively high temperatures at germination may inactivate some enzymes and denature proteins, upsetting the anabolic activities involved in producing the seedlings.

Once the seeds are imbibed, the respiratory activities within the embryo provide energy for metabolic processes and the oxygen supply to the seed can be a critical governing factor in the process of germination. Excess water supply or anatomical obstructions such as seed coat impermeability or adhering floral structures may limit the oxygen supply, impairing the oxidative phosphorylation and resulting in impaired germination together with abnormal physiological conditions such as 'hollow heart' in peas. Depressed dehydrogenase activity has been observed in water-sensitive seeds and other respiratory enzymes are probably affected too. Mechanical injuries may also affect respiration levels when seeds become imbibed.

The role of membranes in the regulation of germination is not fully understood but increased uncontrolled permeability does allow leakage of soluble sugars, amino acids, inorganic solutes and fatty acids that can adversely influence the germination of the seeds. Grunwald (1971) added free sterols, esters and glycoside, and showed that the free sterols control the permeability of membranes of germinating barley seeds, while Glinka and Reinhold (1971) indicated that when abscisic acid (ABA) is added it raises the permeability of plant cells to water, making them more leaky. Pea seeds that have been bleached at maturation show increased permeability and leakage of soluble sugars and other materials, and with round-seeded cultivars, as we have seen, this condition results in loss of germination and vigour. The ability of the embryos to utilise these substrate metabolites rapidly and effectively is apparently more important than the actual amount of leakage, and effects on viability may be governed by the efficiency of seeds in the utilisation of substrate during germination, rather than by changes in membrane permeability *per se*.

As indicated by Malhotra and Spencer (1970), seeds of *Phaseolus vulgaris* exert control of their respiration and enzymic activities in conjunction with the swelling and contracting of the mitochondria. After the first two days of imbibition the seeds lose control for three to five days, finally recovering it on the seventh day. Leaching of metabolites may occur during this lag period. Inherent differences may exist between cultivars and between seeds stored under different environmental conditions. Berjak and Villiers (1972) have shown that germinating seeds of maize are able to repair membrane damage resulting from seed deterioration. These seeds compensate for delay in metabolic activities by a temporary increase in the production of mitochondria, endoplasmic reticulum (ER), RNA and protein, along with stimulated DNA replication and division. The polar lipids found in cell membranes may serve as structural and functional barriers to leaching (Koostra and Harrington, 1969). Any condition that reduces the phospholipid content adversely affects membrane permeability. An interesting hypothesis on the structure of cell membranes has been postulated by Fox (1972) who suggests that the proteins may act as 'gate keepers' and active carriers, determining what passes through the membrane; the enzymes provide distinct functional properties while the lipids govern gross structural properties. It may well be that permeability of the functioning organelles like the mitochondria is the key to the metabolic activities in the seeds. With further understanding of the structure and functions of seed membranes in response to environmental changes it may be possible to elucidate the biochemical events that occur during germination and thus ascertain the specific causes of physiological disorders in germinating seeds.

REFERENCES

ABDUL-BAKI, A. A. (1969). 'Relationship of glucose metabolism to germinability and vigor in barley and wheat seeds', *Crop Sci.*, **9**, 732–736

ABDUL-BAKI, A. A. and ANDERSON, J. D. (1970). 'Viability and leaching of sugars from germinating barley', *Crop Sci.*, **10**, 31–34

ANDERSON, J. D. (1970). 'Physiological and biochemical differences in deteriorating barley seed', *Crop Sci.*, **10**, 36–39

ANDERSON, J. D. (1973). 'Metabolic changes associated with senescence', *Seed Sci. Technol.* (in press)

ASPINALL, D. and PALEG, L. G. (1971). 'The deterioration of wheat embryo and endosperm function with age', *J. exp. Bot.*, **22**, 925–935

AUSTIN, R. B. (1972). 'Effects of Environment Before Harvesting on Viability', in *Viability of Seeds*, (ed. Roberts, E. H.), 114–119, Chapman and Hall, London

BAIN, J. M. and MERCER, F. V. (1966). 'Subcellular organization of the developing cotyledons of *Pisum sativum* L.', *Aust. J. biol. Sci.*, **19**, 49–68

BERJAK, P. and VILLIERS, F. A. (1972). 'Ageing in plant embryos II. Age-induced damage and its repair during early germination'. *New Phytol.*, **71**, 135–144

BIDDLE, A. J. (1970). 'Pea and bean disorders', *Conf. Pea and Bean Crop Protection*, Peterborough, England

BRADBEER, J. W. and COLEMAN, B. (1967). 'Metabolic changes in seeds during chilling', in *Physiologie, Ökologie und Biochemie der Keimung*, (ed Borriss, H.) Part 1, 473–482 Ernst-Moritz-Arndt-Universität, Greifswald

CHING, T. M. (1973). 'Biochemical aspects of seed vigor', *Seed Sci. Technol.* (in press)

DENT, J. W. (1963). 'Soluble carbohydrate determination', *J. natn. Inst. agric. Bot.*, **9**, 282–291

DUNCAN, A. A., SIDOR, T., VITTUM, M. T., OHLING, H. and PUMPHREY, F. V. (1965). 'Cultural studies on blonding of peas', *Oregon Vegetable Digest.*, **14**, 9–11

FLENTJE, N. J. (1964). 'Pre-rotting of peas I. Factors associated with seed', *Aust. J. biol. Sci.*, **17**, 643–650

FOX, C. F. (1972). 'The structure of cell membranes', *Scient. Am.* **226**, 30–47

GABER, S. D. and ROBERTS, E. H. (1969). 'Factors controlling water sensitivity in germination of barley', *J. Inst. Brew.*, **75**, 103–106

GLINKA, Z. and REINHOLD, L. (1971). 'Abscisic acid raises the permeability of plant cells to water', *Pl. Physiol.*, **48**, 103–105

GRABE, D. F. (1964). 'Glutamic acid decarboxylase activity as a measurement of seedling vigor, *Proc. Ass. off. Seed Analysts. N. Am.*, **54**, 100–109

GRUNWALD, C. (1971). 'Effects of free sterols, steryl ester and sterol glycoside on membrane permeability', *Pl. Physiol.*, **48**, 653–655

HARRINGTON, J. F. (1973). 'Biochemical basis of seed longevity', *Seed Sci. Technol.* (in press)

HENDRICKS, S. B. (1971). Personal communication

HEYDECKER, W. (1967). 'Critical factors in the germination of non-dormant seeds', in *Physiologie, Ökologie and Biochemie der Keimung*, (ed. Borriss, H.) Part 1, 65–73, Ernst-Moritz-Arndt-Universität, Greifswald

HEYDECKER, W. (1972). 'Vigour', in *Viability of Seeds,* (ed. Roberts, E. H.), 209–252, Chapman and Hall, London

HEYDECKER, W. and FEAST, P. (1969). 'Studies of the "hollow heart" condition of pea (*Pisum sativum* L.) seeds', *Proc. int. Seed. Test. Ass.*, **34**, 319–327

JANSSEN, G. (1962). 'Studies on water sensitivity in barley seeds', *Svensk kem. Tidsk.*, **74**, 181–193

JONES, R. L. and ARMSTRONG, J. E. (1971). 'Evidence for the osmotic regulation of hydrolytic enzyme production in germinating barley seeds', *Pl. Physiol.*, **48**, 137–142

KOOSTRA, P. T. and HARRINGTON. J. F. (1969). 'Biochemical effects of age on membranal lipids of *Cucumis sativus* L.', *Proc. int. Seed Test. Ass.* **34**, 329–340

KROPH, J. P. (1971). 'Pea seed vigor in relation to bleaching', *M.S. research study,* Wash. St. Univ.

LANG, A. (1965). 'Effects of some internal and external conditions on seed germination', in *Encyclopedia of Plant Physiology*, Vol. 15 (ed. Ruhland, W.), 850–993, Springer-Verlag, Berlin

LARSON, L. A. (1968). 'The effect soaking pea seeds with or without coats has on seedling growth', *Pl. Physiol.*, **43**, 255–259

LYONS, J. M., WHEATON, T. A. and PRATT, H. K. (1964). 'Relationship between the physical nature of mitochondrial membranes and chilling sensitivity in plants, *Pl. Physiol.*, **39**, 262–268

MCDANIEL, R. G. (1969). 'Relationship of seed weight, seedling vigor and mitochondrial metabolism in barley', *Crop Sci.*, **9**, 823–827

MACKAY, D. B. (1972). 'The measurement of viability', in *Viability of Seeds*, (ed. Roberts, E. H.), 114–149, Chapman and Hall, London

MAGUIRE, J. D. (1969). 'Seed dormancy, germination and seedling vigor of some Kentucky bluegrass (*Poa pratensis* L.) varieties as affected by environmental and endogeneous factors', *PhD. Thesis*. Oregon State University

MAGUIRE, J. D. and KROPF, J. P. (1972). 'Pea seed vigor in relation to bleaching', *Crop Sci.*, (in press)

MAGUIRE, J. D. and STEEN, K. M. (1971). 'Effects of potassium nitrate on germination and respiration of dormant and non-dormant Kentucky bluegrass (*Poa pratensis* L.) seed', *Crop Sci.*, **11**, 48–50

MALHOTRA, S. J. and SPENCER, M. (1970). 'Changes in the respiration, enzymology swelling and contractal properties of mitochondria from cotyledons of *Phaseolus vulgaris* L. during germination', *Pl. Physiol.*, **46**, 40–44

MANOHAR, M. S. and HEYDECKER, W. (1964). 'Effects of water potential on germination of pea seeds', *Nature*, **202**, 22–24

MATTHEWS, S. and BRADNOCK, W. T. (1967). 'The detection of seed samples of wrinkle-seeded peas (*Pisum sativum* L.) of low potential planting value', *Proc. int. Seed Test. Ass.*, **32**, 553–563

MAYER, A. M. and POLJAKOFF-MAYBER, A. (1963). *The Germination of Seeds*, Pergamon, Oxford

MOORE, R. P. (1963). 'Previous history of seed lots and differential maintainace of seed viability and vigor in storage', *Proc. int. Seed Test. Ass.*, **28**, 691–697

MOORE, R. P. (1966). 'TZ tests for diagnosing causes for seed weakness and for predicting and understanding performance', *Proc. Ass. off. Seed Analysts. N. Am.* **56**, 70–73

MOORE, R. P. (1968). 'Seed deterioration symptoms as revealed by tetrazolium and growth tests', *Proc. Ass. off. Seed Analysts N. Am.*, **58**, 107–110

NAWA, Y. and ASAHI, T. (1971). 'Rapid development of mitochondria in pea cotyledons during the early stage of germination', *Pl. Physiol.*, **48**, 671–674

OBENDORF, R. L. and HOBBS, P. R. (1970). 'Effect of seed moisture on temperature sensitivity during imbibition of soybeans', *Crop Sci.*, **10**, 563–566

ORPHANOS, P. I. and HEYDECKER, W. (1968). 'On the nature of the soaking injury of *Phaseolus vulgaris* seeds', *J. exp. Bot.*, **19**, 770–784

OVCHAROV, P. I. (1969). 'The physiology of different quality seeds', *Proc. int. Seed Test. Ass.*, **34**, 305–313

PALEG, L. (1961). 'Physiological effects of gibberellic acid. III Observations on its mode of action on barley endosperm', *Pl. Physiol.*, **36**, 829–837

PERRY, D. A. (1969). 'Seed vigour in peas (*Pisum sativum* L.), *Proc. int. Seed Test. Ass.*, **34**, 221–232

PERRY, D. A. and HARRISON, J. G. (1970). 'The deleterious effect of water and low temperature on germination of pea seed', *J. exp. Bot.*, **21**, 504–512

PERRY, D. A. and HOWELL, P. J. (1965). 'Symptoms and nature of "hollow heart" of pea seed', *Pl. Path.*, **14**, 111–116

POLLOCK, B. M. (1969). 'Imbibition temperature sensitivity of lima bean seeds controlled by initial moisture', *Pl. Physiol.*, **44**, 907–911

POLLOCK, B. M. (1972). 'Effects of environment after sowing on viability', in *Viability of Seeds*, (ed. Roberts, E. H.), 150–171, Chapman and Hall, London

POLLOCK, B. M. and TOOLE, V. K. (1964). 'Seed vigor in lima beans', *Proc. Ass. off. Seed Analysts N. Am.*, **54**, 26

POLLOCK, B. M. and TOOLE, V. K. (1966). 'Imbibition period as the critical temperature sensitive stage in germination of lima bean seeds', *Pl. Physiol.*, **41**, 221–229

PRITCHARD, G., QUARTLEY, C. E. and TURNER, E. R. (1967). 'The effects of high oxygen pressure on the respiratory metabolism of pea seeds', in *Physiologie, Ökologie und Biochemie der Keimung*, Part 1, 519–529, Ernst-Moritz-Arndt-Universität, Greifswald

ROBERTS, E. H. (1972). 'Cytological, genetical and metabolic changes associated with loss of viability', in *Viability of Seeds*, (ed. Roberts, E. H.), 253–306, Chapman and Hall, London

ROOS, E. E. and POLLOCK, B. M. (1971). 'Soaking injury in lima beans', *Crop Sci.*, **11**, 78–80

RULES FOR TESTING SEEDS (1965). *Proc. Assoc. off. Seed Analysts N. Am.*, **55**, 112

SIMON, E. W. (1972). 'Leakage from seeds during early imbibition', *J. exp. Bot.* (in press).

STREET, H. E. and ÖPIK, H. (1970). *The Physiology of Flowering Plants*, Arnold, London

TAYLORSON, R. B. (1969). 'Photocontrol of rough cinquefoil seed germination and its enhancement by temperature manipulation and KNO$_3$', *Weed Sci.*, **17**, 144–148

THOMAS, H. (1972). 'Control mechanisms in the resting seed', in *Viability of Seeds*, (ed. Roberts, E. H.), 360–396, Chapman and Hall, London

TOOLE, E. H., HENDRICKS, S. B., BORTHWICK, H. A. and TOOLE, V. K. (1956). 'Physiology of seed germination', *A. Rev. Pl. Physiol.*, **7**, 299–324

WOODSTOCK, L. W. (1969). 'Seedling growth as a measure of seed vigor', *Proc. int. Seed Test. Ass.*, **34**, 273–280

WOODSTOCK, L. W. and COMBS, M. F. (1967). 'Application of the respiration test for seed vigor in peas', *Proc. Ass. off. Seed Analysts N. Am.*, **57**, 144–148

WOODSTOCK, L. W. and FEELEY, J. (1965). 'Early seedling growth and initial respiration rate as potential indicators of seed vigor in corn', *Proc. Ass. off. Seed Analysts N. Am.*, **55**, 131–139

WOODSTOCK, L. W. and POLLOCK, B. M. (1965). 'Physiological predetermination: imbibition, respiration and growth of lima bean seeds', *Science*, **150**, 1031–1032

DISCUSSIONS

Moore: Are the plants, or parts of plants, in which bleached seeds occur characterised by any particular feature?

Maguire: Bleaching of pea seeds occurs at random. The chlorophyll can disappear from seeds anywhere on the plant and anywhere in the pod provided the temperature is high and the pods have been rained on. There is no difference in the pattern of distribution of bleached seeds between plants staked upright and plants lying on the ground, although the latter may have a higher incidence.

Storey: Is this bleaching identical with the disorder known as 'hollow heart'?

Maguire: No. It is not identical with hollow heart, although both troubles originate during the maturation of the seeds.

Norton: From our examination of the dry seeds it appears that, apart from very minor differences, the chemical composition of bleached seeds is identical with that of green seeds.

17

INTERACTING EFFECTS OF
SEED VIGOUR AND ENVIRONMENT
ON SEEDLING ESTABLISHMENT

D. A. PERRY

Scottish Horticultural Research Institute, Invergowrie, Dundee

INTRODUCTION

The germination capacity of a seed lot is determined by the proportion of seeds capable of producing a normal shoot and roots under conditions designed to ensure maximum germination. Deviations from optimal conditions usually result in reduced germination, provided there are no complicating factors such as dormancy present. Conditions for germination in the soil are rarely optimal because parasitic biotic factors are present even when physical factors are good. As soil conditions deteriorate the percentage emergence from seed lots of similar germination capacity within any crop may differ. It is largely the absence of a consistent relation between germination in the laboratory and emergence in the field which has been responsible for the development of the concept of seed vigour. No definition of vigour has been generally accepted; some workers consider that all factors which affect emergence should be included, e.g. seed size, mechanical and insect damage and seed-borne pathogens (Isely, 1957; Schoorel, 1957), whereas others prefer to restrict the use of the term to a description of the physiological condition (Woodstock, 1969) and to exclude characteristics, such as those mentioned above, which can be readily observed.

The following definition of vigour has recently been proposed: 'A physiological property determined by the genotype and modified by the environment, which governs the ability of a seed to produce a seedling rapidly in soil, and the extent to which that seed tolerates a range of environmental factors. The influence of seed vigour may

persist through the life of the plant and affect yield' (Perry, 1972).
This definition applies to a single seed but may also refer to the average within a bulk. The term 'vigour' will be used in this restricted,
physiological sense throughout this paper.

The International Rules for Seed Testing (Anon., 1966) state that
the object of the germination test is to indicate the planting value of
a seed lot under *good soil conditions*, and Essenburg and Schoorel
(1962) have reviewed the ample evidence supporting the validity of
the test procedure. However, there have been few studies of the
relationships between germination, vigour, and emergence under
conditions which deviate from the optimum. Schoorel (1957)
showed that the relation between germination and emergence of
peas and beans was poorer in adverse soil conditions of low moisture
content than in a more favourable environment.

This paper discusses the implications of the results obtained in
emergence trials on peas at Invergowrie (Perry, 1967) and at
several other sites (Perry, 1970), on carrots (Perry and Hegarty,
1971; Hegarty, 1971), and on sugar beet.

PEAS

The wide variation in emergence found when the same seed lots are
sown at different sites (Heydecker, 1969) has demonstrated that the
emergence of peas is very sensitive to the soil environment. Failure
of the germination test to indicate the percentage survival from a
given seed lot, particularly in early spring sowings, has led to the
development of the vigour concept for this crop (Perry, 1967) and to
a realisation of the need for vigour tests to detect seeds liable to fail
in the field.

In 1968, five different seed lots of the cultivar Kelvedon Wonder
and five of the cultivar Lincoln were included in field trials in which
there were a total of 27 different 'sowing treatments' comprised of
different sowing dates at several sites in Scotland and England. At
some sites the effect of captan seed dressing on emergence was
investigated. Full details of the methods and the sites have been
described by Perry (1970). A critical assessment of seedling growth
(Perry, 1969a) and the measurement of the electroconductivity of
the water in which peas had been soaked for 24 h (Matthews and
Bradnock, 1967) were employed as vigour tests and examined for
their applicability to emergence at all the sowing treatments.

This discussion will refer mainly to the results obtained at Cambridge, where seed treated with captan was sown at two sites on

three dates, and at Colworth House near Bedford and Invergowrie near Dundee where there were three sowing dates and where both treated and non-treated seeds were used.

EFFECT OF SOWING DATE

Table 17.1 shows that the mean emergence percentage of the 10 lots improved with later planting, except at Invergowrie where high soil temperatures and drought followed the third sowing. The soil factors limiting emergence in early sowings were probably low

Table 17.1 MEAN EMERGENCE OF 10 PEA SEED LOTS AT 3 SITES ON 3 SOWING DATES

Site	Sowing date	Mean percentage emergence
Cambridge	20 Feb	29·4
	19 Mar	37·0
	22 Apr	64·5
	S.E. (mean)	± 1·15
Colworth House	29 Feb	40·4
	21 Mar	56·0
	22 Apr	78·2
	S.E. (mean)	± 0·82
Invergowrie	7 Mar	48·4
	22 Apr	68·3
	4 June	63·2
	S.E. (mean)	± 0·81

temperatures and high moisture content. Within the mean emergence figures, individual lots varied considerably and, furthermore, reacted differently to changes in the environment, as shown by a significant interaction between lots and date of sowing at each site. On the basis of their performance in the laboratory vigour tests, the seed lots could be divided into four categories of vigour: high, medium, low and very low (Perry, 1970) and these groupings were in good agreement with emergence in the field. *Table 17.2* shows the performance of the 10 lots separated on the basis of the results of a vigour test at one site on three sowing dates. A similar pattern was observed at the other sites. The emergence of high vigour seeds was better than that of the other grades and increased with later sowing. The improvement in emergence in the medium

314

Table 17.2 PERCENTAGE EMERGENCE OF 10 PEA SEED LOTS ON 3 SOWING DATES AT CAMBRIDGE

Sowing date	Seed lot	Vigour grade									
		High				Medium		Low			Very low
		Lin 4	Lin 2	Lin 5	KW2	KW3	Lin 1	Lin 3	KW5	KW4	KW1
20 Feb		54·5	38·5	44·0	30·0	18·5	17·0	14·0	5·0	6·5	1·5
19 Mar		62·5	44·5	58·5	41·0	34·0	29·5	31·0	13·0	31·0	4·5
22 Apr		97·5	78·0	96·5	75·5	71·0	62·8	62·5	42·5	66·0	15·0

S.E. for differences of means within the table ±4·84

Table 17.3 MEAN PERCENTAGE EMERGENCE OF PEA SEED LOTS IN THE HIGH, MEDIUM AND LOW VIGOUR GRADES ON THREE SOWING DATES AT CAMBRIDGE

Sowing date	High vigour (4 lots)		Medium vigour (3 lots)		Low vigour (2 lots)	
	Emergence	Reduction*	Emergence	Reduction	Emergence	Reduction
20 Feb	41·8	52·0	16·5	74·8	5·7	89·4
19 Mar	51·6	40·6	31·5	51·9	22·0	59·4
22 Apr	86·9	—	65·4	—	54·3	—

* Seeds which failed to emerge expressed as a percentage of those emerging from 22 April sowing

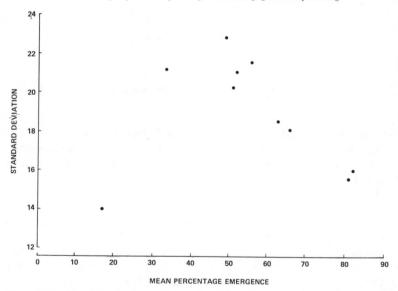

Figure 17.1   The relationship between mean soil emergence of 10 pea seed lots and the standard deviations of the means from 27 sowing 'treatments' in 1968

and low grades in later sowings was, however, proportionately greater than that shown by the high vigour seeds. Conversely, medium and low vigour seeds were more sensitive to an adverse environment than high vigour seeds, as shown by the percentage reduction in mean emergence from the first and second sowings compared with that at the third sowing (*Table 17.3*).

The sensitivity of a seed lot to the environment may be expressed by the variability of its emergence under different soil conditions. The standard deviation of emergence is a measure of variability and in *Figure 17.1* the values for the 10 lots are plotted against the mean percentage emergence over all sites and sowing dates. Although the standard deviations appear to follow a normal distribution with a maximum at around 50 per cent emergence, only the very low vigour sample which emerged poorly at all times, contradicts the alternative hypothesis that variability declines as vigour increases.

EFFECT OF ENVIRONMENT ON CORRELATION BETWEEN GERMINATION CAPACITY AND FIELD EMERGENCE

Mean emergence of all the seed lots sown on any particular occasion is a measure of the effect of the environment on the seed on that occasion, i.e. a low mean emergence implies a severe stress, while a high mean emergence indicates favourable soil conditions. Correlation coefficients were calculated between the germination capacity (laboratory germination) of the 10 seed lots used in these trials and their field emergence at each of the 27 'sowing treatments'. The relation between mean emergence and the magnitude of the correlation coefficient demonstrates the effect of the environment on the correlation for that sowing. Information on the influence of differing levels of stress in the soil environment on the relation between germination capacity and field emergence was obtained by calculating a second correlation coefficient, this time between the mean emergence and the correlation coefficients obtained for each sowing treatment. As the correlation coefficient, r, is not normally distributed, it was transformed to a value z ($\tanh^{-1} r$) before analysis. The correlation coefficient for the 27 sowing treatments was $+ 0.53$, which is significant at the 1 per cent level. Because the experimental technique and environment were similar, the 12 sowing treatments at Cambridge and Colworth House were analysed together and a correlation coefficient of $+ 0.75$ was obtained. From the regression line and individual points shown in *Figure 17.2* it is clear that the best relation between laboratory germination

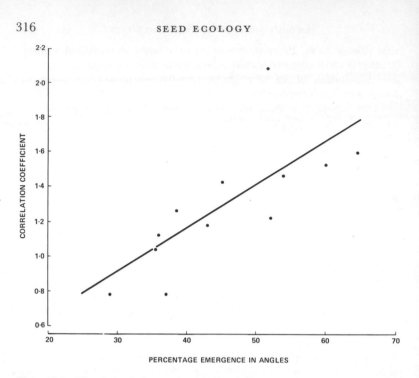

Figure 17.2   The relationship between the correlation coefficient (between laboratory germination and field emergence) and mean emergence on each of 12 sowing 'treatments' of peas at Cambridge and Colworth House in 1968

and field emergence was obtained when emergence was high and, by implication, when the environment was favourable.

INTERACTION BETWEEN VIGOUR AND SOIL-BORNE PATHOGENS

Soil-borne pathogens are an important part of the soil environment and *Pythium ultimum* Trow. in particular is found on seedlings which fail to emerge. Fungicides applied to seeds modify the environment by protecting against fungal pathogens and their effect on emergence and their interaction with seed vigour and with soil conditions were studied in some trials.

Treatment with captan usually improved emergence in the 1968 trials and the greatest response was shown in the adverse conditions of the earliest sowings where emergence from untreated seeds was least (*Table 17.4*). Protection was also provided in the third sowing at Invergowrie when the seeds were exposed to moisture stress.

The response of seeds to the fungicidal seed dressing was also affected by their vigour and *Table 17.5* shows that the increase in

Table 17.4   THE RESPONSE OF PEA SEEDS TO TREATMENT WITH CAPTAN FUNGICIDE

Site	Sowing date	Mean percentage emergence (10 seed lots)		Percentage improvement following treatment
		Not treated	Treated	
Colworth House	29 Feb	33·9	46·9	38·3
	21 Mar	50·2	61·8	23·1
	22 Apr	74·9	81·4	8·7
		S.E. ±1·19		
Invergowrie	7 Mar	42·0	54·7	30·2
	22 Apr	64·2	72·4	12·8
	4 June	56·2	70·3	25·0
		S.E. ±1·11		

emergence of high vigour seeds was less than that of medium and low vigour seeds when averaged over 3 sowing dates at both Colworth House and Invergowrie.

In 1964 the effect of seed dressing on 24 lots of several pea varieties was investigated (Perry, 1967). The pattern of response was similar to that in 1968 and, because there were more seed lots, a valid regression analysis could be carried out. *Figure 17.3* shows the emergence of each lot without treatment plotted against the improvement obtained with a fungicidal treatment (ie. percentage emergence of treated seeds minus percentage emergence of untreated seeds) for the first sowing on 3 March. A correlation coefficient of -0.83 was

Table 17.5   RESPONSE OF SEEDS OF DIFFERENT VIGOUR GRADES TO FUNGICIDE TREATMENT

Vigour grade	Site	Mean percentage emergence		Percentage improvement following treatment
		Not treated	Treated	
High	Colworth House	71·2	79·2	11·2
(4 lots)	Invergowrie	75·1	83·0	10·0
Medium	Colworth House	49·9	60·3	20·8
(3 lots)	Invergowrie	52·5	67·2	28·0
Low	Colworth House	40·9	56·8	38·9
(2 lots)	Invergowrie	34·8	49·4	42·0

calculated from the data. If emergence without treatment is taken as a measure of seed vigour, *Figure 17.3* shows that low vigour seeds responded more to treatment than high vigour seeds. In one case no response was obtained with high vigour seeds. If fungicide treatment restored emergence to its potential maximum, which should approximate to the germination capacity of the samples (see Perry, 1967),

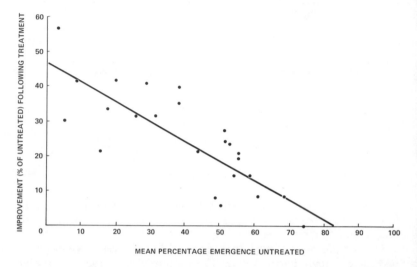

Figure 17.3   The relationship between emergence of untreated pea seed and the improvement due to fungicide treatment after sowing on 3 March, 1964 at Invergowrie

by protecting the seeds against pathogens the regression coefficient would be close to $-1 \cdot 0$. However, the calculated value of $-0 \cdot 55$ shows that this was not achieved and a physiological factor may have contributed to the poor emergence of the low vigour samples.

CARROTS

There are insufficient data on emergence of other crops on which to test the hypotheses presented above for peas. Recently, Hegarty (1971) has studied some of the factors which influence the relation between germination and emergence in carrots at Invergowrie and further work on this subject is given in Chapter 23.

The correlation coefficients between the results of the laboratory germination test as prescribed in the International Rules for Seed

Testing (Anon., 1966) and field emergence on 3 sowing dates (1 April, 29 April and 27 May) for 11 seed lots were 0·56, 0·65 and 0·39, respectively, and the mean percentage emergence figures were 63·8, 65·2 and 54·4 (Hegarty, 1971). The decrease in the value of the correlation coefficient as mean emergence fell suggests that the two may be related as they are in peas, but more comparisons are needed to establish a definite relation. Hegarty further found that the laboratory germination of some seed lots was depressed when the test temperature was lowered to 10°C while other lots were un- affected. As the results of these tests correlated better with field emergence than those obtained under the International Rules it is likely that a similar temperature sensitivity occurred in the emer- gence trials where mean soil temperatures for the first 2 sowings were about 10°C. Thus there is an interaction between germination temperature and seed quality which—in the absence of any recog- nisable abnormality in the seed—is evidence of the existence of vigour differences which only become recognisable at low tempera- tures.

SUGAR BEET

Preliminary experiments at Invergowrie have indicated that the emergence of sugar beet seeds, in contrast to carrot seeds, is not reduced by low temperature but is sensitive to high soil moisture. Four samples of commercial, pelleted, monogerm seeds were sown at weekly intervals for 10 weeks commencing 7 April 1970, using the techniques described by Perry and Hegarty (1971) for carrots. The results are depicted in *Figure 17.4*. Emergence of the sowings on the various dates fluctuated, but there was no trend towards increasing emergence as soil temperatures rose, although the rate of emergence was greater. The temperature data presented in *Figure 17.4* are daily means from a mercury-in-steel thermograph with the bulb 25 mm below soil level. The emergence of two of the seed lots was markedly depressed following sowing on 4 May. Rainfall data show that 10·2 mm rain fell on the day following sowing and a total of 40·6 mm between sowing and emergence 15 days later. No other sowing received as much rain.

It has been established that sugar beet seeds are sensitive to high substrate water level in laboratory gemination tests (Chetram and Heydecker, 1967; Lazar, 1969) and the results presented in *Figure 17.4*, suggest that a similar sensitivity may occur in the field, al- though not all seed lots are affected to the same extent. A differ- ential effect of an adverse environment on different seed lots is

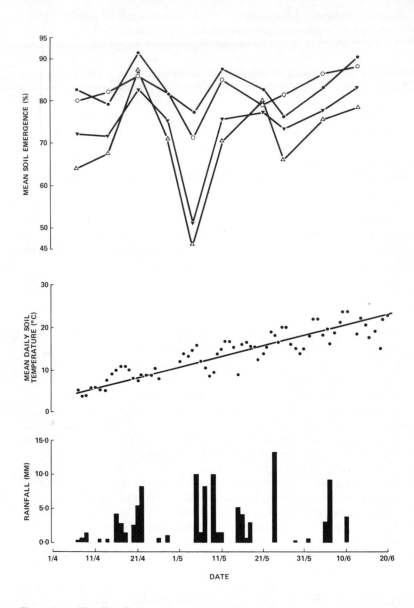

Figure 17.4   The effect of soil temperature and rainfall on emergence of 4 lots of sugar beet sown on 10 occasions at Invergowrie

therefore evident in sugar beet, and the existence of vigour differences in this crop may be surmised.

DISCUSSION

Some generalisations about vigour may be made from the results of the extensive trials on peas. The manifestation of differences in vigour depend on an interaction between the seed and its environment. When conditions are favourable field emergence approaches, and is well related to, germination in the laboratory, but when they impose a stress on the seed the relationship declines. Low vigour seeds are more sensitive than high vigour seeds to stress factors and the predictability of plant populations from calculated seed rates becomes uncertain. Provided tests for vigour are available, high vigour seeds which are more tolerant of adverse circumstances should be selected for sowing in poor soil conditions. Evidence from the 27 sowing treatments in the 1970 trials showed that high vigour seeds of peas retained their superiority and emerged better than all lower grades at all sites despite the wide range of soil types and conditions encountered (Perry, 1970).

Pea seeds which fail to emerge in field soil become colonised by soil-borne pathogens such as *Pythium ultimum* and an improvement in emergence is usually obtained by treating them with a fungicide. In these studies the improvements were greatest under stress conditions and in low vigour seeds, but they were not as great as theoretically possible if pathogenic attack was the sole cause of death. The failure to restore emergence to its theoretical maximum may be due to incomplete protection by the fungicide used. However, germination of low vigour lots even in a sterile system is adversely affected by low temperature, by anaerobiosis and by soaking in water (Caldwell, 1960; Perry, 1969b; Perry and Harrison, 1970). Thus, the failure to emerge probably results from stress factors which either kill the seeds or increase their susceptibility to attack by soil-borne pathogens.

The evidence included in this paper shows that vigour differences also exist in carrots and sugar beet, although these seeds are less sensitive to stress, and the specific factors within the environment which affect them differ. Carrot seeds are sensitive to low temperature, while sugar beet seeds are affected by high moisture content, and from the limited data available it seems possible that there are interrelations between germination, emergence and the environment similar to those found in peas.

Other stress factors such as low moisture and mechanical impedance may induce differential sensitivity, and assessments based on field observations are necessary to establish which of these are most commonly encountered by growers. It is possible that a seed lot distinguished by high vigour under one stress condition may show a different response under another.

Although the concept of seed vigour is now well established, little is known of the factors that cause variation in vigour. There may be several causes within a crop; for example, in peas rapid drying of immature seeds in the field or artificially and deterioration of seeds in store result in low vigour. But it is frequently impossible to determine the cause of low vigour in a commercial seed lot because its history is unknown. Studies of the incidence and frequency of low vigour, investigation into its causes and into the conditions under which it becomes manifest, should result in the production of consistently reliable seeds which are capable of tolerating the normal range of environmental conditions expected at sowing time, and which are so necessary for modern techniques of precision sowing.

REFERENCES

ANON. (1966). 'International rules for seed testing', *Proc. int. Seed Test. Ass.*, **31**, 1–152

CALDWELL, W. P. (1960). 'Laboratory evaluation of vigor of garden peas', *Proc. Ass. off Seed Analysts N. Am.*, **50**, 130–136

CHETRAM, R. S. and HEYDECKER, W. (1967). 'Moisture sensitivity, mechanical injury and gibberellin treatment of *Beta vulgaris* seeds', *Nature*, **215**, 210–211

ESSENBURG, J. F. W. and SCHOOREL, A. F. (1962). 'Het verband tussen de kiemkrachts-bepaling van zaaizaden in het laboratorium en de opkomst te velde', *LitOverzicht, Centrum voor Landbouwpublikaties en Landbouwdocumentatie, Wageningen*, **26**, 42–68

HEGARTY, T. W. (1971). 'A relation between field emergence and laboratory germination in carrots', *J. hort. Sci.*, **46**, 299–305

HEYDECKER, W. (1969). 'Report of the Vigour Test Committee, 1965–1968', *Proc. int. Seed Test. Ass.*, **34**, 751–773

ISELY, D. (1957). 'Vigor tests', *Proc. Ass. off. Seed Analysts N. Am.*, **47**, 176–182

LAZAR, O. (1969). 'Le freinage de la germination des glomérules de la betterave sucrière par l'excès d'eau', *J. int. Inst. Sugar Beet Res.*, **4**, 49–61

MATTHEWS, S. and BRADNOCK, W. T. (1967). 'The detection of seed samples of wrinkle-seeded peas (*Pisum sativum* L.) of potentially low planting value', *Proc. int. Seed Test. Ass.*, **32**, 553–563

PERRY, D. A. (1967). 'Seed vigour and field establishment of peas', *Proc. int. Seed Test. Ass.*, **32**, 3–12

PERRY, D. A. (1969a). 'A vigour test for peas based on seedling evaluation', *Proc. int. Seed Test. Ass.*, **34**, 265–270

PERRY, D. A. (1969b). 'Seed vigour in peas (*Pisum sativum* L.)', *Proc. int. Seed Test. Ass.* **34**, 221–232

PERRY, D. A. (1970). 'The relation of seed vigour to field establishment of garden pea cultivars', *J. agric. Sci., Camb.*, **74**, 343–348

PERRY, D. A. (1972). 'Seed vigour and field establishment', *Hort. Abstr.*, **42**, 334–342

PERRY, D. A. and HARRISON, J. G. (1970). 'The deleterious effect of water and low temperature on germination of pea seed', *J. exp. Bot.*, **21**, 504–512

PERRY, D. A. and HEGARTY, T. W. (1971). 'Effect of captan seed dressing on carrot emergence', *Pl. Path.*, **20**, 29–32

SCHOOREL, A. F. (1957). 'The use of soil tests in seed testing', *Proc. int. Seed Test. Ass.*, **22**, 287–301

WOODSTOCK, L. W. (1969). 'Seedling growth as a measure of seed vigor', *Proc. int. Seed Test Ass.*, **34**, 273–280

DISCUSSIONS

Hudson: Why do you relate sugar beet germination with rainfall rather than soil water?

Perry: The effect *is* one of excessive soil water and the rainfall distribution was found to indicate this best.

Heydecker: We too have found that what happens in the shallow top layer of the soil where the seeds are sown and through which all rain must pass is what mainly matters.

Heydecker: If fungi are not chiefly responsible for the discrepancy between laboratory germination and field stands, what is?

Perry: Vigour is a question of 'whole seed physiology'. In peas one can distinguish seeds of low and high vigour by deliberately attempting to reduce their germination capacity, e.g. by letting seeds imbibe under sterile conditions at 4°C or less, or immersing them in water for a period, and then placing them on moist paper in 20°C at a high substrate water content. Such treatments weaken the seeds and render them susceptible to ubiquitous soil pathogens, especially *Pythium ultimum*, which exploit the areas of weakness. I believe that edaphic factors are the primary causative agents of failure.

Storey: Dr. Perry's figures show a surprising correlation between temperature and emergence of peas which are usually much more water than low temperature sensitive.

Perry: Temperature and water are closely linked. In early spring when the temperatures are low the soil is wet because the mean evaporation rate is lower than the mean precipitation rate.

Heydecker: Peas belong to the group of seeds which are peculiarly susceptible to microbial attack because they are bred not only to be sown but also to be eaten.

Matthews: The effect of pathogens should not be underestimated.

Perry: I fully agree; for instance, in early sowings the emergence of peas may take 28 days, compared to 9–10 days later on, and the conditions favour the pathogens. Adverse physical conditions when reproduced in the laboratory may reduce germination by 10 to 15 per cent, but in the field in the presence of pathogens 90 per cent of the seeds may fail to emerge.

18

SEED-BORNE DISEASES AND THEIR CONTROL

R. B. MAUDE

National Vegetable Research Station, Wellesbourne, Warwick

INTRODUCTION

Many serious plant diseases arise from infection carried by the seeds. Initially, plant establishment may be reduced by pre-emergence losses and death of seedlings; later, spread of disease may cause the loss of all or part of a crop. Where seed production is required, infection of the inflorescences may result in contamination, deterioration and loss of seeds. Seed-borne fungi, bacteria and viruses are implicated as the causal pathogens of such diseases. In this paper, the biology and control of seed-borne organisms are discussed with special reference to vegetable diseases.

THE IMPORTANCE OF SEED-BORNE DISEASES

The spread of many plant pathogens is dependent on the transmission of their spores through the air for long or short distances and the alighting of such spores upon suitable host plants. Seed-borne pathogens, however, are subject to none of these restrictions. Not only are they spread through the natural seed-dispersal mechanisms of plants but they are carried over great distances by man, to wherever the crop is grown. When they arrive, moreover, so does their natural host, ready to develop from the seed which carries them.

The world-wide transportation of infected seeds is well authenticated. For example, certain stocks of celery seed imported into Britain from America have been shown to bear viable infections of the fungus *Septoria apiicola* (leaf blight) (Hewett, 1968). This disease is already present on locally raised seeds; on the other hand, French bean (*Phaseolus vulgaris*) seeds, from the same country, are accepted as the main source of introduction of the non-indigenous race 2 of *Pseudomonas phaseolicola* (halo blight) (Wharton, 1967). On a more local scale, outbreaks of carrot leaf blight (a disease more typical of warmer climates) which occurred in East Anglia, the important carrot growing region in Eastern England, in 1958 (Hawkins, Ives and Storey, 1959) most probably resulted from the introduction of seed samples from the continent of Europe, many of which have been shown to carry the disease (Hewett, 1964; Maude, 1966b). Certainly, the recent increases in the incidence of *Ascochyta fabae* (leaf spot) in field beans in Britain (Hewett, 1966) are attributable to seed samples obtained from a continental source.

Once introduced, a seed-borne disease may persist in plant debris. Some, e.g. *Septoria apiicola* on celery, may survive only long enough to infect one subsequent crop (Maude and Shuring, 1970), while others, e.g. *Stemphylium radicinum*, can persist for as long as eight years in contaminated soil and annually reinfect the roots of carrots raised for seed production (Maude and Moule, 1972). Infected seeds are clearly an important source of introduction of a pathogen into a crop. For a disease to develop within a crop by the sowing of infected seeds, however, there are two main requirements: (a) transfer of the disease from seeds to seedlings, resulting in outbreaks of primary infection, and (b) the spread of infection through the crop from these sources.

TRANSFER OF DISEASE

The transmission of disease from seeds to seedlings is influenced by many factors, some of which are discussed below.

SURVIVAL OF THE PATHOGEN

Some seed-borne pathogens are carried superficially, whereas others are more deeply seated. In the latter group, the bacteria and fungi mainly invade the seed coat and cotyledon tissues and are only rarely found deep in the embryos. The reverse is true of viruses, however, and seed coat infection is less common (Dykstra, 1961). The position of the pathogen on or in the seed may affect its survival, since superficial pathogens are usually lost more rapidly than deep-seated ones from seeds during storage or after planting in the soil (presumably by the action of the soil microflora).

Thus, lettuce mosaic virus, situated within the embryos of lettuce seeds, survives for the life-span of the seeds (3–5 years); *Ascochyta pisi*, within the cotyledons of peas, persists for at least six years (Butler and Jones, 1949), *Phoma betae* on red beet seeds and *Alternaria brassicicola* on cabbage seeds for at least four years (Maude, unpublished) and *Phoma lingam* on swede seeds for up to two years (Lloyd, 1959). Some fungi, although deep-seated, survive less well. The fact that *Septoria apiicola*, on celery, is relatively short lived has led to the recommendation that the seeds should be kept for two or more years to remove infection (Chupp and Sherf, 1960). Interacting conditions of temperature and humidity so affect the longevity of the pathogen, however, that such a method is likely to be unreliable. Conditions such as low temperature and humidity which are favourable for the maintenance of seed vigour will often also prolong the life-span of the pathogen, and under favourable normal warehouse storage conditions (at

Table 18.1   THE EFFECT OF DIFFERENT TEMPERATURE AND HUMIDITY CONDITIONS UPON THE VIABILITY OF *SEPTORIA APIICOLA* SPORES FROM STORED CELERY SEEDS

Storage temperature	Percentage viability of spores after storage								
	25% R.H.			50% R.H.			75% R.H.		
	Storage period (years)								
	1	2	3	1	2	3	1	2	3
5°C	68·3	53·8	59·3	74·2	69·0	37·5	45·3	1·5	0
15°C	64·2	45·5	41·2	67·4	49·4	14·9	13·5	0	0
25°C	51·7	18·9	1·3	31·1	0	0	1·1	0	0

ambient temperatures, in ventilated stores) the longevity of most pathogens of seeds is at least sufficient to ensure carry-over of disease from year to year (*Table 18.1*).

TRANSMISSION OF PATHOGENS FROM SEEDS TO SEEDLINGS

When seeds are planted in the soil and their tissues imbibe water, the organisms present with them resume growth and can infect the developing seedlings. Many factors, however, influence the extent to which this occurs. Antibiotic effects of seed or soil microflora, for instance, may suppress the pathogen. Thus, pea seeds superficially infected (35 per cent) with *A. pisi* may fail to give rise to diseased seedlings in field soil (Wharton, personal communication). The positions of the infective seed tissues in relation to the developing seedling are also of considerable importance. Thus, where the infected seed cotyledons remain below soil level, transfer of infection—e.g. of *Ascochyta* in peas—from cotyledon to plumule depends on contact between them during the process of seed germination. Once emerged, the young plumule can no longer become infected from the seed source. Rate of germination and degree and location of cotyledon infection in relation to the germinating plumule are also important factors. The pea disease caused by *Mycosphaerella pinodes* presents an entirely different situation, however. The fungus grows directly from the seed into the hypocotyl of the seedling. The chances of plant escape are therefore small and many seedlings become infected (Maude, 1966a).

Yet another effect is seen when, as with many plants, the young cotyledons emerge above the soil, often bearing the remains of an infected seed coat. In such cases seedling infection may depend upon the humidity of the environment. Thus, the infection of celery grown from infected seeds was increased from 0·7 per cent in a ventilated glasshouse compartment to 20 per cent when grown in a relative humidity of 100 per cent (Maude, 1964).

The fate of the seedlings which contract infection from the seeds may vary considerably. Some die quickly and may be of no further consequence but, on others, the pathogen may develop to a stage when it can be transmitted to other plants and the second and more serious phase of crop infection may ensue.

SPREAD OF DISEASE

Under suitable climatic conditions, very small numbers of infected seeds and seedlings can cause major outbreaks of disease. Thus, two diseased seeds in 10 000 (Walker and Patel, 1964) or one in 16 000 (Guthrie, Huber and Fenwick, 1965) can lead to serious outbreaks of halo blight in dwarf bean crops. Similarly, more than one infected lettuce seed in 30 000 infected with mosaic virus has been found to represent a considerable hazard to crops in America (Zink, Grogan and Welch, 1956); and in Britain, Tomlinson (1962) has demonstrated that levels of seed infection greater than 0·1 per cent can result in serious outbreaks of the disease. As already mentioned, under normal celery-raising conditions the proportion of *Septoria apiicola* transferred from seeds to seedlings is small. From such a level (i.e. 0·07 per cent infected seedlings), however, a complete celery crop can become diseased (Maude, 1964).

Ultimately, spread of disease within a crop results in the reinfection of the seeds. In many cases this occurs in the first year of growth, when the seed-bearing organs, e.g. pea and bean pods, become infected. Some crops, e.g. celery, become infected in their first year of growth, but flower and produce seed inflorescences in their second year. In such cases, carry-over of disease on the older leaves results in infection of the flowering stems and finally the seeds. With both types of crop remedial sprays can be used to prevent spread of disease in the field and reduce the chances of infection to the seed inflorescences.

The seed-borne fungi *Stemphylium radicinum* (black rot of carrots), *Alternaria brassicicola* (dark leaf spot of cabbage) and the bacterium *Corynebacterium betae* (silvering of red beet), on the other hand, all cause diseases which are latent in the 'ware' crop (one grown for market) but become widespread and produce characteristic disease symptoms on plants in their second year of growth. Under these conditions, where initial spread of infection is apparently symptomless, e.g. in silvering of beet (Keyworth and Howell, 1961), remedial field spray programmes are difficult to prescribe and such diseases therefore become well established by the second year of crop growth.

CONTROL OF DISEASE

The seed is the focal point of the disease cycle and most methods for the control of seed-borne diseases aim either to prevent infection of the seed crop, or to disinfect the diseased seeds.

PREVENTIVE MEASURES

Considerable reduction in the incidence of seed-borne pathogens such as *Septoria apiicola* on celery seed and *Pseudomonas phaseolicola* on French beans (*Phaseolus vulgaris*) have been achieved by the multiplication of seed stocks in warm dry climates unsuitable for the spread of disease as in California and Idaho in America. In addition, field inspection schemes normally used to ensure varietal purity have also been used for crop health estimations. Thus, in New South Wales a bean crop inspection scheme has for the past twenty years been used to check not only for trueness to varietal type but also for levels of halo blight (*Pseudomonas phaseolicola*) (Wilson, 1947). The laboratory testing of seeds can be used by official seed testing stations to confirm field inspection declarations. The effectiveness of such tests, however, is directly proportional to the number of seeds tested. As in most cases only 400 seeds per sample are examined, low levels of seed-borne infection may well escape detection. For instance, less than 0·1 per cent of the seeds in many commercial French bean stocks are infected with halo blight. To determine whether stocks are carrying the disease, it has been necessary to devise a method which will detect one infected seed in 10 000 (Taylor, 1970).

The ease with which a pathogen can be eradicated from seeds largely depends on the type of the organism, i.e. virus, bacterium or fungus, and its position on or in the seed tissues. Thus, the majority of virus pathogens which inhabit the embryos of seeds do not readily respond to therapy. Only in a relatively few cases, e.g. tomato mosaic virus, where external infection of the seed coat occurs, can disease control be achieved by various seed treatments (Broadbent, 1965). Bacterial and more especially fungal diseases are more responsive to treatment. Where fungal infection is relatively superficial on seeds, treatment by means of organomercury or dithiocarbamate fungicides have been effective. Where, however, the bacterial and fungal pathogens are more deeply seated, e.g. *Pseudomonas phaseolicola* beneath the seed coat of beans, or even *Ascochyta pisi* in the tissues of pea cotyledons, normal seed dressings are not effective and penetrative treatments are required. One such method is heat treatment. Thus, dry heat using hot carbon tetrachloride (Cruickshank, 1954) or wet heat using hot water (Bant and Storey, 1952) or steam–air mixtures (Baker, 1962) have been employed against some seed-borne fungi. The heat treatments have the disadvantages that they are not always completely effective and that depression of germination sometimes results from their use. For these reasons, Maude (1966a; 1966b) studied methods of introducing fungicides into the seed tissues by letting pea seeds infected with *Ascochyta* imbibe in water suspensions of thiram or captan. Complete disease eradication occurred if the suspensions were maintained at 30°C for 24 h, and when dried after treatment the seeds germinated normally. The method finally adopted for use consisted of soaking seeds in a 0·2 per cent water suspension of thiram at 30°C for 24 h, after which they were dried in a rapid air stream at 25°C (Maude and Keyworth, 1967). This treatment was found to be effective against over twenty seed-borne fungal pathogens (Maude, Vizor and Shuring, 1969—see *Table 18.2*). Because of the soaking and drying involved, the method has proved of commercial use mainly against pathogens of small, high value seeds such as *Septoria apiicola* on celery, *Phoma lingam* and *Alternaria brassicicola* on brassicas and *Phoma betae* on red beet, but is not commercially practicable for the treatment of bulky seeds such as peas and beans. The disinfection achieved by thiram soaking results from the uptake of the fungicide in solution (30 ppm) by the seed tissues in the course of imbibition. The treatment eradicates not only internal and external pathogens but also contaminant fungi, and thiram from the suspension which remains as a dry dressing on the surfaces of seeds after drying, gives protection against soil-borne fungi. It has been found that many other fungicides have the same effect on internal pathogens but many damage the seeds. It is probable that the low solubility of thiram in water ensures that although sufficient fungicide is

Table 18.2 SEED-BORNE DISEASES CONTROLLED BY THE THIRAM SOAK TREATMENT

Host	Disease	Pathogen
Peas	Leaf and pod spot	*Ascochyta pisi*
Peas	Foot rot	*Mycosphaerella pinodes*
Celery	Leaf spot	*Septoria apiicola*
Celery	Root rot	*Phoma apiicola*
Brassicas	Canker	*Phoma lingam*
Brassicas	Dark leaf spot	*Alternaria brassicicola*
Brassicas	Leaf spot	*Alternaria raphani*
Red beet	Black leg	*Phoma betae*
Red beet	Anthracnose	*Colletotrichum spinaceae*
Carrot	Leaf blight	*Alternaria dauci*
Carrot	Black rot	*Stemphylium radicinum*
Trefoil	Spring black stem	*Ascochyta imperfecta*
Flax	Grey mould	*Botrytis cinerea*
Oats	Leaf spot	*Pyrenophora avenae*
Wheat	Glume blotch	*Septoria nodorum*
Wheat	Bunt	*Tilletia caries*
Lobelia	Damping-off	*Alternaria spp.*
Zinnia	Blight	*Alternaria zinniae*

taken into seeds to control fungal infection this amount is insufficient to harm germination. The germination of red beet seed is actually stimulated by the treatment. This is due to the washing out of a germination inhibitor from the fruit coats of beet combined with the eradication of *Phoma betae* from the seeds (*Table 18.3*). Some *Brassica* seed lots have, however, shown a germination loss as a result of thiram soaking. This has occurred with seed samples which have been removed from temperature- and humidity-controlled stores

Table 18.3 STIMULATION OF SEEDLING EMERGENCE IN
FIELD SOIL BY THIRAM SOAKING OF *PHOMA BETAE*
INFECTED RED BEET SEEDS

Seeds soaked in	Emergence (*seedlings/1 000 'seeds'*)
0·2% thiram for 24 h at 30°C	1 027
Water for 24 h at 30°C	552
Untreated seeds	290

after 6–24 months' storage and appears to be an entirely random effect which is difficult to explain.

Although the thiram soak treatment was available for the control of a variety of fungal pathogens of small seeds, there was, until the late 1960s, no practicable method for use against similarly infected samples of large seeds. A treatment was needed which allowed deep fungicide penetration of the seed system, as did thiram soaking, but did not require the soaking and subsequent drying of the seeds. With the development of systemic fungicides this need has been largely fulfilled. The discovery, in 1966, by von Schmeling and Kulka that dry dressings of the systemic fungicide carboxin applied to the surface of barley grains could rid their embryos of the loose smut fungus (*Ustilago nuda*), was strong presumptive evidence of the penetration of this fungicide or a fungitoxic by-product deep into the seed tissues. Unfortunately, carboxin proved to be effective almost exclusively against Basidiomycete fungi and was of no use for the treatment of the majority of fungal seed-borne pathogens of vegetables, many of which are caused by Ascomycetes and *Fungi Imperfecti*.

Table 18.4 PEA SEED DRESSING EXPERIMENT WITH BENOMYL/CAPTAN FUNGICIDE MIXTURES

Fungicide	Application rate (g a.i./kg seed)	Percentage Ascochyta pisi infection		Percentage germination	
		Laboratory	Field	Laboratory	Field
Nil	—	25·6	5·7	74·8	38·9
Captan	0·625	7·5	0·4	85·0	59·0*
Benomyl	1·875	0	0	83·0	40·8
Benomyl + captan	1·875 + 0·625	0	0	84·0	60·5*
LSD (P = 0·05)		3·0	0·9	6·6	—

* Results significantly better than the nil treatment (analysed as angles)

The development of other systemic compounds, particularly the benzimidazole fungicides, benomyl (Delp and Klopping, 1968) and thiabendazole (Weinke *et al.*, 1969) which are toxic to many more of these fungi (Edgington, Khew and Barron 1971; Bollen and Fuchs, 1970) has made the treatment of many vegetable seed diseases possible. As a result a number of those fungi which cause losses to crops raised from large seeds such as peas and beans can now be treated by dressings of systemic fungicides.

It appears that these, when applied dry to the surface of seeds, control internal pathogens by penetrating deeply within the seeds after entering the seed tissues when these imbibe water. This is in contrast to the mode of action of non-systemic seed dressings, e.g. thiram and captan, where there is little evidence of tissue penetration unless a 24 h fungicidal seed soak is used, e.g. during thiram soaking.

Seed surface applications of the systemic benzimidazole fungicides benomyl (Maude and Kyle, 1970) or thiabendazole (Maude and Kyle, 1971) have eliminated *Ascochyta pisi* from seedlings raised from treated infected peas. Dry or slurry dressings have been effective. Slurry dressings of these fungicides also rid field beans of *Ascochyta fabae*, and similar dressings of both fungicides and of methyl thiophanate can control *Colletotrichum lindemuthianum* (anthracnose) infections in French bean seeds (Maude *et al.*, 1972).

Not all fungi are as easily controlled, however, and *Mycosphaerella pinodes*, responsible for foot rot of peas, is not eliminated from seeds by dressings of benomyl or thiabendazole. Although lack of control of *Mycosphaerella pinodes* in infected pea seed samples may in part be due to the mode of transmission of the fungus, it also reflects the specificity of these new systemic compounds. The benzimidazoles, in particular, though of wider toxicity than most systemic fungicides, are still very limited in this respect and though effectively inhibiting the growth of *Colletotrichum lindemuthianum* and *Phoma lingam* they have little effect, even at high concentrations, on *Colletotrichum spinaceae* and *Phoma betae*. Furthermore, most systemic fungicides are not toxic to fungi in the genera *Alternaria* and *Stemphylium*. Thus, seed-borne diseases of carrots caused by *Alternaria dauci* and *Stemphylium radicinum* and of brassicas caused by *Alternaria brassicicola* cannot be controlled with these compounds. A further problem is that although the thiram soak method adds thiram to the seed surface and is effective against soil-borne damping-off fungi, the systemic fungicides do not have this effect. Thus, peas dressed with benomyl (*Table 18.4*), although rid of *Ascochyta pisi*, are not protected from the Phycomycete fungi which cause them to damp-off in the soil. Such fungi, however, can be combated by mixing non-systemic fungicides, e.g. captan or thiram, with benomyl (Maude and Kyle, 1971), and a commercial formulation of benomyl with thiram (Benlate T) is now available for the same purpose.

Thus, although there are limitations, the use of systemic fungicides as seed dressings offers a simple and practical solution to certain previously intractable diseases of large seeds. What is clearly required, however, is a fungicide which is not only systemic but also has the fungitoxic range of activity of a dithiocarbamate compound such as thiram.

REFERENCES

BAKER, K. F. (1962). 'Thermotherapy of planting material', *Phytopathology*, **52**, 1244

BANT, J. H. and STOREY, I. F. (1952). 'Hot-water treatment of celery seed in Lancashire', *Pl. Path.*, **1**, 81

BOLLEN, G. J. and FUCHS, A. (1970). 'On the specificity of the *in vitro* and *in vivo* antifungal activity of benomyl to *Phoma betae*', *Neth. J. Pl. Path.*, **76**, 299

BROADBENT, L. (1965). 'The epidemiology of tomato mosaic XI. Seed-transmission of TMV', *Ann. appl. Biol.*, **56**, 177

BUTLER, E. J. and JONES, S. G. (1949). *Plant Pathology*, Macmillan, London

CHUPP, C. and SHERF, A. F. (1960). *Vegetable diseases and their control*, Constable, London

CRUICKSHANK, I. A. M. (1954). 'Thermo-chemical seed treatment', *Nature*, **173**, 217

DELP, C. J. and KLOPPING, H. L. (1968). 'Performance attributes of a new fungicide and mite ovicide candidate', *Pl. Dis. Reptr.*, **52**, 95

DYKSTRA, T. P. (1961). 'Production of disease free seed', *Bot. Rev.*, **27**, 445

EDGINGTON, L. V., KHEW, K. L. and BARRON, G. L. (1971). 'Fungitoxic spectrum of benzimidazole compounds', *Phytopathology*, **61**, 42

GUTHRIE, J. W., HUBER, D. M. and FENWICK, H. S. (1965). 'Serological detection of halo blight', *Pl. Dis. Reptr.*, **51**, 544

HAWKINS, J. H., IVES, J. V. and STOREY, I. F. (1959). 'New or uncommon plant diseases and pests, *Alternaria* leaf blight of carrots', *Pl. Path.*, **8**, 76

HEWETT, P. D. (1964). 'Testing carrot seed infected with *Alternaria porri* f. sp. *dauci*', *Proc. int. Seed Test. Ass.*, **29**, 463

HEWETT, P. D. (1966). '*Ascochyta fabae* Speg. on tick bean seed', *Pl. Path.*, **15**, 161

HEWETT, P. D. (1968). 'Viable *Septoria* spp. in celery seed samples', *Ann. appl. Biol.*, **61**, 89

KEYWORTH, W. G. and HOWELL, J. S. (1961). 'Studies on silvering disease of red beet', *Ann. appl. Biol.*, **49**, 173

LLOYD, H. B. (1959). 'The transmission of *Phoma lingam* (Tode) Desm. in the seeds of Swede, Turnip, Chou Moellier, Rape and Kale', *N.Z. J. agric. Res.*, **2**, 649

MAUDE, R. B. (1964). 'Studies on *Septoria* on celery seed', *Ann. appl. Biol.*, **54**, 313

MAUDE, R. B. (1966a). 'Pea seed infection by *Mycosphaerella pinodes* and *Ascochyta pisi* and its control by seed soaks in thiram and captan suspensions', *Ann. appl. Biol.*, **57**, 193

MAUDE, R. B. (1966b). 'Studies on the etiology of black rot, *Stemphylium radicinum* (Meier, Drechsl and Eddy) Neerg., and leaf blight, *Alternaria dauci* (Kühn) Groves & Skolko on carrot crops; and on fungicide control of their seed-borne infection phases', *Ann. appl. Biol.*, **57**, 83

MAUDE, R. B. and KEYWORTH, W. G. (1967). 'A new method for the control of seed-borne fungal disease', *Seed Trade Rev.*, **19**, 202

MAUDE, R. B. and KYLE, A. M. (1970). 'Seed treatments with benomyl and other fungi-cides for the control of *Ascochyta pisi* of peas', *Ann. appl. Biol.*, **66**, 37

MAUDE, R. B. and KYLE, A. M. (1971). 'Leaf and pod spot (*Ascochyta pisi*) of peas', *Rep. natn. Veg. Res. Stn* (1970), 106

MAUDE, R. B., KYLE, A. M., MOULE, C. G. and DUDLEY, C. (1972). 'Anthracnose (*Colleto-trichum lindemuthianum*) of French beans'. *Rep. natn. Veg. Res. Stn* (1971), 75

MAUDE, R. B. and MOULE, C. G. (1972). 'Black rot (*Stemphylium radicinum*) of carrots', *Rep. natn. Veg. Res. Stn* (1971) (in press)

MAUDE, R. B. and SHURING, C. G. (1970). 'The persistence of *Septoria apiicola* on diseased celery debris in soil', *Pl. Path.*, **19**, 177

MAUDE, R. B., VIZOR, A. S. and SHURING, C. G. (1969). 'The control of fungal seed-borne diseases by means of a thiram seed soak', *Ann. appl. Biol.*, **64**, 245

TAYLOR, J. D. (1970). 'The quantitative estimation of the infection of bean seed with *Pseudomonas phaseolicola* (Burkh.) Dowson', *Ann. appl. Biol.*, **66**, 29

TOMLINSON, J. A. (1962). 'Control of lettuce mosaic by the use of healthy seed', *Pl. Path.*, **11**, 61

VON SCHMELING, B. and KULKA, M. (1966). 'Systemic fungicidal activity of 1,4 oxathiin derivatives', *Science N.Y.*, **152**, 125

WALKER, J. C. and PATEL, P. N. (1964). 'Splash dispersal and wind as factors in epi-demiology of halo-blight of bean', *Phytopathology*, **54**, 140

WEINKE, K. E., LAUBER, J. J., GREENWALD, B. W. and PREISER, F. A. (1969). 'Thiabend-azole, a new systemic fungicide', *Proc. 5th Br. Insectic. Fungic. Conf.*, 340

WHARTON, A. L. (1967). Sources of infection by physiological races of *Pseudomonas phaseolicola* (Burkh.) Dowson in dwarf beans', *Pl. Path.*, **16**, 27

WILSON, R. D. (1947). 'Rainfall in relation to the production of bean seed free of the bacterial blight diseases', *Agric. Gaz. N.S.W.*, **58**. 15

ZINK. F. W., GROGAN, R. G. and WELCH, J. E. (1956). 'The effect of the percentage of seed transmission upon subsequent spread of lettuce mosaic virus', *Phytopathology*, **46**,662

DISCUSSIONS

The discussions for Chapter 18 are combined with those at the end of Chapter 19.

19

SAPROPHYTIC FUNGI AND SEEDS

G. J. F. PUGH

Department of Botany, University of Nottingham

INTRODUCTION

A seed may be defined as the product of the ovule after fertilisation, containing the embryo and surrounded by a testa (simplified after Willis, 1919). In the Angiosperms, seeds are produced inside ovaries and in many plants the propagule consists of the true seed plus the ovary wall, so that many commercial 'seeds' are strictly fruits. This situation occurs, for instance, in the cereals, in beet, carrots, lettuce and spinach (see Boswell, 1961). It also occurs in non-crop plants such as sycamore and *Salsola kali*, while in *Halimione portulacoides* there are also fleshy bracts surrounding and attached to the fruit.

When considering the role of fungi on 'seeds' this botanical distinction is important, because many common fungi may arrive on the surface of a fruit. Of these, only a comparatively small number are able to survive and grow there. Growth on the true seeds before they are dispersed is limited to those micro-organisms which can penetrate the ovary wall and grow to the seed. Samish, Tulczynska and Bick (1961) have pointed out that although the tissues of normal, healthy, undamaged fruits are generally considered to be sterile, there is little proof that this is so. Their own work showed that bacteria could frequently be isolated from within healthy tomatoes, and they discussed earlier work which showed the presence of bacteria in healthy cucumbers (Samish and Dimant, 1959). While various plant pathogenic fungi are well known on seeds, other species which are normally regarded as saprophytic can be regularly isolated from dispersal units, such as the fruits of sycamore, and also from the seed within the pericarp (Pugh and Buckley, 1971).

The purpose of this paper is to review briefly the presence of fungi and other micro-organisms in seeds *sensu lato*, and to examine the possible interactions between the micro-organisms and the seeds and seedlings.

FUNGI ON THE SEEDS

The fungi which can normally be found on seeds can, for convenience, be divided into three main categories (see *Table 19.1*). These broadly represent the predominant characteristics of each group, but are not mutually exclusive.

PLANT PATHOGENIC SPECIES

These fungi are dealt with in detail in Chapter 18. They can be separated into those fungi which cause immediately obvious infections on the plant and those which produce systemic diseases and which only become apparent when the plant is mature, such as the smut diseases.

FIELD FUNGI

These species can readily be isolated from cereal grains and many plant propagules in the field. They can be divided into the transient or casual contaminants, which are of little importance, and the members of the phylloplane flora, which are micro-organisms adapted to survive and grow on the surfaces of living aerial parts of plants. They are best known as leaf-surface micro-organisms, but fruits and leaves have often been shown to have a common flora, consisting of both fungi and bacteria. Amongst the filamentous fungi which are common on leaves, *Alternaria* spp., *Aureobasidium pullulans*, *Cladosporium* spp., *Epicoccum purpurascens* (syn. *E. nigrum*) and *Fusarium* spp. can be isolated easily and frequently from fruits and grains (see Gambogi, 1960). Similarly, yeasts such as species of *Candida, Cryptococcus, Rhodotorula, Sporobolomyces*, and *Torulopsis* (see Carmo-Sousa, 1969) and a range of bacteria, including species of *Pseudobacterium, Pseudomonas* and *Sarcina* (see Klincare, Kreslina and Mishke, 1971), can be found on both fruits and leaves.

The field fungi, together with the bacteria, are present on the cereal grain and other 'seeds' at harvest, or at dispersal in the case of non-crop plants. Many of the fungi can be isolated from freshly collected

Table 19.1 SOME EXAMPLES OF THE POTENTIAL EFFECTS OF FUNGI ON SEEDS

Fungus	Effect on plant	Effect on seedling	Effect on man and animals
Plant pathogenic species			
Claviceps purpurea	Ergot	No viable seed	St. Anthony's Fire
Field fungi			
Alternaria tenuis	Phylloplane fungus		
Cladosporium herbarum	Phylloplane fungus	Auxin producer	Allergy in man
Aureobasidium pullulans	Phylloplane fungus	Auxin producer	Allergy in man
Epicoccum purpurascens	Phylloplane fungus	Auxin producer	
Sporobolomyces roseus	Phylloplane fungus	Auxin producer	
Fusarium spp.	Some root pathogens		Can cause infertility in cattle
Storage fungi			
Absidia corymbifera	Detericration of grain		Mycotic infections
Aspergillus flavus	Deterioration of grain	Enhances germination	Liver disease
Aspergillus fumigatus	Deterioration of grain	Enhances germination	Pulmonary infection
Penicillium cyclopium	Deterioration of grain	Enhances germination	Nervous disorder and death of sheep
Penicillium citrinum	Deterioration of grain		Kidney disease

seeds both immediately after dispersal and later throughout the winter.

During the overwinter storage of cereal grain the incidence of the field fungi gradually diminishes, and the storage fungi become more abundant. Thus, Mulinge and Chesters (1970) found that after two months of storage of barley grain, the fungi present were mainly species of *Absidia*, *Aspergillus* and *Penicillium*. The most abundant species of these genera were *Absidia corymbifera*, *Aspergillus flavus*, *A. fumigatus*, *A. versicolor*, *Penicillium cyclopium*, *P. decumbens*, *P. expansum*, *P. frequentans* and *P. roqueforti*.

Warnock (1971) studied the extent of the mycelium of *Penicillium cyclopium* in barley grains and estimated that the dry weight of mycelium of this species was of the order of 1·5 to 1·9 μg in individual grains. Other seeds undergoing storage also become contaminated by a similar range of fungi and sometimes by the same species. Jackson (1965), for example, discussed the role of *Aspergillus flavus*, *A. niger*, *Rhizopus stolonifer* and *Sclerotium bataticola* on peanuts.

Another group of fungi can be isolated from stored grain which is incubated at 37°C and 45°C. Mulinge and Apinis (1969) isolated a range of thermophilic species from barley grains, including species of *Absidia*, *Aspergillus* and *Monascus*, *Dactylomyces crustaceus*, *Eurotium amstelodami* and *Mucor pusillus*. These fungi were mostly present in the husk early in storage, but after long storage, and particularly if self-heating occurred, they invaded the grain tissues.

The growth of fungi on grain before and during storage must lead to a reduction in the value of the crop by loss of weight and by deterioration of the quality of the grain. More importantly, this contamination may result in a health hazard to man and animals in three ways.

1. Toxic metabolites are produced by many of the storage fungi (e.g. Aflatoxin by *Aspergillus flavus*; the F2 factor which causes infertility in cattle and other animals by *Fusarium* spp., especially *F. graminearum* and *F. moniliforme*).
2. Fungal spores are well known as allergens, and species of *Alternaria* and *Cladosporium* are particularly associated with allergic responses in man.
3. Mycotic diseases may be caused by *Aspergillus fumigatus* (pulmonary aspergillosis), while *Mucor pusillus* and species of *Absidia*

have been shown to cause ulcers and even death in young pigs (Gitter and Austwick, 1959).

ROLE OF FUNGI ON SEEDS

SAPROPHYTIC ACTIVITY

Many of the fungi present on seeds are known to be cellulose decomposers (species of *Alternaria*, *Cladosporium*, *Cephalosporium*, *Epicoccum* and *Fusarium*, for example), while *Aureobasidium pullulans* produces pectinolytic enzymes (Smit and Wieringa, 1953). Collectively, these saprophytic fungi help in the breakdown of the pericarp and testa. They may also exert some influence on potentially pathogenic fungi; thus Tveit and Wood (1955) have demonstrated that *Chaetomium cochloides* and *C. globosum*, isolated from oat seeds, are antagonistic to species of *Fusarium* and particularly to *F. nivale*.

EFFECT OF METABOLITES

A number of workers have shown that micro-organisms present on seeds may produce metabolites which stimulate or inhibit the germination of seeds and the subsequent growth of seedlings. Krasil'nikov (1958) separated soil organisms into 'microbial activators' and 'microbial inhibitors'. The former group consisted mainly of nodule bacteria and mycorrhizal fungi, but he also included various free-living bacteria, filamentous fungi and yeasts which can cause stimulation of growth. His account of microbial inhibitors is mainly concerned with bacteria, although he mentioned toxins produced by *Fusarium*, and actinomycete suppression of seed germination.

More recently, Abdulla (1970) has shown that exudates from seed-borne fungi—both field and storage species—can either stimulate or reduce the amount of germination of oats. Exudates from *Alternaria tenuis*, *Cladosporium herbarum* and *Fusarium culmorum* tended to show no effect, or a slight reduction in germination. On the other hand, those of *Epicoccum purpurascens*, *Penicillium cyclopium*, *P. expansum*, *Aspergillus flavus*, *Eurotium amstelodami*, *E. chevaliera*, *E. repens* and *E. rubrum* normally showed a marked stimulation of germination under most of the test conditions. Abdulla grew his test fungi in Czapek's solution and in potato dextrose solution, but he did not analyse the growth factor. However, Valadon and Lodge (1970), working with *Cladosporium herbarum*, showed that this species could produce indole-3-acetic acid (IAA) and indole-3-acetonitrile (IAN) as well as other growth substances. The same auxins were reported by Buckley and Pugh (1971) to be produced by both *Aureobasidium pullulans* and *Epicoccum purpurascens*. The use of tryptophane as nitrogen source produced a marked effect by the culture filtrates on the growth of *Avena* coleoptiles. Other phylloplane organisms have also been investigated: Diem (1971) showed the production of IAA by two yeasts, *Candida muscorum* and *Sporobolomyces roseus*; Klincare, Kreslina and Mishke (1971) demonstrated that epiphytic bacteria could produce auxins, gibberellin-like substances and various vitamins. Inoculation of barley seeds with these bacteria enhanced germination and aided the growth and development of the plants.

Pugh and Buckley (1971) found A. *pullulans* in all freshly collected sycamore fruits. It can be seen inside the hairs which line the pericarp

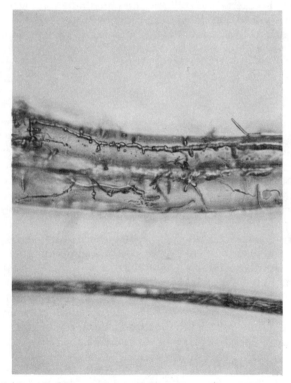

Figure 19.1   Aureobasidium pullulans *visible inside the hairs which line the pericarp of* Acer pseudoplatanus

(*Figure 19.1*) and it can be isolated from surface-sterilised seeds. In order to kill the endogenous fungus and then assess the growth of

Acer pseudoplatanus in the absence of the fungus, seeds collected in January, 1972 were given a heat treatment (immersed in water kept at 54°C for 10 min). Subsequent growth of the seedlings in sterile soil was noted and compared with untreated controls. Germination

Figure 19.2   Germination of seeds of Acer pseudoplatanus *in sterile soil, after being collected in January, 1972. The treated seeds (on the right) were immersed in water at 54°C for 10 min to kill the endogenous* Aureobasidium pullulans *(control plants on the left)*

of the treated seeds was delayed, and subsequent growth was retarded compared with the controls (*Figure 19.2*). When this was repeated with seeds collected on the point of germination in March, 1972, there was no longer any obvious difference between the treated and control seedlings. While much more work needs to be done, it appears that the January heat treatment may either have had a direct effect on germination, or, possibly, that it had an indirect effect through killing *Aureobasidium pullulans*. The lack of difference produced by the March treatment may imply either that the embryo after-ripened between January and March or that the substances required for germination and supplied by the fungus were now present and were not affected by the treatment.

Once the seed has germinated, the seedling becomes colonised by a range of fungi and other micro-organisms, some of which may have been stimulated by exudates from the developing seed (see Schroth and Snyder, 1961). The ability of many of the organisms to act as 'microbial activators' in the laboratory indicates that the potential is there for stimulation in the field. It is also known that leachates from leaves may contain auxins, and that leachates can be

344   SEED ECOLOGY

absorbed by both leaves and roots (Tukey, 1971). Thus, not only
the saprophytic fungi on the propagule, but also those on the leaves
of trees and other plants in the neighbourhood, may play a part in
enhancing the germination of seeds in nature.

ACKNOWLEDGEMENT

I would like to thank Dr. N. G. Buckley for allowing me to use the
photograph in *Figure 19.1.*

REFERENCES

ABDULLA, M. H. (1970). 'Preliminary study on the influence of fungal metabolites on
germination of barley grains', *Mycopath. Mycol. appl.* **41**, 307–313
BOSWELL, V. R. (1961). 'What seeds are and do: an introduction', in *Seeds, The Yearbook
of Agriculture*, 1–10 U.S.D.A., Washington D.C.
BUCKLEY, N. G. and PUGH, G. J. F. (1971). 'Auxin production by phylloplane fungi',
Nature, Lond., **231**, No 5301, 332
CARMO-SOUSA, L. D. (1969). 'Distribution of yeasts in nature', in *The Yeasts*, 79–105,
Academic Press, London
DIEM, H. G. (1971). 'Production de l'acide indolyl-3-acétique par certaines levures
epiphylles', *C. r. hebd Séanc Acad. Sci.*, **272**, 941–943
GAMBOGI, P. (1960). 'Seconda nota sulla microflora presente nelle cariossidi di *Avena*,
Annali. Fac. Agr. Univ. Pisa., **21**, 3–17
GITTER, M. and AUSTWICK, P. K. C. (1959). 'Mucormycosis and moniliasis in a litter of
suckling pigs', *Vet. Rec.*, **71**, 6–11
JACKSON, C. R. (1965). 'Peanut kernel infection and growth *in vitro* by four fungi at
various temperatures', *Phytopathology*, **55**, 46–48
KLINCARE, A. A., KRESLINA, D. J. and MISHKE, I. V. (1971). 'Composition and activity of
the epiphytic microflora of some agricultural plants', in *Ecology of Leaf-Surface
Micro-organisms*, 191–201, Academic Press, London
KRASIL'NIKOV, N. A. (1958). *Soil Micro-organisms and Higher Plants*, Academy of Sciences
of the U.S.S.R. Moscow
MULINGE, S. K. and APINIS, A. E. (1969). 'Occurrence of thermophilous fungi in stored
moist barley grains', *Trans. Br. mycol. Soc.*, **53**, 361–370
MULINGE, S. K. and CHESTERS, C. G. C. (1970). 'Ecology of fungi associated with moist
stored barley grain', *Ann. appl. Biol.*, **65**, 277–284
PUGH, G. J. F. and BUCKLEY, N. G. (1971). '*Aureobasidium pullulans*: an endophyte in
sycamore and other trees,' *Trans Br. mycol. Soc.*, **57**, 227–231
SAMISH, Z. and DIMANT, D. (1959). 'Bacterial population in fresh, healthy cucumbers',
Fd Mf., **34**, 17–20
SAMISH, Z., TULCZYNSKA, R. E. and BICK, M. (1961). 'Microflora within healthy tomatoes',
Appl. Microbiol., **9**, 20–25
SCHROTH, M. N. and SNYDER, W. C. (1961). 'Effect of host exudates on chlamydospore
germination of the bean root rot fungus, *Fusarium solani, F. phaseoli.*', *Phytopathology*,
51, 389–393
SMIT, J. and WIERINGA, K. T. (1953). 'Microbiological decomposition of litter', *Nature*,
Lond., **171**, 794–795

TUKEY, H. B. JNR (1971). 'Leaching of substances from plants', in *Ecology of Leaf Surface Micro-organisms*, 67–80, Academic Press, London

TVEIT, M. and WOOD, R. K. S. (1955). 'The control of *Fusarium* blight in oat seedlings with antagonistic species of *Chaetomium*', *Ann. appl. Biol.*, **43**, 538–552

VALADON, L. R. G. and LODGE, E. (1970). 'Auxins and other compounds of *Cladosporium herbarum*', *Trans. Br. mycol. Soc.*, **55**, 9–15

WARNOCK, D. W. (1971). 'Assay of fungal mycelium in grains of barley, including the use of the fluorescent antibody technique for individual fungal species', *J. gen. Microbiol.*, **67**, 197–222

WILLIS, J. C. (1919). *A Dictionary of the Flowering Plants and Ferns*, Cambridge University Press

DISCUSSIONS

E. H. Roberts: It is well known that hot water treatment can delay the germination and ultimately decrease the viability of seeds. In Dr. Pugh's experiment on sycamore seeds the direct effect of the heat treatment on the seeds is inextricably bound up with the effect of the heat on the fungus.

Pugh: Only seeds collected in January did not germinate normally; on those collected in March the heat treatment did not have any detrimental effects. Abdalla's work, which showed an increase in germination of barley with some culture filtrates, did not make use of any heat treatment.

E. H. Roberts: Sycamore seeds have a long after-ripening period and this may well account for the greater resistance of the seeds in the hot water treatment at the later date; so the role of the fungus is still not proven.

Pugh: I entirely agree that it is not proven in sycamore. I feel that the whole question of the effect of fungal metabolites on seed germination needs to be studied more deeply.

Hegarty: Is the occasional depression of germination, experienced after the warm water thiram soak, due to germination of seeds before they have completely dried back?

Maude: No, in brassica seeds, where some radicles occasionally emerge by the 24th hour of soaking, the proportion is so small that, in the seed trade's point of view, it makes no difference. Premature germination during soaking is not the primary cause for the depression of germination. Where it does occur, the depression is usually associated with a long storage period after treatment.

20

TETRAZOLIUM STAINING FOR ASSESSING SEED QUALITY

R. P. MOORE

Department of Crop Science, North Carolina State University, U.S.A.

INTRODUCTION

The seed production and marketing industries, educational establishments and legal control agencies frequently need a method that yields faster and more comprehensive results concerning seed quality than those provided by germination or growth tests. The tetrazolium test has been developed and refined during the past 30 years to fulfil some of these needs. The test is sound and can serve many useful purposes.

BASIS OF TEST

The tetrazolium test makes use of a colourless testing solution that produces colour differences between normal, weak and dead embryo tissues. These colour differences, together with other insights, permit an assessment of the presence, location and nature of disturbances within embryo tissues. From these assessments an analyst can obtain a rapid estimate of potential germination capacity and embryo soundness.

The reaction of tetrazolium molecules with hydrogen atoms released by the dehydrogenase enzymes which are involved in the respiration processes of living tissues, results in the production of a water-insoluble, oil-soluble red pigment, formazan. The staining reaction is as follows:

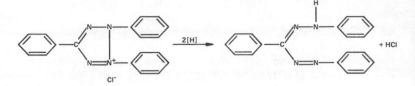

Various aspects of the test are presented in considerable detail in a handbook edited by Grabe (1970).

TEST PROCEDURE

TESTING SOLUTION

The testing solution is prepared by dissolving the tetrazolium salt (TTC) in distilled or tap water of, approximately, pH 6·5–7·0. Solutions of 0·1–1·0 per cent are commonly used. The unused portion of a stock solution will remain effective for several months when protected from excessive light, but the solution should only be used once.

EQUIPMENT

The equipment needed to attain the most consistent and accurate results includes a temperature-controlled cabinet running at between 25°C and 30°C, in which the seeds are moistened and stained, a refrigerated environment capable of maintaining 5°C for the preservation of stained samples, apparatus to provide magnification of approximately 5–15 ×, and a good light. Miscellaneous equipment includes dissecting needles, razor blades, tweezers, petri dishes and beakers.

SAMPLE SIZE

Since the testing environment can be precisely controlled, the need for large samples is not as great as in growth tests when 4 × 100 seeds are commonly used (ISTA, 1966). Representative samples of 50–100 seeds are usually adequate for practical tests. A 25-seed sample will often suffice. Random variations and the precision of results desired are the major concerns in selecting sample size.

MOISTENING OF SEEDS

Unless already moist, seeds should ordinarily be slowly and fully moistened with water before staining. This preliminary moistening activates metabolic activities and conditions embryonic tissues for an even absorption of the tetrazolium solution. It also minimises the occurrence of undesirable injuries during seed preparation. However, unless dry and brittle, many kinds of seeds, except large-seeded legumes, can be soaked by immersion in warm water without serious consequences. Decisions as to acceptable short-cuts in moistening will come with experience (Moore, 1970a).

PREPARATION FOR TETRAZOLIUM ABSORPTION

To assure satisfactory absorption of the TTC solution, physical barriers to penetration must be overcome (Moore, 1964; 1970a). This is accomplished by cutting, puncturing or removal of seed coats or by exposing cut surfaces of embryos. Cuts should preferably be made with a single sliding stroke of a sharp razor or other suitable blade. A sliding cut reduces pressure damage and ensures a smooth surface that facilitates evaluation. Certain kinds of seed absorb tetrazolium adequately without special preparation.

When suitably moistened, most kinds of seed can be adequately prepared by one or more of the following methods, or by slight modifications of these methods:

Method 1 *Moistening or preparation unnecessary*
Small-seeded legumes: alfalfa—*Medicago sativa,* clovers—*Trifolium* spp., and *Lespedeza* spp.
Seed coats can be nicked to break hardseededness. Hardseededness can also be reduced by brief alternate exposures to hot and cold water. Although not necessary, pre-moistening of the seeds can improve staining quality, especially if the seeds are aged.

Method 2 *No preparation except moistening*
Large-seeded legumes: beans—*Phaseolus* spp., peas—*Pisum sativum* and soya beans—*Glycine max.*
Staining can be accelerated by slitting the seed coats.

Method 3 *Puncturing or cutting pericarp and endosperm near top of embryos*
Small-seeded grasses: bent grasses—*Agrostis* spp., meadow grass (bluegrass)—*Poa* spp. and timothy—*Phleum pratense.*

Method 4 *Cut off cotyledon end of seed including tips of cotyledons*
Tobacco—*Nicotiana tabacum.*
To minimise pressure injury, cuts should be made with a sliding stroke of a sharp razor blade. The cut should be a

sufficient distance from the end of a seed to permit easy removal of embryos after staining.

Method 5   *Cut longitudinally through seed coats and into or through underlying cotyledons (or endosperm)*

Brassica spp., celery—*Apium graveolens*, endive—*Cichorium endivia*, lettuce—*Lactuca sativa*, French and African marigold—*Tagetes erecta*, and sesame—*Sesamum orientale*.

The cut should involve about three-quarters of the length of the seed and be at the cotyledon end. Guard against radicle and plumule injury.

Method 6   *Cut laterally entirely or partly through endosperm*

Small-seeded grasses: Fescues—*Festuca* spp., cocksfoot (orchard grass)—*Dactylis glomerata* and canary-grass—*Phalaris* spp.

Cut or dissect slightly beyond top of embryos.

Method 7   *Cut longitudinally through centre of embryo and endosperm*

Cereals, corn—*Zea mays*, large-seeded grasses and *Sorghum* spp. Expose leaves, scutellar node, roots, etc. For increased precision in evaluation, seed halves should be left connected by an uncut section of the pericarp on the side or at the end of the seed, away from the embryo.

Method 8   *Cut parallel to narrow dimension*

Egg plant—*Solanum melongena*, onion—*Allium cepa*, pepper—*Capsicum* spp. and tomato—*Lycopersicon esculentum*.

Keep seed halves attached by an uncut section to prevent separation of embryo parts.

Method 9   *Remove seed coats*

Beets—*Beta* spp., cotton—*Gossypium* spp., *Cucurbita* spp. and peanuts—*Arachis hypogaea*.

Avoid injuries during preparation to radicles and other highly important embryo parts. It is sometimes expedient to delay removal of coats until radicles enlarge within the partly opened seed coats.

STAINING

Adequate staining is accomplished by adding sufficient staining solution to ensure coverage of all seeds. Large imbibed seeds can be dropped directly into a beaker of solution as prepared. A sample of small seeds can be conveniently and efficiently moistened, prepared, stained, and evaluated on the same filter paper in the same petri dish. The filter pad can be folded over the prepared seeds to prevent them from floating when the TTC solution is added.

The staining time is dependent upon methods of seed preparation, kind of crop and temperature. Altering the temperature can alter the duration of staining required from approximately 30 min at 40°C to 24 h at 15°C. To make assessment possible it is only essential for the extent of staining to be sufficiently advanced to permit a clear distinction between normal, weak and dead tissues. In any case the duration of staining should not be prolonged to the extent that healthy tissues deteriorate as a result of the chemical treatment and the staining solution becomes discoloured. An experienced analyst can often detect normal, weak and dead embryo tissues shortly after the red colour starts to appear and before interior tissues have had time to stain completely.

When prolonged staining time is unavoidable, the beginning and advance of harmful effects can be delayed by the use of low temperatures or fungicides. Within the acceptable temperature range of approximately 15°C to 40°C, each 5°C reduction in temperature tends to cut the rate of deterioration in half, but also to double the required staining time. When time permits, staining at temperatures near the optimum for growth tests tends to improve differences in staining and to ease the evaluation of normal and weak tissues.

PRESERVATION OF STAINED SAMPLES

When adequately stained, seeds should be transferred to a refrigerator or a dark cool area to retard deterioration. A fungicide can be used for restricting infection and thus prolonging the reading quality of stained seeds. If refrigeration is used, it will not be necessary to replace the testing solution with water. Samples should, however, be evaluated within 24 h of staining whenever possible, and shortly after staining when temperature and facilities are not available for preservation of quality.

EVALUATION OF TESTS

Colour intensity and staining patterns are of major importance in evaluation, but not of sole importance. Correct evaluation requires a commonsense appraisal of the presence, location and nature of normal, weak, dead and fractured tissues. Condition and colour of tissues must sometimes be used jointly in evaluation. An analyst must determine by adequate examination of the individual embryo structures of each seed whether it possesses the minimum 'normality' of structures and soundness of tissues needed for the production of seedlings that could be regarded as 'germinable' under favourable growth test conditions. It is now necessary to recognise and interpret the tissue conditions revealed by individual staining patterns.

Requirements for precision in interpretation include:

1. An analyst endowed with common sense and a desire to make the test perform correctly.
2. Familiarity with embryo structures and with their relevance to seedling structures and seedling development.
3. Knowledge of differences between sound ('countable') and unsound ('non-countable') seedlings in growth tests.
4. Understanding of the nature, limitations, and use of TTC and growth tests, respectively, and of their results.
5. An appreciation of differences between potential germination percentages of a seed lot as revealed by TTC tests and actual germination percentages obtained from growth tests with and without fungicide treatment, and under conditions favourable and less favourable for germination.
6. Good light and magnification of approximately 5–15 ×.
7. Training and experience.

A single tetrazolium test can simultaneously and rapidly provide estimates for: (a) potential germination percentage, (b) appraisal of soundness of individual embryos (Moore, 1961; 1962) and (c) diagnosis of causes for embryo disturbances (Moore, 1966).

POTENTIAL GERMINATION PERCENTAGE

Potential germination percentage provides an estimate of seeds that are sufficiently sound to produce 'countable' seedlings in a favourable growth test. The appraisal is made by observing the soundness of each major embryo structure and the connections between structures. The soundness of the endosperm or perisperm is also important (Moore, 1964) where these tissues are present.

The determination of the germinable or non-germinable condition of a seed is based largely upon an appraisal of the presence, nature, and location of normal, weak, fractured, diseased and dead tissues and of abnormalities. By relating embryo conditions observed in TTC tests to seed and seedling conditions observed in favourable growth tests, an analyst can gradually learn to appreciate which embryo conditions lead to which seedling conditions.

Experienced analysts can frequently gain decisive information at a glance as to whether a TTC-treated seed is obviously germinable or not. When required, additional information concerning the nature of a disturbance can be quickly obtained by cutting additional tissues with sliding strokes of a razor blade.

With adequate light and magnification it is possible to evaluate with an acceptable degree of accuracy a high percentage of seeds of soya beans, peas, clovers, alfalfa and small-seeded grasses, without even removing the structures that cover the embryos. When pigments within seed coats or lemmas prevent clear vision, a few drops of lactophenol clearing solution can be used. The solution consists of 2, 2, 2, 4 parts, respectively, of lactic acid, phenol, water, and glycerin. Since phenol is toxic, the solution should be used with suitable precautions to avoid excessive inhalation or physical contact. Removal of pigmentation from the covering structure requires approximately 10–30 min.

Discrepancies may occur between the potential germination percentages of TTC tests and actual germination percentages in growth tests. These discrepancies do not necessarily indicate that either test is in error. In striving to understand why results of the two tests differ it is helpful to realise that inadequately controlled environmental conditions play an important role in growth test results. These results reflect an interaction between seed and environment and not a simple straightforward measurement of seed conditions. Especially with weak seed lots, a minor change in the environment (temperature or moisture) can alter growth test results, though clearly not influencing the initial germination potentialities of the seed lots. With weak seed lots also, the absence or ineffective use of a fungicide may cause a reduction in total germination. On the other hand, an excessive application of a phytotoxic fungicide likewise does not necessarily influence the potential germination capabilities of dry seeds, but it can prevent normal seedling development in growth tests when seeds become moistened and the chemical is not sufficiently diluted. Furthermore, it is important to realise that dormancy does not alter tetrazolium test results but can be of importance in growth tests.

An experienced analyst will agree that unsuspected errors creep into a growth test much more frequently than into TTC tests. Unsuspected sources of discrepancies between results of TTC and growth tests can frequently be traced by repeating growth tests. At least one commercial seed testing laboratory in America regularly conducts TTC tests on samples submitted for growth test so as to verify the validity of the growth test results, and to permit intelligent shortening of the duration of these tests.

EVALUATION OF EMBRYO SOUNDNESS

The degree of soundness of germinable embryos, which must be ascertained in the assessment of potential germination under field conditions, is determined largely from an appraisal of the presence, location and nature of normal, weak and dead areas, of missing tissues and of fractures. The extent and location of imperfections have a considerable influence upon the degree to which seed quality is impaired. Imperfections of a given dimension that involve or adjoin vital regions of the embryo must be considered more destructive than if they were present in less important parts of the embryo. As an example, a small necrosis or a fracture in a clover seed that penetrates partially into the stele near the attachment of the embryonic axis to the cotyledons is more disturbing to seed quality than a similar or even larger imperfection on the end of a cotyledon away from the radicle, or near the tip of the radicle.

A working knowledge of developmental patterns and germination behaviour of the kind of seed tested serves as a guide for evaluating the relative influence of specific disturbances on seed quality. Comparative tetrazolium and growth test studies can be helpful in developing and refining the ability to assess the quality of tetrazolium treated seeds (Moore. 1968a).

Experience reveals that embryo disturbances, unless minor, will cause seeds to be sensitive to adverse storage and germination conditions. Seeds with extensive or critically located disturbances tend to depend strongly upon a favourable environment for acceptable performance. Such seeds do not tolerate much diversity of environ-

Table 20.1 USE OF THE TETRAZOLIUM TEST AS A MULTIPLE-SCALE QUALITY EVALUATION SYSTEM FOR REVEALING CHANGES IN SOUNDNESS OF GERMINABLE EMBRYOS DURING STORAGE

| Crop | Seed lot | Relative humidity (%) | Tetrazolium ratings (1 = best) | | | | |
| | | | 1 | 2 | 3 | 4 | 5 |
			(Percentage of 100 germinable seeds)				
Maize	A	65	0	40	52	4	4
		78	0	0	56	40	4
		85	0	0	36	36	28
	B	65	0	8	64	28	0
		78	0	0	20	64	16
		85	0	0	8	36	56
Wheat	C	65	24	64	4	4	4
		78	4	52	28	0	16
		85	0	16	68	12	4
	D	65	36	63	1	0	0
		78	12	32	44	12	0
		85	8	32	20	36	4

ment either in storage (Moore, 1963a) or during germination and early seedling development. On the other hand, seeds with sound, or essentially sound, embryos tend to perform normally under a wide range of environmental conditions. Seed lots with a high percentage of sound embryos can be considered as being stable and thus more predictable in performance than the more 'labile' seed lots with a high percentage of deteriorated seeds.

Differences between high- and low-quality seed lots of similar germination percentages are frequently due to shifts in the percentage of seeds that are of sound, intermediate, and poor quality rather than to distinct differences in the nature of disturbances present. There are usually some high-quality seeds present in low-quality seed lots, and low-quality seeds present in high-quality lots. *Table 20.1* illustrates these differences according to a classification scheme commonly used by the writer. The seed lots were stored for 24 days at room temperature at three different levels of relative humidity in order to vary physiological ageing.

DIAGNOSING CAUSES OF EMBRYO DISTURBANCES

Differences observed in staining patterns and tissue soundness are the results of specific causes. The tetrazolium test is unexcelled for revealing the major causes for variations in symptoms (Moore, 1966; 1970b). Different seeds in a sample may well reveal symptoms originating from different causes. Each sample, however, tends to reflect the influence of one major, and perhaps one or more minor, causes. The most common disturbances vary widely with the crop and the regions of seed production. The initial symptoms of a specific factor are frequently altered by ageing and may be somewhat obscured by accelerated deterioration within, and outward from, the initially disturbed tissues.

Causes commonly at work in reducing seed soundness include mechanical and water damage, freeze, heat, ageing, insects and diseases. Recognition of major causes of disturbances can be helpful in quality evaluation and control programmes, and in obtaining new insights into seed life. The following brief comments will do no more than illustrate a few examples of the numerous symptoms encountered.

MECHANICAL INJURY

Mechanical injuries are revealed as fractures and/or bruises (*Figure 20.1c*). Injuries that appear somewhat similar can result from forces

Figure 20.1A   Tetrazolium staining of Phaseolus vulgaris—*natural crushing of plumule and radicle*

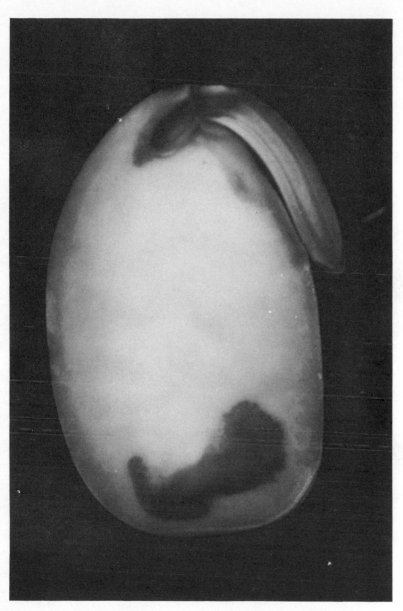

Figure 20.1B   Tetrazolium staining of Glycine max—*water damage*

N

358

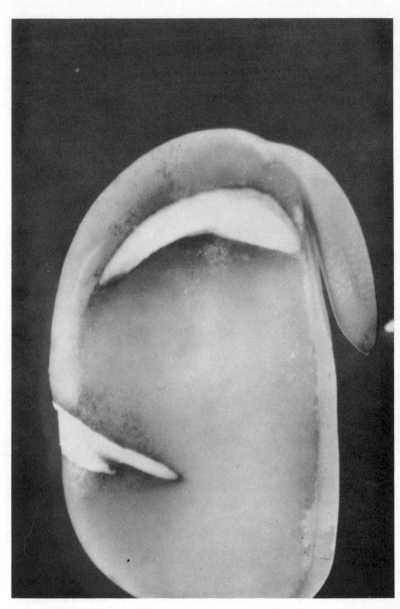

Figure 20.1C   Tetrazolium staining of Glycine max—mechanical damage

359

Figure 20.1D   Tetrazolium staining of Glycine max—*water damage and ageing*

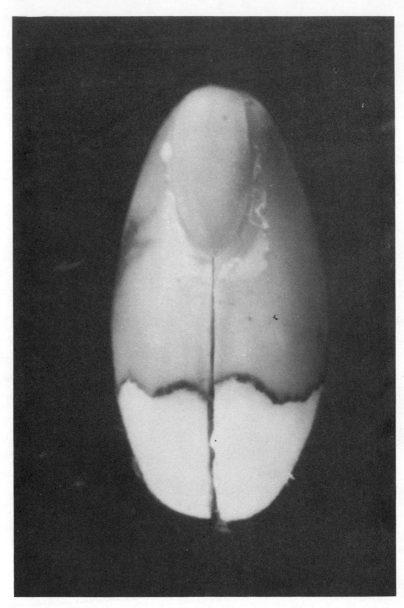

Figure 20.1E   Tetrazolium staining of Glycine max—*water damage and infection*

361

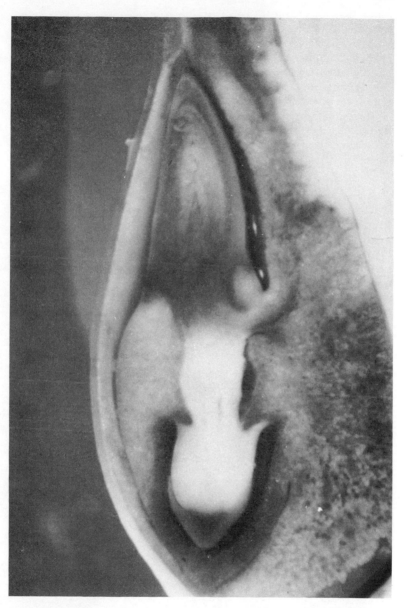

Figure 20.1F   Tetrazolium staining of Zea mays—*severe freeze injury*

other than mechanical. Water damage, especially in large-seeded legumes, results in symptoms that may tend to resemble mechanical injuries. Circumstantial evidence, when used along with direct evidence, usually provides the needed guidance for distinguishing between the two types of injury. An analyst must strive to visualise the circumstances that could possibly result in the nature of the symptoms observed.

Mechanical bruises of recent occurrence present a deep-red and liquid-logged appearance which changes on ageing to the white colour and flaccid appearance of dead tissues.

WATER DAMAGE

Water damage (*Figure 20.1B, D, E*) is especially prevalent in seeds of large-seeded legumes that are exposed to alternate wet and dry weather conditions while awaiting harvest after reaching maturity (Moore, 1965). Rapid uptake of water by localised embryo tissues caused by rain after the original drying phase results in various patterns of weakened and dead tissues on the embryo surfaces and internally. Fractures may also occur. These are common in wrinkled cultivars of peas and in certain varieties of snap beans (*Phaseolus vulgaris*) (*Figure 20.1A*).

Rapid swelling and wrinkling of seed coats upon initial moistening often account for various patterns of surface necroses and mottling. On being treated with TTC such disturbed areas, when of recent origin, tend to stain faster and darker red than normal ones. Given time for ageing, the disturbed tissues die and will then no longer stain. In addition, the areas of disturbance gradually enlarge and may eventually join together to form extensive patches of surface necroses.

Water that accumulates between cotyledons tends to move quickly by surface film attraction to open spaces in the vicinity of where the embryonic axis is attached to the cotyledons. Considerable disturb-

Figure 20.2   Tetrazolium stained embryos of wrinkled pea seeds showing water-disturbed tissues that may result in 'hollow heart'

ances can thereby be initiated within hypocotyls, epicotyls, cotyledon attachments, and plumules.

In wrinkled peas, especially those matured under water stress, rapid water uptake by the cotyledons causes fractures or cracks with special consequences. These fractures initiate the displacement of adjoining and nearby cells which often die (*Figure 20.2*). The discoloured mid-sections reflect displaced groups of cells. This displacement of cells results in accelerated deterioration and early death. Such seeds may produce seedlings with the 'hollow heart' abnormality of cotyledons. The disturbance is initiated by tissue stresses and fractures or cracks during the early phases of wa   uptake. The localised release of natural pressure which results causes the nearby cells, and then indirectly a sequence of adjoining, closely compacted cells, to expand on the low-pressure side and to be forced away from their original close contact with adjoining cells. It is probably the progressive displacement and early death of cells in the cotyledons which causes the hollow heart condition to develop that has been of concern to Perry and Howell (1965), Heydecker and Kohistani (1969) and Heydecker and Feast (1969). More detailed insights concerning various aspects of water damage have been published elsewhere (Moore, 1965; 1970b).

PRE-HARVEST FREEZE

Symptoms of freeze injury vary extensively with the crop, with dryness of the seeds when exposed, temperature and duration of frost period, etc. The symptoms of freezing may even become partially obscured by ageing.

In monocotyledons such as maize (*Figure 20.1F*) or small grains, the stained cut surfaces of freeze injured seeds will vary in colour. These tissue colours vary from slightly darker than the normal red in tissues of near-normal cell turgidity to brownish-red and blurred and in extreme cases to partially or completely white (Moore and Goodsell, 1965; Moore, 1967). Slight degrees of freeze injury do not necessarily prevent germination. An experienced analyst can successfully distinguish between germinable and non-germinable freeze injured seeds.

Dicotyledonous seeds do not clearly reveal the various levels of freeze injury that are commonly encountered in monocotyledonous seeds. Immature seeds are generally killed; mature seeds tend to resist damage. Damaged tissues are revealed by embryos being darker red than normal and by a tendency to release leachate that produces a red precipitate. This precipitate accumulates within seed coats and in the testing solution. Freeze injured tissues may also tend to develop a bluish-red, liquid-logged appearance, especially if mechanical injury is also present.

HEAT

Heat damage, as other types of damage, results in various levels of deterioration. In cereals the cut surfaces of the embryonic leaves and roots may at first tend to be slightly flaccid and pale or deep red with a brownish tint. With advanced stages of injury the tissue may even fail to stain and may reveal a heat-toughened, rather firm embryo with a light brownish tinge. Between these two extremes, seeds are to be found with various intermediate grades of colour and tissue condition.

Since the seeds of a heat injured sample do not necessarily all reveal obvious symptoms, an analyst must frequently draw conclusions on the basis of evidence provided by a sufficient number of seeds. Seeds with intermediate levels of damage are usually the most informative for providing evidence for diagnosing causes of the deterioration of seeds with less obvious symptoms.

INFECTION BY MICRO-ORGANISMS

Infections of concern are commonly reflected by enlarged necrotic embryo tissues that are soft in texture and of dull appearance. Often the necrotic areas are surrounded by dark-red, liquid-logged tissues (*Figure 20.1E*).

Necrotic tissues, whether initially infected or not, tend to provide colonising bases for pathogens which start as saprophytes but, once established, can cause extensive seed and seedling decay. When infections are deep-seated and extensive, the commonly used fungicides may fail to provide the desired eradication and protection even though the seeds may remain germinable, and diseased seedlings may result.

AGEING

Two types of ageing must be taken into consideration in tetrazolium testing, namely chronological and physiological. Physiological ageing is of much greater economic importance than chronological ageing. It is, in fact, strictly speaking, only physiological ageing that causes deterioration which is made recognisable by the tetrazolium stain and by tissue flaccidity. Seeds that are chronologically young may nevertheless reveal extensive physiological ageing; whereas seed lots that are much older chronologically, if stored under cool dry conditions, often reveal very little physiological ageing. Differences in the rate of physiological ageing are influenced by both genetic and environmental factors.

Physiological ageing is commonly accelerated by embryo injuries. The injured tissues tend to deteriorate rapidly and this in turn promotes accelerated and progressive deterioration of adjoining tissues. Initial injuries that adjoin important tissues can affect these and become critical in storage with only a minor degree of ageing (Moore, 1963a; 1963b). Leachates from the deteriorated tissues and from minor infections can likewise readily block the germination of such seeds.

In dicotyledonous seeds the early stages of ageing are largely reflected by the enlargement and smoothing of the edges of initially disturbed areas and by the deterioration of radicle tips. Monocotyledonous seeds that are cut longitudinally through the germs commonly reveal their level of ageing by the extent of paleness and flaccidity along the cut surfaces. Not all embryo structures, or parts of structures, age uniformly. The tissues within and surrounding injured areas are usually much more advanced in ageing than are further-off undisturbed tissues. Aged living tissues react more adversely than sound tissues to injuries and also to exposure during cutting in preparation for tetrazolium testing. Advanced stages of ageing tend to mask the symptoms of initial injuries. In all cases, seeds that show clearly distinguishable patterns of strong, weak and dead tissues are the most informative for diagnosing causes for loss of seed quality.

REFERENCES

GRABE, D. F. (ed.) (1970). 'Tetrazolium testing handbook for agricultural seeds', Contribution No. 29 to the Handbook on Seed Testing, Ass off. Seed Analysts
HEYDECKER, W. and FEAST, P. M. (1969). 'Studies of the hollow heart condition of pea (Pisum sativum) seeds', Proc. int. Seed Test. Ass., 34, 319–328
HEYDECKER, W. and KOHISTANI, M. R. (1969). '"Hollow Heart" and poor stands of peas (Pisum sativum L.)', Ann. appl. Biol., 64, 153–160
ISTA (1966). 'International rules for seed testing', Proc. int. Seed Test. Ass., 31, 1–152
MOORE, R. P. (1961). 'Tetrazolium evaluation of the relationship between total germination and seed quality', Proc. Ass. off. Seed Analysts N. Am., 51, 127–130
MOORE, R. P. (1962). 'Tetrazolium as a universally acceptable quality test of viable seed', Proc. int. Seed Test. Ass.', 27, 795–805
MOORE, R. P. (1963a). 'The dying seed', Proc. Ass. off. Seed Analysts N. Am., 53, 190–193
MOORE, R. P. (1963b). 'Previous history of seed lots and differential maintenance of seed viability and vigor in storage', Proc. int. Seed Test. Ass., 28, 691–699
MOORE, R. P. (1964). 'Tetrazolium testing of tree seed for viability and soundness', Proc. Ass. off. Seed Analysts N. Am., 54, 66–72
MOORE, R. P. (1965). 'Natural destruction of seed quality under field conditions as revealed by tetrazolium tests', Proc. int. Seed Test. Ass., 30, 995–1004

MOORE, R. P. (1966). Tetrazolium tests for diagnosing causes for seed weaknesses and for predicting and understanding performance', *Proc. Ass. off. Seed Analysts N. Am.*, **56**, 70–73

MOORE, R. P. (1967). 'Freeze injury to seed corn as evaluated in tetrazolium and growth tests', *Proc. Ass. off. Seed Analysts N. Am.*, **57**, 138–140

MOORE, R. P. (1968a). 'Seed deterioration symptoms as revealed by tetrazolium and growth tests', *Proc. Ass. off. Seed Analysts N. Am.*, **58**, 107–110

MOORE, R. P. (1968b). 'Merits of different vigor tests', *Proc. Ass. off. Seed Analysts N. Am.*, **58**, 89–94

MOORE, R. P. (1969). 'History supporting tetrazolium seed testing', *Proc. int. Seed Test. Ass.*, **34** No. 2, 233–242

MOORE, R. P. (1970a). 'Tetrazolium seed testing aids', *Proc. Ass. off. Seed Analysts N. Am.*, **60**, 100–103

MOORE, R. P. (1970b). 'Tetrazolium for diagnosing causes for disturbances in seed quality', in *Hundert Jahre Saatgutprüfung*, 1869–1969 (ed. Ader, F.), 104–109, Sauerländer, Frankfurt am Main

MOORE, R. P. and GOODSELL, S. F. (1965). 'Tetrazolium tests for predicting cold test performance of seed corn', *Agron. J.*, **57**, 489–491

PERRY, D. A. and HOWELL, P. J. (1965). 'Symptoms and nature of "hollow heart" of pea seed', *Pl. Path.*, **14**, 111–116

DISCUSSIONS

Bekendam: Is not careful moistening, with sufficient oxygen supply, e.g. between filter papers, always necessary to prevent damage to the imbibing seeds?

Moore: Slow seed-softening is usually beneficial with large-seeded legumes and with very dry or aged seeds. Most other kinds of seed which come immediately from the field are usually not too dry, and small grains especially can be suitably prepared and placed directly into the TTC solution. With experience one soon learns to determine whether a staining pattern is man-made or is a result of an inherent seed condition.

Hudson: What are the limitations in seed size for the TTC test?

Moore: Essentially none; even tobacco seeds can be tested. After presoftening of tobacco seeds the cotyledon ends are cut off with a razor blade. After staining the embryos are squeezed out of the testas under a microscope. With experience, one can evaluate 100 tobacco seeds within 5–6 min.

Gordon: Can the TTC test be used on seeds with pigmented embryos?

Moore: Certain of such seeds if dry (e.g. wild oats, *Citrus*) may take longer, say 48 h, to pre-soften before they can be stained. The green cotyledon colour of some (e.g. forest tree) seeds is no obstacle to the experienced eye. Frequently it is unnecessary to wait until the seeds are fully stained; they can be judged from the way they start to stain. On the other hand, overstained tissues become deteriorated and blurred. The detrimental influence of prolonged staining can be delayed by adding a fungicide to the staining solution. If reading is to be delayed the solution can be replaced with water and the seeds kept in a refrigerator for assessing later. The pH should be 6·5 to 7·0.

21

THE IMBIBITION PROCESS

M. SPURNÝ*

Institute of Experimental Botany, Czechoslovak Academy of Sciences, Brno, Czechoslovakia

PERMEABILITY OF THE SEED COAT

The occurrence of seeds with hard seed coats, in particular seeds of several kinds of grasses and of leguminous plants, has led to a deeper study of the structural composition of the integumentary system and of the factors controlling the permeability of plant membranes in general. The seed coat has thus become a model object for studying the semipermeability and polar permeability of plant membranes (Brown, 1907; Schröder, 1911; Shull, 1913; Denny, 1917; Collins, 1918; van der Marel, 1919; Rudolfs, 1925; Kotowski, 1927; Gurewitsch, 1929; Brauner, 1928; 1930; Muhlack, 1929; Úlehla, 1930; Brown, 1931; 1932; Pugh, Johann and Dickson, 1932; Kisser and Schmid, 1932; Singh and Tandon, 1935; Gregor and Sollner, 1946; Zentsch, 1962; Spurný, 1954a; 1954b; 1954c; 1962).

Crocker and Barton (1957) considered the question of the occurrence of semipermeable phytomembranes on the basis of data obtained by Skene (1943) and Gregor and Sollner (1946) and concluded that from the evidence available on the submicroscopic structure of the cell membranes (Mühlethaler, 1955 and 1956) one cannot assume that the cellulose membranes possess pronounced semipermeable properties. The permeability of membranes for

---

* This communication was illustrated by three time-lapse photomicrographic films produced by Dr. Spurný, 'Perméabilité du tégument séminal pour l'eau et pour l'air chez le petit pois (*Pisum sativum* L.)', 'Cell wall structure of epidermal cells of the pea seed coat (*Pisum sativum* L.)' and 'Phases of the swelling mechanism during the germination of pea seeds (*Pisum sativum* L.)'.

dissolved substances is closely connected with the type of impregnation of the cellulose membranes. Substances regulating the penetration of water through the seed coat have been looked for, and explanations have been sought in the special structure of cellulose (Shull, 1913; Brauner, 1928; Stein von Kamienski-Jancke, 1957;

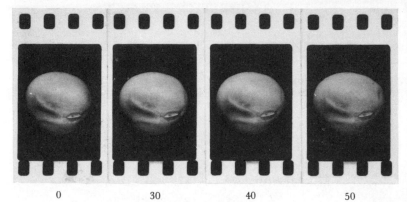

0   30   40   50

phase (min)

60   70   80   90

phase (min)

Figure 21.1   Localisation of the primary water influx through the pea seed coat (Pisum sativum L.). *Characteristic corrugation of the seed coat in the area of the chalaza indicates the spot of the primary water influx. Phases taken after 0, 30, 40, 50, 60, 70, 80 and 90 min of swelling*

Frey-Wyssling, 1959; Cavazza, 1950a; 1950b)—e.g. Frey-Wyssling reports that the cell membrane has no interfibrillar system but is built compactly of cellulose microfibrils, which determine its greater density. Further an explanation has been sought in the impregnation of the cell membranes with lignin (Küster, 1935), with fats (Denny, 1917), with suberin (Hamly, 1832; 1935; Scott,

1948; Spurny, 1954a; 1963) or with tannin (Denny, 1971). On the other hand, Watson (1948) has not found in the hard seed coat of the tribes *Trifoliae* and *Lotae* any structures which hinder water intake. In some more recent papers (Frey-Wyssling, 1959 and Cavazza, 1950b) the permeability of the cellulose membranes of the seed coat has been associated with their changes in the course of dehydration (Frey-Wyssling uses the term *'verhornen'*). It is a matter of irreversible changes of the micellar structure of the membranes, manifest by changes in the swelling potential of gels. The structural changes in the tissue of the seed coat, which result in non-permeability, have also been considered to be due to a genetic factor that can be modified by selection and by improvement (White and Stevenson, 1948; Hübner, 1938; James, 1949).

As to methods, the problem of the permeability of the seed coat and the primary influx of water has been studied by various modifications of micropermeameters functioning on the principle of osmosis (Muhlack, 1929; Kisser and Schmid, 1932; Spurný, 1954c; Manohar and Heydecker, 1964), by dyeing methods (Muhlack, 1929; Pringsheim, 1930; Lakon, 1949), by volumetric and gravimetric methods (Pringsheim, 1930; Stiles, 1949), by electric conductivity (Kotowski, 1927; Brauner, 1928), and by cinematography (Spurný, 1961).

With the problem of permeability are connected some findings of non-uniform permeability of the seed coat in particular areas. There exist data indicating that the most permeable spot of the testa is the hilum (Eberhart, 1906), an opening comprising the outer portion of the micropyle (Detmer, 1880; Manohar and Heydecker, 1964), while in the case of pea seeds it has been established that water penetrates through the chalaza tissue twice as fast as through the smooth part of the testa (Spurný, 1954c; 1961; 1962) (see *Figure 21.1*).

CELL WALL STRUCTURE OF THE MACROSCLEREIDS (PALISADE CELLS) OF THE PEA SEED COAT (*PISUM SATIVUM* L.)

In previous papers (Spurný, 1954a; 1954b; 1954c) it has been shown that the water permeability of the pea seed coat may be influenced by specific adcrustations which reinforce the walls situated on the outer ends of the epidermal palisade cells, immediately under the cuticle

It has been shown by microtechnical methods that this adcrustation is present in the form of a suberised starlike formation lining the outward facing part of the narrow lumen of the epidermal cells. This adcrustation was detected on cells isolated in macerating agent after treating them with a solution of COOXAM (copper oxide–ammonia or Schweitzer's reagent) (Müller, 1929). The structure of the suberised formation remained intact after the cellulose cell wall had been completely dissolved. These findings must have been unknown to some later investigators, e.g. to Stein von Kamienski-Jancke (1957), who made a detailed study of the submicroscopic structure of the epidermal layer of the seed coat of Leguminosae (including *Pisum sativum* L.), but must have overlooked the existence of the suberised formation. (In addition to the classical papers pertinent to this problem referred to by Stein von Kamienski-Jancke (1957), mention should be also made of papers by Rowson (1952) and Reeve (1946a; 1946b).)

The technique of isolating the epidermal cells in the macerating solution has been reported in Spurný (1954a). The process of swelling and dissolution of the isolated epidermal cell walls was followed on specially fixed preparations for the purpose of microcinematographic studies. The isolated cells were fixed in such a manner as to enable microreactions to be carried out. The procedures employed were as follows:

(a) From a suspension of isolated epidermal cells a film was spread on a clean cover glass, and after drying it was overlain with a solution of 6 per cent collodion. The dry collodion film was then gently removed and placed on a slide and overlain with water, and a microscopic slide was prepared in the normal way. It was then carefully dehydrated and two opposite sides of the preparation were framed with a mixture of paraffin and wax. The collodion film with the attached epidermal cells was thereby held in position in such a way that the COOXAM solution could be applied from the two free sides (*Figure 21.2*(a)). A 10 per cent solution of methyl metacrylate in chloroform was sometimes used to good advantage as an alternative to collodion.

(b) A number of slides were carefully cleaned in tetrachloroethane and a thin layer of 8 per cent Umaplex solution in tetrachloroethane was spread on them (care being taken to avoid microbubble formation). The slides prepared in this manner were then dried in a dust-free room for 2 h over an infra-red lamp. (The solvent must be thoroughly evaporated, otherwise there is danger that after subsequent contact with water an emulsion may arise which produces optical interference.) Meanwhile, films from the suspension of isolated epidermal cells were spread on cover glasses, evaporated and dried.

The principle of preparation then consisted in transferring the isolated cells from the dry suspension of the cover glasses into a

371

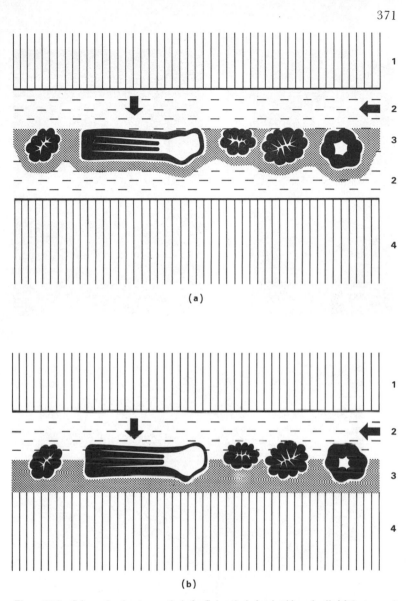

Figure 21.2   Scheme showing two methods for fixing the isolated epidermal cells of the pea seed coat (Pisum sativum *L.). (a) collodion film method: 1, cover glass; 2, water film (arrow indicates attackable areas of cell walls after treatment with a solution of COOXAM); 3, collodion film with isolated cells; 4, microslide. (b) Umaplex adhesive method: 1, cover glass; 2, water film (arrow indicates the attackable areas of cells after treatment with a solution of COOXAM); 3, Umaplex layer with impressed epidermal cells; 4, microslide*

372

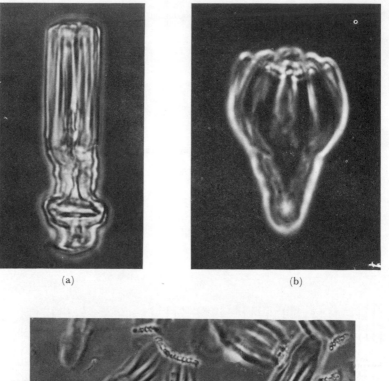

(a)   (b)

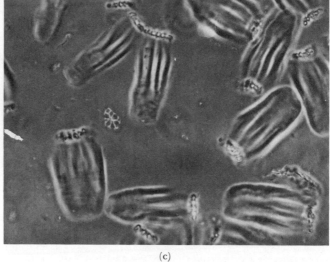

(c)

Figure 21.3 (a–h)   Isolation of the suberised star-shaped adcrustation H *of the macrosclereids of the pea seed coat after treatment with* COOXAM.

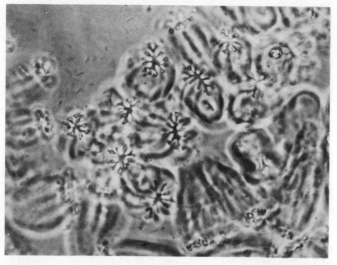

(d)

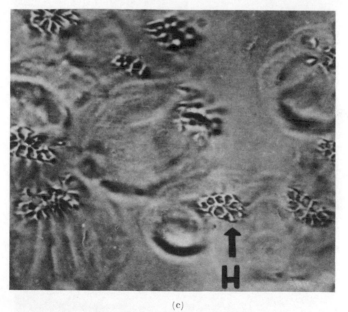

(e)

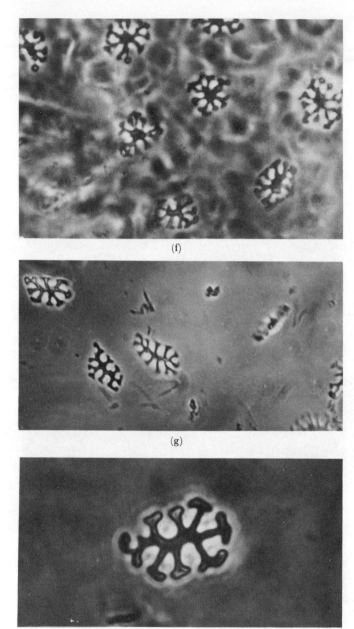

(f)

(g)

(h)

Umaplex layer, the surface of which had been softened with tetra-chloroethane (*Figure 21.2(b)*). Some practice and experience is required to judge with certainty whether the surface of the slides is properly 'adhesive'.

Owing to the fact that not all epidermal cells could be transferred from the cover glasses on to the adhesive Umaplex layer at once, this technique enabled the transfer of cells from one cover glass on to as many as five 'adhesive' slides. The solvent was then thoroughly removed by drying over an infra-red lamp, a normal water prepara-tion was made, and the opposite sides were fixed, as described in pro-cedure (a), for the purpose of carrying out microreactions.

Prior to the reaction proper, the preparations were, in both cases, treated with water to remove all insufficiently fixed cells. For taking film shots, a Mark MKU-1 Soviet microcinematographic equipment for normal 35 mm film was used. The prepared slides were then firmly fixed to the revolving cross table of the microscope by means of clamps, and at the time indicating the start of the microreaction a few drops of COOXAM solution were added and drawn through the preparation by means of blotting paper. Concurrently, the mechanism of the time-lapse equipment and the optical timing device were set in operation. As a result of this, the beginning and the end of the performed microreaction could be realiably located on the film recordings. The phase contrast system was used with lenses 20 × and 40 × magnification in combination with Homal 3·5 × and green filter. The light source used was a high pressure XBO 250 xenon discharge lamp. Under these conditions, it was possible to achieve sufficient exposures even at a frequency of 12 frames/second on Agfa Superpan 19/10° DIN films. The pro-gramme control device of the equipment used made it possible to set this frequency, while some partial sections of the microreaction were even taken satisfactorily with a frequency of 5 frames/second. Copies of the film recordings were then evaluated on an ordinary magnifying glass microscope specially adapted for this purpose.

The microcinematographic recordings of the microreaction have shown that the suberised starlike formation is released during the dissolution of the cellulose membrane of the epidermal cells from the outer part of the cellulose wall (*Figure 21.3*).

From the film recordings which registered the freely drifting isolated starlike formations, after the cellulose membrane had been dissolved, it was possible to reconstruct the space structure of these adcrustations (*Figure 21.4*). The centre of these membranes is slightly depressed in the middle so that the margins of the arms reach beyond the centre. The margins of the arms have the shape of thornlike processes. The number of arms of the starlike formation corresponds with the shape of the starlike lumen of the epidermal cells. The suberised formation is released as soon as the thin cellulose membrane on the outer part of the epidermal cell wall is dissolved.

Both the above-described methods of fixing isolated epidermal cells have proved to be effective. In the case of collodion preparations, the cells were fixed on the film during the whole course of the micro-reaction until the dissolution of the membrane. The percentage of

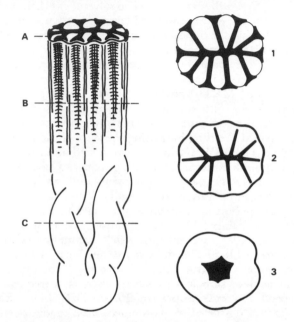

Figure 21.4 Suberised star-shaped adcrustation on the apical part of the macrosclereid of the pea seed coat (Pisum sativum *L.*). *1 (top right)*, *suberised structure viewed from above (black) in plane A, covering the narrow cell lumen; 2, section through the cell in plane B—narrow cell lumen (black); 3, section in plane C—wide open cell lumen*

badly fixed cells, i.e. of those washed away, was small. One of the disadvantages of this technique is that even after the swelling and dissolution of the cell wall of the epidermal cells, its imprinted relief remains preserved in the collodion matrix, and this produces optical interference when phase contrast is used for observation.

Conversely, when Umaplex film was used, the cells were not so well fixed, and consequently the percentage of badly fixed and washed away cells was much higher. As the microreaction went on, this percentage increased. On the other hand, the optical quality of the film recordings was much better.

DEVELOPMENT OF THE STARLIKE STRUCTURE OF THE PEA SEED MACROSCLEREIDS (PALISADE CELLS) DURING THEIR ONTOGENY

From the above review it follows that a number of factors regulating the permeability of the seed coat to water have been ascertained. Relevant is the fact that suberin adcrustations of the secondary membrane of the macrosclereids have been discovered in the seed coat of the pea by Spurný (1963), but were overlooked both by Reeve (1946a; 1946b) and Stein von Kamienski-Jancke (1957), although the former made a detailed study of the macrosclereid structure of the pea seed coat and of its ontogeny. In order to verify the occurrence of suberin formations, the author has made a study of the development of the starlike structure forming in the apical portion of the macrosclereids and has considered the results in relation to the permeability of the seed coat of pea seeds from samplings at various stages of seed ripeness. The results can be summarised as follows:

1. From seeds planted and harvested in 1960 and 1961, samples were collected during the vegetation period (June 30, July 17, August 1—1960; June 15, July 17, August 1—1961), for the

Figure 21.5   Secondary thickening of the cell wall in the macrosclereids of the pea seed coat (Pisum sativum L.). Cross-section scheme of the macrosclereids of seeds collected on May 1, June 1 and August 1, 1961

purpose of determining the rate of primary water influx through the seed coat (for methods see Spurný, 1961) and for microtechnical studies of the development of suberin adcrustations (*Figure 21.5*). For methods see Spurný (1954a; 1954b; 1954c; 1963).

2. The suberin adcrustation on the apical part of the cell wall of the macrosclereids begins to develop about 14 days prior to full seed maturity; fully developed suberin structures have been found only in the seed coats of fully matured seeds.

3. Corresponding with these structural changes of the macrosclereid walls are changes in the permeability of the seed coats; the resistance of the smooth parts of the seed coat (those which do not comprise the hilum, the chalaza and the space of the radicular pocket) to water penetration increases with the ripening of the seed. The cells in the area of the chalaza, which may be regarded as the water inlet region (Spurný, 1962), bear practically no continuous suberin adcrustations and consequently they allow the water to penetrate (in mature seeds) approximately twice as fast as the cells of the smooth part of the seed coat (Spurný, 1954c).

4. From the examination of seeds (selected from the proximal (old) and the distal (young) ends of pods) for the existence of suberin adcrustations, the author was unable to infer any relationship between seed coat permeability and degree of seed maturity. Consequently, it has not been possible to confirm the view that seeds with less permeable seed coats are, in fact, not fully matured seeds, where the cellulose of the cell walls of macrosclereids becomes irreversibly changed (Cavazza, 1950a; 1950b; Frey-Wyssling, 1959). The results have shown that the largest number of delayed swelling seeds are found amongst seeds which have been produced near the ends of pods, proximal or distal, more or less regardless of their maturity.

5. It appears that in addition to proven factors that may regulate the water permeability of the pea seed coat, such as suberin adcrustations of the cell walls on the apical part of the macrosclereids and changes in the cellulose membranes of not yet matured seeds, such as were postulated by Cavazza (1950a; 1950b), there must exist some differences in the development, structure and composition of the integumentary system of seeds growing at either end of a pod. The nature of changes in the metabolism of seed coat cells is hitherto not known.

EXISTENCE OF THE SUBERISED ADCRUSTATIONS IN THE TESTA OF CERTAIN LEGUMINOUS SEEDS

In previous papers (Spurný, 1963; 1964), the author has described the occurrence of specific starlike formations in the cell walls of the epidermal cells in the seed coat of *Pisum sativum* L. These adcrustations, probably composed of suberin or cutin, partly fill the outward facing portion (the top end) of the narrow cell lumen of the epidermal cells. In this way the permeability of the integumentary system of the seed to water and solutions can be controlled.

The question arose whether, by the use of a similar microchemical technique, analogous adcrustations could be discovered in the cell walls of the seed coats of other leguminous seeds.

Seeds of *Tetragonolobus purpureus, T. conjugatus, T. requieni* and *T. siliquosus, Trifolium incarnatum* and *T. hybridum, Medicago denticulata, M. lappacea* and *M. sativa* and of *Melilotus albus,* harvested in 1966, were used in the experiments carried out in 1967. Only the smooth part of the testa, without chalaza, hilum and radicular pocket was used for microchemical operations. Portions of the swollen seed coat were suspended in a macerating solution, shaken strongly, then separated by centrifuge and washed thoroughly with distilled water. Normal microscopic slides were then prepared from the suspension of isolated epidermal cells. The following microchemical reagents were used; a mixture of 10–50 per cent chromic and nitric acids for tissue maceration, Schweitzer's reagent (COOXAM) for swelling and dissolution of cellulosic cell membranes, solution of chlor-zinc-iodide as well as iodine in combination with potassium iodide and 60–72 per cent sulphuric acid for identification of cellulosic cell membranes and a saturated solution of Sudan IV in 70 per cent ethanol as used by Spurný (1954a; 1963) for staining of suberised cell wall adcrustations. However, a modification of the staining procedure with Sudan IV was used as follows: the suspension of the isolated epidermal cells was spread on a cover glass and after drying it was overlain with a saturated alcoholic solution of Sudan IV; it is recommended to repeat this staining operation three times using drops of fresh solution and then contrast by mounting in glycerol.

The results may be summarised as follows:

The basic structure of the epidermal cells of the seed coat of the leguminous seeds under examination is very similar, although small variations in the length and in the length/width ratio of the cell could be ascertained. In most cases, differences were found in the type of the terminal thickening of the wall just under the cuticle; formations analogous to the adcrustations described in macrosclereids of the pea seed coat could be discovered in two cases only, namely *Tetragonolobus conjugatus* and *T. requieni* (*Figure 21.6*). In the first case, the isolated adcrustations seem to be built of a suberin-like substance, since they are resistant to attack by COOXAM and 72 per cent sulphuric acid as well as to a mixture of 10 per cent chromic and nitric acids, giving at the same time a positive test with

380

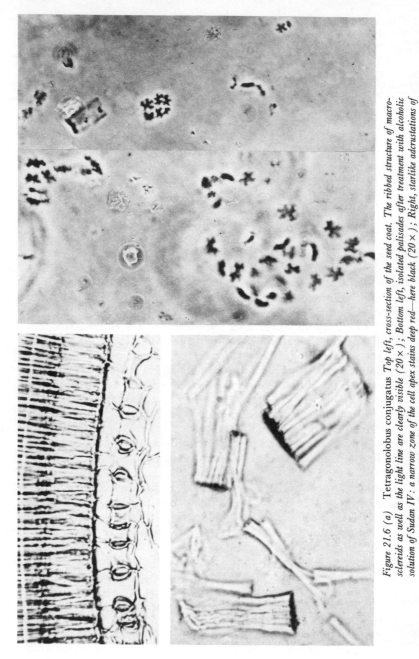

Figure 21.6 (a)   Tetragonolobus conjugatus   Top left, cross-section of the seed coat.   The ribbed structure of macro-sclereids as well as the light line are clearly visible (20 ×); Bottom left, isolated palisades after treatment with alcoholic solution of Sudan IV: a narrow zone of the cell apex stains deep red—here black (20 ×); Right, starlike adcrustations of the cell wall isolated after dissolving the cellulose membrane in COOXAM solution (phaco 20 ×)

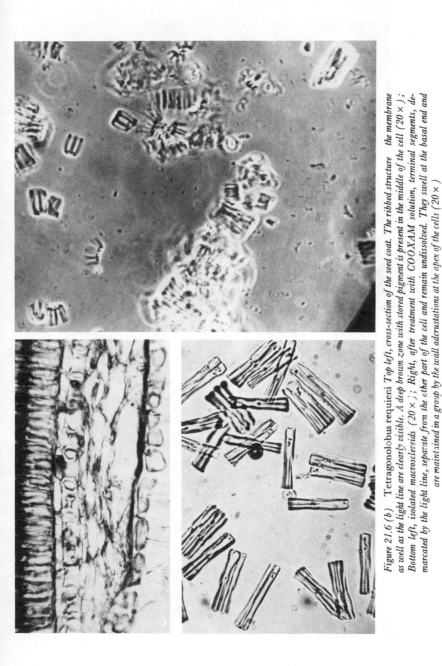

Figure 21.6 (b) Tetragonolobus requieni Top left, cross-section of the seed coat. The ribbed structure the membrane as well as the light line are clearly visible. A deep brown zone with stored pigment is present in the middle of the cell ($20 \times$); Bottom left, isolated macroscleroids ($20 \times$); Right, after treatment with COOXAM solution, terminal segments, demarcated by the light line, separate from the other part of the cell and remain undissolved. They swell at the basal end and are maintained in a group by the wall adcrustations at the apex of the cells ($20 \times$)

382

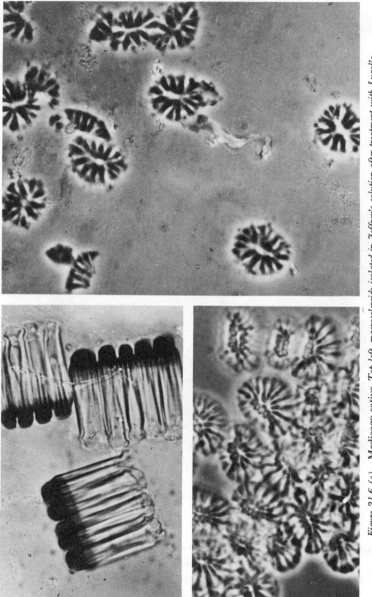

Figure 21.6 (c) Medicago sativa *Top left, macrosclereids isolated in Jeffrey's solution after treatment with Lugol's solution: terminal thickenings of the cell ('caps') stain yellowish brown (40 ×); Bottom left, macrosclereids swell spherically after trehtment with COOXAM (phaco 40 ×); Right, after the cells have dissolved completely, isolated caps resist the effect of COOXAM (phaco 40 ×)*

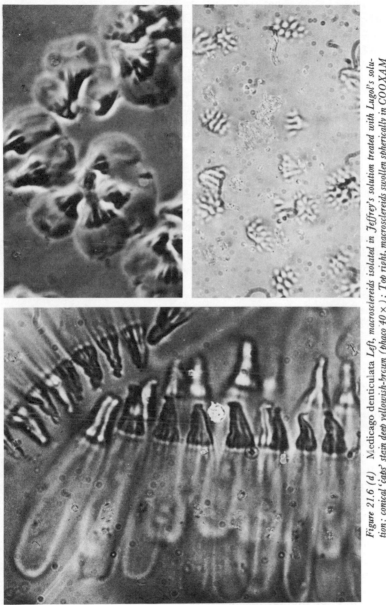

Figure 21.6 (d) Medicago denticulata Left, macrosclereids isolated in Jeffrey's solution treated with Lugol's solution: conical 'caps' stain deep yellowish-brown (phaco 40 ×); Top right, macroscleroids swollen spherically in COOXAM solution (40 ×); Bottom right, 'caps' isolated after dissolution of the cellulose membrane of the palisade (phaco 40 ×)

Sudan IV. Concerning adcrustations in the epidermal cells of
T. requieni, all tests indicate a non-cellulosic membrane. The test with
Sudan IV could not be carried out here since the structure of this
formation is so fine that after the swelling of the thickened walls it
disintegrates into fragments, the staining of which is then technically
extremely difficult.

Macrosclereids of the other species of *Tetragonolobus* and of
Trifolium, Medicago and *Melilotus* (with the exception of those of
Trifolium incarnatum) present characteristic thickening of the cell
wall on the cell apex. These formations could be considered as
structures analogous to the adcrustations of the membranes in the
two species mentioned above. They are developed especially in the
form of terminal caps, being demarcated from the other part of the
cell by the light line. The caps are resistant to attack by COOXAM,
72 per cent sulphuric acid and a mixture of 10–50 per cent chromic
and nitric acid, and this is strong evidence for their non-cellulosic
nature. It seems probable that they consist instead—like other
tertiary thickened membranes—of suberised or cutinised substances.
However, their negative response to Sudan IV argues against this,
the evidence is therefore not unequivocal (Scott, 1948; 1950).

PHASES OF THE SWELLING MECHANISM DURING THE
GERMINATION OF PEA SEEDS (*PISUM SATIVUM* L.)

The process of germination can be studied from different points of
view; the fundamental approach is the biochemical one, no doubt.
But too little attention is given to the role of the anatomical and
morphological background in the germination process which
enables the successive co-ordination—algorithm—of the developing
tissues and organs. Important phenomena in this respect seem to be,
for example, changes in the elasticity of the seed coat, its permeability to air and water as a whole, the spatial anisotropy of the
swelling capacity of the cotyledons, as well as the time co-ordination
of all these processes in relation to the beginning of the zonal swelling
and the subsequent elongation of the hypocotyl-with-radicle.

On the basis of data obtained by the quantitative evaluation of
the film recordings, a diagram of the whole process could be constructed on a time–space matrix (*Figure 21.7*) (under standardised
conditions of light, temperature and RQ), within which the seed
passes from inactivity into growth and development.

1. After 45 min in contact with water, the water penetrates
   through the chalaza tissue into the seed; this marks the starting

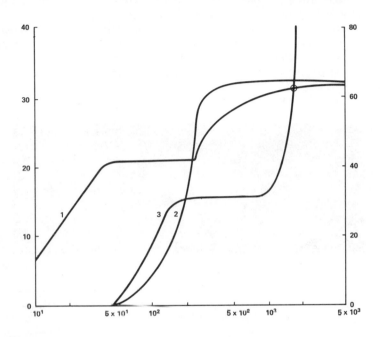

Figure 21.7   Time–space co-ordinates limiting the swelling process of the germinating pea seed (Pisum sativum *L.*). *Abcissa: time in log min, ordinate (left)—size of the swollen seed coat and that of cotyledons in per cent of their original values in direction A ; ordinate (right)—length of the swollen and elongated hypocotyl-with-radicle in per cent of its original value. Curve 1, seed coat; 2, cotyledons; 3, (lower half), swelling, (upper half), elongating growth of the hypocotyl-with-radicle*

point in the swelling of the cotyledons and the hypocotyl with the radicle.

2. After 3 h the seed coat is maximally swollen in direction C (first maximum) (*Figure 21.8*), the length of the hypocotyl is increased by about 30 per cent, particularly in the zone adjoining the first cotyledon buds.

3. After 4–5 h, a plastic elongation of the cotyledons takes place in direction *A*; simultaneously, a slight decrease in the size of the cotyledons can be established in direction *C*. The seed coat is stretched by the swelling cotyledons and its surface increases by about 25 per cent of the value determined for a 'freely' (cotyledon-free) swollen seed coat (*Figure 21.7*—curve 1).

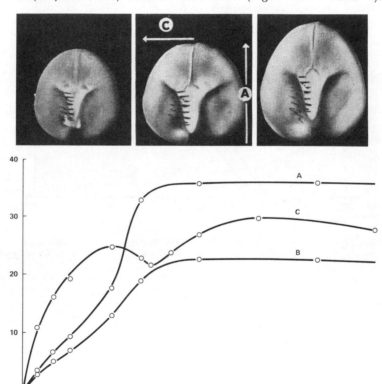

Figure 21.8 Spatial anisotropy of the swelling capacity of the pea seed cotyledons (Pisum sativum *L.*). *Abscissa: swelling in hours, ordinate: swelling capacity of cotyledons in per cent of their original values in respect to the swelling direction A, B (perpendicular to A and C) and C. Pictures taken 0, 100, 500 min after beginning of the swelling*

Owing to its elasticity, the seed coat can stand this pressure without being torn. The cotyledons, still slightly swelling, increase their size in direction C after 8 h of swelling (second maximum).

4. After 15 h of swelling, the growth of the hypocotyl with the radicle commences in the zone adjoining the first cotyledon buds; after 25 h, the root tip grows through the seed coat at a rate of 0·1 to 0·5 mm h^{-1} (*Figure 21.9*). When the hypocotyl begins to grow, the cotyledons are practically swollen to their full extent (200 per cent of their original volume); and the

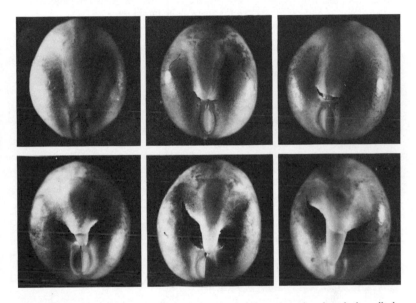

Figure 21.9   Phases of the elongation of the primary pea root growing through the radicular pocket of the seed coat (Pisum sativum L.)

radicular pocket, enclosing the hypocotyl and the radicle, increases in dimension along the longitudinal axis by about 20 per cent so that it presents no obstacle to the hypocotyl in the first growing phase.

REFERENCES

BRAUNER, L. (1928). 'Untersuchungen über das geoelektrische Phänomen II. Membranstruktur und geoelektrischer Effekt. Ein Beitrag zum Permeabilitäts-problem', *Jahrb. wiss. Bot.*, **68**, 711–770

BRAUNER, L. (1930). 'Über polare Permeabilität', *Ber. schweiz. bot. Ges.*, **48**, 109–118

BROWN, A. J. (1907). 'On the existence of a semipermeable membrane enclosing the seed of the Gramineae,' *Ann. Bot.*, **21**, 79–87

BROWN, R. (1931). 'The absorption of water by seeds of *Lolium perenne* L.', *Ann. app. Biol.*, **18**, 559

BROWN, R. (1932). 'The absorption of the solute from aqueous solutions by the grain of wheat,' *Ann. Bot.*, **46**, 571

CAVAZZA, L. (1950a). 'Recherches sur l'imperméabilité des graines dures chez les légumineuses', *Ber. schweiz. bot. Ges.*, **60**, 596

CAVAZZA, L. (1950b). 'Beitrag zur Erkennung von hartschaligen Leguminosensamen,' *Schweiz. landw. Mh.*, **28**, 378

COLLINS, E. J. (1918). 'The structure of the integumentary system of the barley grain in relation to localised water absorption and semipermeability,' *Ann. Bot.*, **32**, 381

CROCKER, W. and BARTON, L. V. (1957). *Physiology of Seeds. An Introduction to the Experimental Study of Seed and Germination Problems*, Waltham, Mass.

DENNY, F. E. (1917). 'Permeability of membranes as related to their composition,' *Bot. Gaz.*, **63**

DETMER, W. (1880). *Vergleichende Physiologie des Keimungsprocesses der Samen*, 1–565, Verlag von Gustav Fischer, Jena

EBERHART, C. (1906). *Untersuchungen über des Vorquellen der Samen*, Diss. Jena

FREY-WYSSLING, A., (1959). *Die pflanzliche Zellwand*, Springer Verlag, Berlin

GREGOR, H. P. and SOLLNER, K. (1946). 'Improved methods of preparation of 'permselective' collodion membranes combining extreme ionic selectivity with high permeability,' *J. Phys. Chem.*, **50**, 53–70

GUREWITSCH, A. (1929). Untersuchungen über die Permeabilität der Hülle des Weizenkorhs', *Jahrb. wiss. Bot.*, **70**, 657

HAMLY, D. C. (1932). 'Softening of the seeds of *Melilotus alba*.', *Bot. Gaz.*, **93**, 345–375

HAMLY, D. C. (1935). 'The light line in *Melilotus alba*', *Bot. Gaz.*, **96**, 755–757

HÜBNER, R. (1938). 'Untersuchungen über die Hartschaligkeit der Zottelwicke und ihre Behebung auf züchterischen Wege', *Landw. Jahrb.*, **85**, 751–789

JAMES, E. (1949). 'The effect of inbreeding on crimson clover seed coat permeability', *Agron. J.*, **41**, 261–266

KISSER, J. and SCHMID, H. (1932). 'Untersuchungen über die Permeabilität der Samenhüllen von *Pisum* und *Triticum* für Wasser sowie die Saugkräfte der Samen,' *Anz. Akad. d. Wiss. Wien, Math. Naturw. Kl.*, **69**, 197–200

KOTOWSKI, F. (1927). 'Semipermeability of seed coverings and stimulation of seeds,' *Plant Physiology*, **2**, 177–186

KÜSTER, E. (1935). *Die Pflanzenzelle*, 484, Jena

LAKON, G. (1949). 'The topographical tetrazolium method for determining the germinating capacity of seed, *Pl. Physiol.*, **24**, 389–394

MANOHAR, M. S. and HEYDECKER, W. (1964). 'Effect of water potential on germination of pea seeds', *Nature*, **202**, 22–24

MÜHLETHALER, K. (1955). 'Submikroskopische Morphologie', *Fortschr. d. Bot.*, **17**, 107–117

MÜHLETHALER, K. (1956). 'Submikroskopische Morphologie', *Fortschr. d. Bot.*, **18**, 51–57

MÜLLER, H. (1929). 'Die quellung von pflanzenfasern in Cuoxam', *Faserforsch.*, **VII**, 205

MUHLACK, E. (1929). 'Zur Keimungsgeschichte der Erbse', *Bot. Arch.*, **26**, 437–485

PRINGSHEIM, E. G. (1930). 'Untersuchungen über Samenquellung. I. Mitteilung. Die Abhängigkeit der Quellung von der Beschaffenheit der Samen und vom Medium', *Ztschr. wiss. Biol. Abt. E—Planta*, **11**, 528–581

PUGH, G. W., JOHANN, H. and DICKSON, J. G. (1932). 'Relation of the semipermeable membranes of the wheat kernel to infection by *Gibberella saubinetii*', *J. agric. Res.*, **45**, 609–626

REEVE, R. M. (1946a). 'Structural composition of the sclereids in the integument of *Pisum sativum* L.', *Am. J. Bot.*, **33**, 191–203

REEVE, R. M. (1946b). 'Ontogeny of the sclereids in the integument of *Pisum sativum* L.', *Am. J. Bot.*, **33**, 806–816

ROWSON, J. M. (1952). 'The hypodermal zone in the testa of certain leguminous seeds', *J. Roy. microsc. Soc.*, **72**, 46

RUDOLFS, W. (1925). 'Effect of seeds upon hydrogen-ion concentration equilibrium in solution', *J. agric. Res.*, **30**, 1021–1026

SCHRÖDER, H. (1911). 'Über die Selektivpermeable Hülle des Weizenkorns,' *Flora*, **102**, 186–208

SCOTT, F. M. (1948). 'Internal suberisation of plant tissue,' *Science*, **108**, 654–655

SCOTT, F. M. (1950). 'Internal suberisation of tissues', *Bot. Gaz.*, **111**, 378

SHULL, C. A. (1913). 'Semipermeability of seed coats', *Bot. Gaz.*, **56**, 169

SINGH, B. N. and TANDON, R. K. (1935).'Temperature-absorption characteristics during germination in seeds of different structure and biochemical constitution under varying concentrations of oxygen and water supply', *Proc. Indian Acad. Sci.*, **1B**, 496–518

SKENE, M. (1943). 'The permeability of the cellulose cell wall', *Ann. Bot.*, **7**, 261–273

SPURNÝ, M. (1954a). 'Cell wall structure of the epidermal cells of the pea seed coat (*Pisum sativum* L.)', *Preslia*, **26**, 139–142

SPURNÝ, M. (1954b). 'The light line of the sclereids in the pea seed coat (*Pisum sativum* L.)', *Preslia*, **26**, 143

SPURNÝ, M. (1954c). 'Water permeability of the pea seed coat. (*Pisum sativum* L.)', *Preslia*, **26**, 239–262

SPURNÝ, M. (1961). 'Das Eindringen von Wasser durch die Samenschale der Erbse (*Pisum sativum* L.)', *Research Film* 4, No 1, 41–47

SPURNÝ, M. (1962). 'Primary water intake into the pea seed coat (*Pisum sativum* L.)', *Preslia*, **34**, 272–276

SPURNÝ, M. (1963). 'Cell wall structure of epidermal cells of the pea seed coat (*Pisum sativum* L.) studied by microcinematography', *Mikroskopie*, **18**, 272–279

SPURNÝ, M. (1964). 'Phasen des Quellungsmechanismus keimender Erbsensamen (*Pisum sativum* L.)', *Flora*, **155**, 167–180

STEIN VON KAMIENSKI-JANCKE, (1957). 'Untersuchungen über Bau, Chemismus und Submikroskopische Struktur der Palisadenschicht von Leguminosensamenschalen,' *Mikroskopie*, **12**, 357–392

STILES, I. E. (1949). 'Relation of water to the germination of bean seeds', *Pl. Physiol.*, **24**, 540–545

ÚLEHLA, V. (1930). 'Permeability of the cell wall. I. Membrane of the Red-Grass (*Arundo phragmites v. Pseudonax* Asch. and Graeb.), its use as osmometer and its permeability for sucrose', *Protoplasma*, **9**, 574–600

VAN DER MAREL, I. P. (1919). 'La perméabilité sélective du tégument séminal', *Rec. Trav. bot. néerl*, **16**, 243

WATSON, D. P. (1948). 'Structure of the testa and its relation to germination in the Papilionaceae tribes Trifoliae and Loteae. *Ann. Bot.*, N.S. **12** (48): 367–409

WHITE, W. J. and STEVENSON, R. M. (1948). 'Permeable seeded strains of sweet clover (*Melilotus alba*), their development and nature', *Sci. Agric.*, **28**, 206–222

ZENTSCH, W. (1962). 'Zur Wasseraufnahme keimender Fichtensamen (*Picea abies* L. Karsten)', *Flora*, **152**, 227–235

O

22

THE RATE OF GERMINATION

A. G. GORDON

Forestry Commission Research Station, Alice Holt Lodge, Farnham, Surrey

INTRODUCTION

This chapter deals with changes in the time which seeds take to germinate when sown in standardised conditions suitable for their germination. The rate of germination is the reciprocal of this 'time to germination'.

The problem discussed may be formulated as follows: A viable seed which does not germinate though given presumably favourable conditions is usually termed 'dormant'. Certain aspects of this concept of dormancy in seeds are confusing, especially as the phenomena of imposed and induced dormancy have to be considered in addition to innate dormancy (Roberts, 1972). Dormancy is not an absolute state, but is in most cases subject to gradual changes. Thus, by changing environmental conditions slightly we can make germinable seeds fail to germinate. This does not normally mean that they have suddenly become 'dormant'. Neither do 'dormant' seeds suddenly become 'non-dormant'. Any term implying an all-or-nothing condition is misleading. For this reason, the concept 'resistance to germination' has been introduced (Gordon, 1971). This may be criticised as splitting hairs but it has led on many occasions to a better understanding of the factors involved. Numerically, the resistance to germination is measured in terms of the time between sowing and germination (radicle emergence). But the importance of the concept lies within the fact that, like dormancy, it gradually changes: for many populations of seeds, the time between sowing and germination first decreases and then increases again.

The situation can be illustrated by visualising innate, post-maturation, dormancy as a projection backward in time of the resistance to germination. Consider the course of events inside the seed of any of the common cereals, and of the new hybrid *Triticale*. An immature caryopsis or seed commonly has an infinitely high resistance to germination. That is to say, no matter what conditions are given to the intact seed, no germination will take place, even though, very soon after anthesis (in barley after seven days), the excised embryo will grow on moist filter paper. Like many properties amongst biological populations germination resistance is not uniform but more or less normally distributed, and, as the seeds mature, a time will be reached where the resistance to germination of the whole population has been reduced to such a degree that a few seeds will germinate. As the resistance of the population to germination decreases, so more and more seeds will be able to germinate, until a time is reached when all seeds will germinate under exactly the same conditions under which they would have remained dormant at the start. This does not mean, however, that the decrease in resistance to germination will cease as soon as a seed is capable of germinating under these conditions. For instance, in one of the experiments to be described, a barley sample which failed to germinate completely for several months after harvest still had a high resistance to germination when it ultimately did germinate, that is, each seed took a long time from sowing to germination.

Improvements in the rate of germination, which were equivalent to further decreases in resistance to germination, will often continue for some time after seeds first become able to germinate and even when all seeds in the population have become germinable. Consequently, in all cereal and conifer studies undertaken, the most rapid germination has always been correlated with the highest germination percentage, although the converse does not necessarily hold. Despite the dissimilarity in the time course of resistance and percentage which militates against a close relationship, correlation coefficients, ranging from -0.863 (P $<$ 0.001) to -0.991 (P $<$ 0.001) have been obtained for ten samples of barley and wheat between the germination resistance values and the total germination percentages. This may not be true for other species but similar patterns of changes in rates of germination have been found in *Brassica napus* (turnip) and *Picea sitchensis* (Sitka spruce), although in the latter the time factor is increased several-fold.

Figure 22.1 shows the pattern of this change in the rate of germination and its continuing change after the germination percentage has reached a maximum. The argument of 'recovery from dormancy' is easier to follow for seeds in which no additional germination control system is known to exist, than when complex control mechanisms are involved, such as embryo dormancy or inhibitor systems. No investigations of factors affecting the germination of the latter types of seed have been included in the present work, but it is possible to formulate a model along the above lines which will fit all kinds of seeds, including light requiring ones. It must, of course, be recognised that where the mean time to visible germination tends to stretch over several weeks or even months, any significance of changes

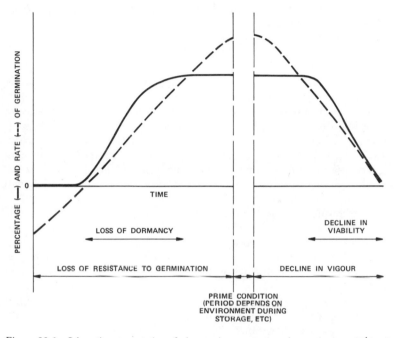

Figure 22.1 Schematic representation of changes in percentage and rate (resistance^{-1}) of germination of any given seed lot

within a population or difference between populations becomes more difficult to establish, because external factors begin to exert a more important influence.

The rate of loss of germination resistance, as well as the final level of germination resistance, can be changed by environmental conditions, including the way in which seeds are handled. Thus, considerable damage to seed quality, not easily demonstrable by normal tests, but expressed by greater resistance to germination, may result from poor harvest conditions. Differences in germination quality may indeed occur within a germination level of 100 per cent.

From the practical and ecological point of view it can be argued that amongst non-dormant seeds those showing least germination resistance are the ones most likely to survive adverse soil conditons.

Heydecker (1969) has reviewed the complex problem of 'vigour' in seeds. But this review shows how limited is the amount of information available on factors which affect germination vigour, other than the well-known one of the micro-flora during storage. The reviewer points out that the rate of germination has often been considered a symptom and index of germination vigour, and with some justification. But he also points out that a clear distinction is needed between part-dormant and non-vigorous seeds, both of which germinate slowly: part-dormant ones because they have not yet shed all their resistance to germination, non-vigorous ones because they have begun to age.

The germination resistance test consists simply of a series of germination tests separated by a suitable interval and each comprising a series of counts within a predetermined time limit (Gordon, 1971), with a view to tracing shifts in the mean germination time of a seed population. An idea of the uniformity of germination time of a group of seeds can be given by the standard deviation of this mean time on each occasion.

The results of the germination resistance test have the same drawbacks as the results of any tests based on a modification of the laboratory germination test (Goodchild and Walker, 1971) in that it is based on a unique set of environmental conditions. On the other hand, the conditions employed are reproduceable and thus able to be used in themselves as direct measures of seed quality. Confusion is avoided by not making a weighting for the total number of seeds germinating (Czabator, 1962; Timson, 1965) but completely separating germination percentage from time to germination (Heydecker, 1966).

FACTORS AFFECTING THE RATE OF GERMINATION OF A GIVEN SAMPLE

The major part of the work reported here was carried out on cereal grains which have a naturally rapid germination and are highly susceptible to minor changes in the germination conditions. However, exploratory work has also been carried out on factors affecting the germination of some conifer seeds, which take some ten times as long to germinate as do cereals, and the results suggest a similar pattern of dependence upon germination conditions.

The factors influencing the germination rate and total germination percentage of cereals in tests have been described elsewhere (Gordon, 1969). They include:

1. The temperature of germination.
2. The initial temperature of the germination water.
3. The amount of water present.
4. The position of the embryo relative to the level of the germination water.
5. The amount of oxygen present in the germination medium.
6. The level of infestation by micro-organisms.
7. The presence of dissolved ions in the germination medium.
8. The number of pre-germinated grains present.
9. The length of storage at high temperatures.
10. The seed moisture content during storage.
11. The moisture content of the grains as they go into the test.

The specifications for the germination resistance test attempt to control as many of the above factors as can still be controlled at this stage, such as water temperature and availability, and, of course, germination temperature (Gordon, 1971). Heydecker (1961) and Orphanos and Heydecker (1967) have observed similar critical relationships concerning the germination of some vegetable seeds.

RELATIONSHIP BETWEEN RATE OF GERMINATION AND TOTAL GERMINATION PERCENTAGE

As mentioned previously, germination resistance, the measure of the rate of germination used here, makes no allowance for the final germination percentage. However, in the species studied—except

Table 22.1 MEAN GERMINATION PERCENTAGE (GP) AND GERMINATION RESISTANCE (GR)* VALUES (IN HOURS) FOR CONQUEST BARLEY AND SELKIRK WHEAT DURING MATURATION AND STORAGE AT 12 PER CENT MOISTURE CONTENT (DATES CORRESPOND TO CURVES ON FIGURE 22.2)

Species	Cultivar		*11/8*	*18/8* (*A*)	*22/8* (*B*)	*25/8* (*C*)	*30/8* (*D*)	*10/9* (*E*)	*30/9* (*F*)	*15/11* (*G*)	*4/4* (*H*)
Barley	Conquest	GP	0	36	65	90	98	100	100	100	100
		GR		65·5	60·9	52·1	40·7	27·6	25·7	26·2	25·6
Wheat	Selkirk	GP	0	0	2	42	70	92	92	98	97
		GR			84·0	74·6	67·4	46·5	38·1	35·5	28·1

* GR = hours to mean germination, based on those seeds which germinate

396

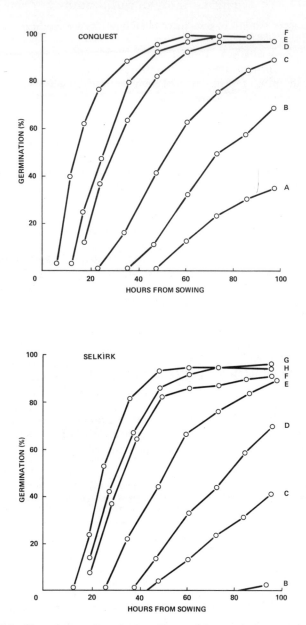

Figure 22.2 Change in rate of germination with age of Conquest barley and Selkirk wheat.
(A. 18.8.69, B. 22.8.69, C. 25.8.69, D. 30.8.69, E. 10.9.69, F. 30.9.69, G. 15.11.69,
H. 4.4.70)

where no seeds or all live seeds germinate—the germination resistance is invariably linked with percentage germination; the percentage is initially low and the resistance high. The resistance then gradually declines as the percentage increases although the resistance is still not at its lowest when the percentage has first reached its maximum (see also Renard, 1967). *Table 22.1* shows the inverse relationship between the germination resistance values and the total germination percentages. Both regressions are significant (P < 0·001), but the position of each differs, indicating that Conquest barley has relatively less germination resistance (a consistently higher rate of germination) than Selkirk wheat (*Figure 22.2*). Germination resistance values of Sitka spruce (*Figure 22.3*) show a similar pattern although they tend

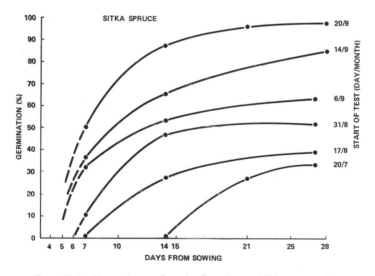

Figure 22.3   Changes in rate of germination with age of Sitka spruce seed

to fall in the ten day range, whereas cereal values are invariably measured in hours.

Table 22.1 again demonstrates clearly that considerable improvement in germination takes place long after the samples have lost 'dormancy' in the narrow sense of the word. It will be shown that use can be made of the knowledge of this continued improvement as well as of the absolute level of germination. However, it must be made clear that these differences do not manifest themselves consistently without careful control of the germination conditions; nor are they normally observable in the routine seed tests which are only made at one stage in time.

Table 22.2 MEAN GERMINATION PERCENTAGE (GP) AND GERMINATION RESISTANCE (GR) VALUES (IN HOURS) FOR SAMPLES OF SEVERAL CEREALS DURING A PERIOD OF 10 MONTHS

Species	Class	Cultivar/Selection	Immediately after harvest 1969		October 1969		January 1970		July 1970	
			GP	GR	GP	GR	GP	GR	GP	GR
Wheat 6N*	Hard Red Spring	Marquis	96	54·7	97	37·9	100	35·5	—	—
		Prelude	94	41·9	97	35·2	98	31·5	—	—
	Hard White Spring	Gabo	91	52·7	—	—	96	46·9	—	—
		Stewart	96	59·9	—	—	98	57·9	—	—
	Semi Hard White Spring	Ramona	99	28·4	—	—	99	25·1	98	32·3
		Gamenya	94	71·1	—	—	94	58·4	94	44·3
	Soft White Spring	Bungalla	100	38·6	—	—	100	30·1	100	35·2
		Chile	98	34·6	—	—	98	32·8	98	33·1
	Soft Brown Spring	Penjamo	98	38·4	—	—	99	36·5	97	36·9
		Pitic	100	30·8	—	—	100	30·6	100	31·1
Wheat 4N†	Spring	Stewart 63	20	62·2	97	57·6	96	57·2	94	50·5
		Ghija	67	62·5	94	55·6	98	34·8	97	32·2
Rye	Spring	Prolific	91	49·1	94	28·7	98	27·0	97	29·0
		Petkus	57	52·5	93	39·6	99	27·7	99	26·9
Triticale	Spring	Rosner	98	50·9	97	49·9	99	41·4	95	38·9
		6408	91	65·8	88	61·5	89	58·6	93	36·5
Barley	Malting Spring	Parkland	78	47·4	100	22·5	99	20·3	99	28·5
	Feed Spring	63–1500	99	24·1	100	22·0	100	22·0	100	26·7

* Hexaploid
† Tetraploid

INTERSPECIFIC DIFFERENCES IN RATE OF GERMINATION

The change in rate of germination of a small cross-section of cultivars from the World Wheat Pool and several cultivars and selections of other cereals was followed from the summer of 1969 through to the summer of 1970. All seed lots were grown in the same place and harvested and handled under identical conditions. The results are shown in *Table 22.2*.

For the hexaploid wheats, the germination percentage at harvest exceeded 94 per cent in all but one cultivar. Except in this one case, 'dormancy' would normally have been regarded as almost non-existent. The germination resistance values ranged however from 28·4 h to 71·1 h. By the next test date there was in most lots an insignificant increase in germination percentage but a significant drop in germination resistance values ($P < 0·001$). However, all lots tended to maintain their ranking relative to one another.

The other species tested show a similar pattern. Although at harvest they showed rather lower total germination percentages, by the second test date these had risen almost to their maximum, yet they still exhibited very different germination resistance values. Over such a narrow range of examples it is not possible to draw hard and fast conclusions on specific differences, but that large differences between species do exist, all within a maximum germination percentage, cannot be doubted. It is perhaps relevant to note that Rosner *Triticale* is a cross between Prolific rye and Stewart 63 *durum* wheat and that its germination resistance values are almost intermediate.

Similar specific differences exist within the conifers. Whereas all normally grown British pines show germination resistance values of below 10 days, the spruces and firs show values between 10 and 20 days.

One further interesting point can be made from *Table 22.2*. It will be seen that in some lots, within a more or less constant maximum germination percentage, the germination resistance values rose again by the July 1970 test, whereas in others the germination resistance values were still declining. If differences in these changes in germination resistance are indeed to be of use in assessing germination quality, it is important to explain these observations. However, it is necessary to show first that individual cultivars have germination resistance values sufficiently constant to be used as selection criteria; in other words, that these differences are genetically based.

INTRASPECIFIC DIFFERENCES IN RATE OF GERMINATION

In the autumn of 1968, 1 kg samples of 16 seed lots in the Western Canadian Co-operative six-row barley trials, were forwarded to Winnipeg for germination resistance analysis. They had been grown in different environmental conditions (at five locations in Western Canada and at one in Eastern Canada), but harvesting and handling procedures had all been as similar as is possible in large-scale field trials. After arrival all lots were handled identically. Complete germination resistance tests were carried out on the 16 lots from all 6 locations on arrival and again 10 weeks later. Analysis of variance showed that on both test dates there were significant differences between cultivars (P < 0·001) and between locations (P < 0·001), but that the interaction between cultivars and locations was also significant (P < 0·001). Because the environment in which they had grown, or perhaps the harvesting method, resulted in significant differences in the absolute germination resistance values, it is interesting to compare the 16 selections on their relative ranking. This comparison is shown in *Table 22.3*. Despite the cultivar × location interaction, and although the rankings tend to over-estimate small differences in the actual values, it is at once clear that the relative positions of the selections are fairly constant and therefore of genetic origin. The significant interaction between cultivars and locations is almost all due to cultivars 1 and 501, although cultivars 104, 301 and 105 also show some interaction with location.

In 1969 a similar study was carried out on a selection of 26 wheats grown at four locations in Western Canada. Once again significant differences (P < 0·001) were established by analysis of variance between selections and between locations and for the interaction between the two.

Similar significant differences (P < 0·001) were found between different selected trees within a registered seed stand of Sitka spruce. In this outbreeding example, assuming that cross-pollination has actually taken place, these differences indicate a maternally transmitted inheritance and are thus, perhaps, even more important.

Reverting to wheat, one further interesting observation can be made from these intraspecific comparisons. *Table 22.2* shows that some white kernelled wheats occur which have a germination resistance value as high as that for red wheats. This observation, confirmed in the above-mentioned tests, is in contrast with the oft-quoted relationship between kernel colour and dormancy, originally postulated by Nilsson-Ehle (1914) (see also Chapter 2). *Table 22.2* also

Table 22.3 RELATIVE RANK OF BARLEY SELECTIONS AT EACH LOCATION BASED ON GERMINATION RESISTANCE VALUES, AND THE RANGE OF GERMINATION RESISTANCE VALUES FOR EACH SELECTION AND AT EACH LOCATION IN HOURS (ALL SAMPLES GERMINATED 98 PER CENT OR MORE; W = WESTERN CANADA, E = EASTERN CANADA)

| | Location | | | | | | | Range | |
	Beaver Lodge (W)	Edmonton (W)	Swift Current (W)	Regina (W)	Portage La Prairie (W)	Guelph (E)	Overall rank	Min	Max
Selection									
301	5	11	1	1	3	1	1	14·5	28·9
308	3	8	2=	3	2	2	2	13·8	25·8
311	4	2	5	2	1	12	3	13·9	22·6
304	1=	4	4	4	6	5	4	13·6	24·3
310	1=	7	2=	7	4	6	5	13·6	25·5
406	10	1	12	5	5	4	6	17·7	22·5
4	6	5	8	6	9	7	7	15·7	25·0
106	7	6	13	8	8	3	8	16·3	25·3
405	9	3	10	11	10	10	9	17·0	25·0
805	8	10	6	13	7	13	10	16·4	28·2
107	11	9	9	9	11	8	11	18·0	27·9
1	13	15	7	11	12=	14	12	17·7	36·8
104	12	12	14	14	16	9	13	18·9	36·5
501	14	14	10	15	15	11	14	21·9	36·4
105	15	16	15	16	12=	15	15	22·3	37·9
605	16	13	16	10	14	16	16	24·9	35·6
Rank by location	1	6	2	4	5	3			
Overall range {Min	13·6	22·5	13·9	15·6	17·7	15·1			
{Max	28·1	37·9	27·3	32·9	36·5	29·8			

indicates a possible relationship between kernel hardness and high initial germination resistance, but enough data for confirmation are not available.

FACTORS AFFECTING INTRASPECIFIC DIFFERENCES IN RATE OF GERMINATION

In the experiments described immediately above, every effort was made to ensure that the test conditions were as uniform as possible so that any differences in rate of germination could be regarded as a true reflection of genetic or production–environmental differences. Significant differences due to the environment were obtained but no attempt was made to establish their origin, in part because of a lack of resources, but mainly because work had already been carried out in Scotland in 1964 on two-row barleys which established reasons for similar germination differences. At this time the actual germination resistance test was not employed. Instead, a standard sand test (ISTA,1966) was carried out at intervals in order to determine the time taken to recover from post-harvest dormancy. The length of the period between harvest and the time when germination first exceeded 90 per cent was regarded as an indication of 'germination quality'. Harvesting and handling conditions were kept as similar as possible. It was impossible, using farmers' crops, to design a factorially arranged experiment in which all treatments were orthogonally represented. Instead all relevant data for 115 different fields of barley were collected and treatments analysed singly, in the hope that the large number of degrees of freedom would offset the high level of experimental error in such a system. Using it thus, the following factors influencing the germination quality of barley were investigated (Gordon, 1968b):

1. The cultivar sown.
2. The soil association in which the field lay.
3. The units of nitrogen, phosphorus and potassium applied per acre.
4. The application of lime.
5. The pH of the soil.
6. The date of sowing.
7. The rate of sowing.
8. The height above sea level.
9. The soil water-holding capacity.
10. The presence of a nurse crop of grass.
11. The previous crop.

12. The month of ploughing.
13. The application of herbicides.
14. The application of insecticides.
15. The date of seed harvest.
16. The moisture content of the seeds at harvest.

Simple analysis of variance for the influence of these factors on the recovery from dormancy revealed:

(a) Significant differences between the various *cultivars* sown. This was later confirmed by the experiments already reported.

(b) Significant differences between barleys grown on different *soil associations*. The soil types were classified from the 1 in 63360 (1 inch to 1 mile) Soil Survey map. Soils derived from alluvium produced barleys of lowest germination resistance, and with very few exceptions the 63 barleys grown on soil associations derived from old red sandstone showed very similar levels of germination resistance (122–126 days to recover from dormancy).

(c) Significant differences between barleys grown after different *previous crops*. Rather surprisingly, fields in which leys and legume crops had been ploughed-under in the previous year, and which were therefore capable of supplying nitrogen throughout the following growing season, were found to produce barleys with significantly lower germination resistances than fields in which turnips and beet crops had been previously sown. Turnip and beet crops are normally sown with high-boron fertilisers to eliminate central root rot and in this survey all crops were so treated. The possible presence of high-boron residues in the fields in which the root crops were grown might be of relevance in explaining the higher germination resistance.

(d) A significant inverse relation between *application of nitrogen* and germination resistance. Non-linear regression analysis, using the square of the number of units of nitrogen per acre applied as the quadratic term, produced a curvilinear regression component significant at the $P < 0.001$ level. The higher (up to 70 units per acre) the number of units of nitrogen fertiliser applied, the lower was the germination resistance. This finding agrees with (c) above and also with Junges and Ludwig (1963), but is in marked contrast to a recent review by Belderok (1968). It should be noted that phosphorus and potassium fertilisers produced no significant effects at all.

(e) Significant differences between barleys grown at different pH levels. Soils with abnormally *high pH levels* (pH 7·5–7·9) produced barleys with significantly higher germination resistance than soils with pH 6·0–6·4.

(f) Significant differences between crops sprayed and not sprayed. Fields of barley *sprayed with DDT against leatherjacket larvae* produced barleys with a significantly (P < 0·01) lower germination resistance than barleys from fields which had not been sprayed. Such sprays were used only on fields of ploughed-under leys when these were badly affected. A significant difference (P < 0·05) was also found between barleys from fields of unsprayed and of sprayed ploughed-under leys. No significant effects were obtained from any form of herbicide sprays.

(g) No significant effects in any other criteria studied.

In further randomised block experiments carried out in 1965 and 1966 throughout the north-east of Scotland, the germination resistance was found to be significantly increased by partial lodging of the seed crop and by the length of time the crop stood in the field after maturity and, within one week after drying, it was found to be affected by the moisture content of the seeds *before* drying (Gordon, 1968b). In another experiment the germination resistance was found to be significantly lowered by increased temperatures of drying and of post-drying storage (Gordon, 1968a).

CHANGES IN THE RATE OF GERMINATION DURING DRY STORAGE OF SEED

In *Table 22.2* it has been shown that some lots of cereals started to exhibit an increased germination resistance after only a few months in storage whereas the resistance of others continued to decrease. It can be seen that in those lots where the germination resistance had been high at the beginning it was almost invariably still decreasing after 10 months, but where germination resistance had been initially low it began to increase after the same time. In an attempt to trace its changes over the storage period, in order to ensure that germination resistance was a more-or-less consistent although gradually changing characteristic, use was made of a group of cereals, the Western Co-operative six-row barleys grown at Portage la Prairie in Manitoba, Canada, in 1968. The germination resistance of all lots was measured four times between October 1968 and August 1970. The analysis of variance showed significant differences between selections and between sowing dates, but it also showed a significant interaction (P < 0·001) between the selections and sowing dates.

This means that selections behave significantly differently with time in respect to changes in their germination resistance values. Again, in those selections where the germination resistance values had been initially high they still tended to decrease after 20 months' storage at 12 per cent moisture, whereas those selections with low initial germination resistance values, tended to show the reverse pattern. Some selections remained virtually constant with respect to their germination resistance values throughout the 20 months. There are, it appears, genetic differences, not previously recognised, in the tendency for germination characteristics to change with time.

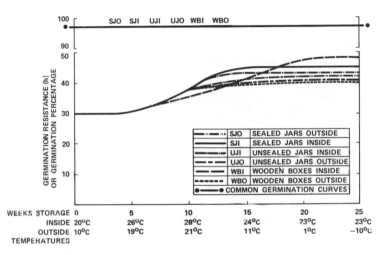

Figure 22.4 Changes in germination resistance values of six differently stored samples of Manitou wheat, within a constant germination percentage. Temperatures given are those at the end of the storage period

Finally, mention should be made of an experiment carried out to trace the change in germination resistance under different storage conditions. Several hundred pounds of Manitou hard red spring wheat (13·8 per cent moisture) were thoroughly mixed and divided into six different lots. Each lot was given a different storage treatment, which can be seen in the key to *Figure 22.4*. Germination resistance tests were carried out at the start and thereafter once every two to three weeks up to 25 weeks. The results are shown in *Figure 22.4*. It is immediately obvious that whereas the total germination percentage remained constant at 97 per cent throughout the 25 weeks, considerable and significant differences occurred in the germination resistance values over the same period of time. It is not intended to discuss here the treatment differences. Suffice it to say that differences in germination resistance values manifested themselves during storage, analogous in this case to what has been termed 'loss of vigour' (Heydecker, 1972). It is interesting to note that similar differences have been observed in Sitka spruce seed in recent experiments.

It is unlikely that the factors responsible for the initial decrease in germination resistance are the same as those responsible for the later increase. No attempt has yet been made to study this aspect, but the decrease in germination resistance has been shown to be related to chemical changes in the germination control mechanisms (Gordon, 1968a). It is not unreasonable to hypothesise that the renewed increase in germination resistance is due to changes brought about by ageing, such as break-down of cell membranes and build-up of microflora, although this remains to be substantiated.

SOME USES OF THE CONCEPT OF GERMINATION RESISTANCE

In practice, the use of batches of seed for sowing is normally based only on the criterion of their attaining a high germination, that is their loss of dormancy. However, it is clear from the evidence above that a large improvement in the germination rate of seeds can take place long after the maximal germination potential has been achieved. No deliberate attempt is usually made to use seeds at a time when their germination resistance has reached a minimum, i.e. when their rate of germination is at its greatest, although sometimes 'vigour' tests are carried out on samples from seed lots previously tested to see how far subsequent deterioration has proceeded, i.e. whether and to what degree germination resistance has increased again.

Attempts to correlate crop yields with seed and seedling vigour have not always been successful. However, other things being equal, one may assume that fast germinators (or seeds with a low germination resistance) will have an advantage in poor soil conditions over slow germinators (or seeds with high germination resistance). An experiment was therefore carried out to test the relationship between germination resistance and subsequent seedling vigour (Boyd, Gordon and Lacroix, 1971). 115 F_3 progenies from a cross between two barleys were used, as well as the original parent types. These latter represented one cultivar with a high, and one with a low, germination resistance. The parent types, plus 54 of their progenies, selected so as to give a range of germination resistance values, were grown in a greenhouse until two weeks after first being tested for germination resistance. After making allowance for differences in seed weights, a significant inverse relationship (P < 0·001) was found to exist between the germination resistance and the dry weight of the seedlings, at any given time after sowing, showing that the advantage gained by early germination was maintained up to the time of sampling. These results offer a good argument for using seeds when their germination resistance has decreased to a minimum.

The germination resistance test can be recommended as a means of assessing the likelihood of sprouting in the ear. If sprouting in the ear is indeed contingent on an unduly small germination resistance value, the optimal germination resistance value would not tend to zero. Selection against sprouting is based at present at many breeding centres on total immersion of the entire ear for 48 h in water followed by warm incubation, to simulate the effect of rain and sunshine. But some selections found to have a high germination resistance when tested by this technique are nevertheless prone to sprouting and vice versa. The method makes no allowance for the tightness of the spikelet or the accumulation of bacteria during immersion, both of which affect the amount of oxygen reaching the embryo and thus the ability of the embryo to germinate. For this reason it is arguable that the germination resistance test, a more natural measure of germination resistance, should be used, possibly in addition to other sprouting tests, to give a more reliable assessment of the overall resistance to sprouting of the grains.

In the malting process a rapid, uniform germination of barley is required. It is suggested that a series of germination resistance tests will give an accurate assessment of exactly when the barley is most likely to produce the most rapid and even germination during malting. They can also be used as a selection criterion for the breeding of malting barley. *Table 22.3* p. 401, shows how effective the test can be in discovering malting barleys. Selections 301, 304, 308, 310 and 311 all have consistently low germination resistance values, more or less independent of the environment in which the grains have matured, and all have at least one parent in common. Some have since been released to commerce as top-quality malting barleys. Tests carried out on as few as 200 grains of barley can thus be used for an initial selection for malting quality and can replace micro-malting techniques requiring many times that number of grains.

Germination resistance values have been found to be significantly correlated with further malting quality characteristics. The water sensitivity test (Essery, Kirsop and Pollock, 1955) is used in malting to give some measure of the improvement in germination quality after maximum germination potential is achieved. It has been shown that for any one cultivar there is a significant positive correlation between high water sensitivity and high germination resistance ($r = 0.963$, $P < 0.001$). This suggests that the water sensitivity and the resistance to germination may well be governed by the same, or related, mechanisms. Nevertheless, the fact that the correlation is reduced to $r = 0.708$ ($P < 0.05$) when different cultivars are studied shows that cultivars differ in their reaction to excess water, and thus makes the water sensitivity test, as a measure of malting quality, less reliable than the germination resistance test.

Furthermore, significant negative correlations have been shown to exist between germination resistance values and the protein content of barley ($P < 0.01$), with the α-amylase content after malting ($P < 0.001$), and the α-amino nitrogen after malting ($P < 0.001$) and a positive correlation with cytase activity after malting ($P < 0.05$). All of these characteristics are used during the selection of barley for malting quality.

CONCLUSIONS

It has been shown that consistent differences in the rate of germination exist even between closely related populations of seeds. Several uses of these findings have been suggested. Some conclusions can be drawn on the importance of these observations.

The investigations described have shown that there is far more to seed quality than simple absence of complete dormancy. They have shown that differences in the rate of germination occur in cereals and in some conifer seeds which are likely to change with time according to a pattern peculiar to individual seed populations. Normal seed-testing techniques fail to identify these important differences in seed quality which can manifest themselves throughout the entire life of a seed. If it is accepted that the highest possible rate of germination represents optimal germination quality, it becomes necessary to be able to recognise this. Unfortunately, single routine seed tests will not allow this to be done. It is arguable whether it is economically justifiable in most cases to perform a series of tests to guarantee optimal germination quality, but in special circumstances it may definitely be justifiable. What is certain is that a knowledge of the factors which affect the rate of germination must lead to a more reliable and meaningful selection for seed germination quality.

From an ecological viewpoint some of the observations on the influence of the environment on rate of germination are interesting. For example, it has been found that where soil conditions are not ideal the germination resistance of the resulting seed crop is high;

and vice versa, better growing conditions of the seed crop lead to lower germination resistance. This is an aspect of seed ecology worthy of receiving more attention.

REFERENCES

BELDEROK, B. (1968). 'Seed dormancy problems in cereals', *Fld. Crop Abstr.*, 21, 203–211

BOYD, W. J. R., GORDON, A. G. and LACROIX, L. J. (1971). 'Seed size, germination resistance and seedling vigour in barley', *Can. J. Pl. Sci.*, 51, 93–99

CZABATOR, F. J. (1962). 'Germination value: An index combining speed and completeness of pine seed germination', *For. Sci.*, 8, 386–396

ESSERY, R. E., KIRSOP, B. H. and POLLOCK, J. R. A. (1955). 'Studies in barley and malt. II. Tests for germination and water sensitivity', *J. Inst. Brew.*, 61, No. 1, 25–28

GOODCHILD, N. A. and WALKER, M. G. (1971). 'A method of measuring seed germination in physiological studies', *Ann. Bot.*, 35, 615–621

GORDON, A. G. (1968a). 'The interaction of dormancy and water-sensitivity of barley with temperature', *J. Inst. Brew.*, 74, No. 4, 355–360

GORDON, A. G. (1968b). 'The variability of barley with special reference to malting quality', *Ph.D. Thesis*, University of Aberdeen

GORDON, A. G. (1969). 'Some observations on the germination energy tests for cereals', *Proc. Ass. off. Seed Analysts N. Am.*, 59, 58–72

GORDON, A. G. (1970). 'Pre-germination in barley', *J. Inst. Brew.*, 76, No. 2, 140–143

GORDON, A. G. (1971). 'Note: The germination resistance test—a new test for measuring germination quality of cereals', *Can. J. Pl. Sci.*, 51, 181–183

HEYDECKER, W. (1961). 'Effects of high and low moisture on seed germination and early growth of vegetables', *Univ. of Nottingham School of Agric. Report*, 1960

HEYDECKER, W. (1966). 'Clarity in recording germination data', *Nature*, 210, 753–754

HEYDECKER, W. (1969). 'The "vigour" of seeds—a review', *Proc. int. Seed Test. Ass.*, 34, No. 2, 201–219

HEYDECKER, W. (1972). 'Vigour', in *Viability of Seeds*, (ed. Roberts, E. II.), 209–252. Chapman and Hall, London

ISTA (1966). 'International rules for seed testing 1966', *Proc. int. Seed Test. Ass.*, 33, No. 1

JUNGES, W. and LUGWIG, H. (1963). 'Einfluss von Wasserversorgung, Düngung und Licht während der Samenbildung auf die Samenruhe bei *Impatiens balsamina* L.', *Proc. int. Seed Test Ass.*, 28, No. 1, 71–96

NILSSON-EHLE, H. (1914). 'Zur Kenntnis der mit der Keimungsphysiologie des Weizens in Zusammenhang stehenden inneren Faktoren', *Z. Pflanzenz.*, 2, 153

ORPHANOS, P. I. and HEYDECKER, W. (1967). 'The danger of wet seedbeds to germination', *Univ. of Nottingham School of Agric. Report*, 1966–67, 73–76

RENARD, H. A. (1967). 'Evolution des propriétés germinatives des semences des graminées pendant et après leur maturation', in *Physiologie, Ökologie und Biochemie der Keimung*, (ed. Borriss, H.). Vol. 1, 347–358. Ernst-Moritz-Arndt-Universität, Greifswald

ROBERTS, E. H. (1972). 'Dormancy: a factor affecting seed survival in the soil', in *Viability of Seeds* (ed. Roberts, E. H.), 321–359, Chapman and Hall, London

TIMSON, J. (1965). 'A new method of recording germination data', *Nature*, 207, 216–217

DISCUSSIONS

E. H. Roberts: The concept of 'resistance to germination' seems unnecessarily complicated. Why not speak of the time it takes a seed to germinate, or of its reciprocal, the rate of germination?

Gordon: The point of the concept of 'resistance to germination' is that it changes: the gradual reduction in the time to germination can be visualised as an extension of the loss of dormancy.

Whittington: In general, white-seeded varieties of wheat lose their dormancy at a faster rate than red ones; is there still a difference in their mean time to germination when the dormancy has disappeared? It may be more logical to count from the date of anthesis rather than the date of sowing.

Gordon: Even when they all have become completely non-dormant, some white varieties germinate more readily than other white ones which themselves may germinate with as much delay as red ones. This is so even in Manitoba, Canada, where maturation conditions eliminate dormancy and all grains are germinable at harvest. The time scale is greatly reduced, but even here a gradual increase in the rate of germination with time from harvest is experienced.

Thompson: The after-ripening process with which Dr. Gordon is concerned may not only involve an increase in the rate of germination but also a widening of the range of temperatures at which the seeds can germinate. Ideally, one needs to look at the germination pattern at the optimal temperature for each batch of seeds, or alternatively, over a range of temperatures.

Gordon: For barley we originally selected the malting temperature. Agronomically, it seems desirable to select the field temperature expected.

Monteith: An agreed nomenclature is desirable in order to foster interdisciplinary studies. Thus, to physicists, 'resistance' has a precise meaning: it is the relation between a flux and a gradient. Similarly, in physical terms 'speed' refers to a distance travelled in unit time whereas 'rate' refers to the occurrence of any event in unit time. We should therefore speak of rate and not of speed of germination (see Glossary).

23

TEMPERATURE RELATIONS OF GERMINATION IN THE FIELD

T. W. HEGARTY

Scottish Horticultural Research Institute, Invergowrie, Dundee

INTRODUCTION

In a number of crop species in which the value of the crop is related in some way to the size of the plant part harvested, optimum yields are often obtained by controlling plant growth through inter-plant competition. Examples are the production of potatoes, carrots or beetroot for canning or bottling whole, and onions, brussels sprouts or cauliflowers for sale on the fresh market or for processing. In a crop such as carrots in which the plant density has to be high when a high yield of small roots is desired (more than 1000 plants per m^2— about 90 plants per ft^2—of bed, using a system of spaced beds) adjustment of the plant stand after emergence is impracticable. Sowing rates must therefore be chosen with the aim of producing the plant population required.

Usually the percentage of seedlings emerging in the field is less than the percentage of viable seeds in a seed lot (the value obtained from the official germination test which is carried out under conditions optimal for germination). The soil is therefore not an ideal environment for germination and emergence, and can be considered to be either more or less adverse at different times. The soil factors affecting emergence are temperature, moisture, aeration, pathogen content and activity and structural properties such as resistance to seedling penetration and liability to crusting (capping). In order to calculate sowing rates it is necessary to predict to what extent these soil factors will affect each seed lot in the period between sowing and emergence, a period that may be of several weeks' duration. At the present state of knowledge this would be an impossible task even if weather conditions could be accurately predicted, and for a crop such as carrots it is generally accepted that wide deviations from the target level of emergence can occur (Bleasdale, 1963; Reynolds, 1968).

Carrots are used as examples throughout this paper for three reasons:

1. The crop is sown 'to a stand' and not thinned.
2. There is a real problem in predicting accurately plant stands of carrots in the field.
3. The cultivated carrot appears to be free from many, although apparently not all, forms of seed dormancy (Dale, 1970).

Since the majority of carrot sowings in East Scotland are made in April and May the effect of temperature, and in particular of low temperature, on the final level and on the rate of germination has been investigated as part of a wider programme of work on the

Table 23.1   MEAN DAILY MAXIMUM AND MINIMUM VALUES OF SOIL TEMPERATURE (°C) AT 2 cm DEPTH FOR WEEKLY PERIODS IN APRIL RECORDED IN 1969, 1970 AND 1971

Week	Mean maximum temperature				Mean minimum temperature			
	1969	1970	1971	Mean	1969	1970	1971	Mean
1	18·0	8·9	7·3	11·4	−1·4	−0·5	3·0	0·4
2	15·0	10·8	18·4	14·8	0·0	0·5	3·0	1·2
3	19·3	14·2	15·4	16·3	−0·6	3·9	4·3	2·5
4	12·2	14·6	11·3	12·7	2·8	1·5	3·0	2·4
			Mean	13·8			Mean	1·6

relation between seedling emergence and the soil environment. Soil temperatures have been recorded at seed depth (2 cm) over 3 years and the results for April are shown in *Table 23.1*. The mean maximum and mean minimum values for the month as a whole are approximately 14°C and 2°C, respectively. The mean maximum temperatures varied considerably, and did not necessarily increase steadily as time advanced. Minimum temperatures varied less. The extreme values recorded in the 3 years were 26°C and −3°C, both in 1969.

The literature on the effects of temperature on both the final level and the rate of germination or emergence dates back to the last century, and in particular to Sachs (1860) and Haberlandt (1874). The concept of cardinal points, i.e. maximum, minimum and optimum temperatures of germination, was proposed at this time by Sachs. In his review on temperature relations of germination, Edwards (1932) described many of the results from earlier papers, although in general he omitted the results for small vegetable seeds. Briefly, it had been shown from an early date that the maximum level of germination tended to occur over a range of temperatures above or below which the level of germination was reduced. The rate of germination was also temperature dependent, normally being fastest in a narrow temperature range above or below which it was progressively reduced. Edwards presented the data given by Atterberg (1928) for oats as an example of a typical temperature response pattern (*Figure 23.1*). It consists of a series of 'inverted-U' curves,

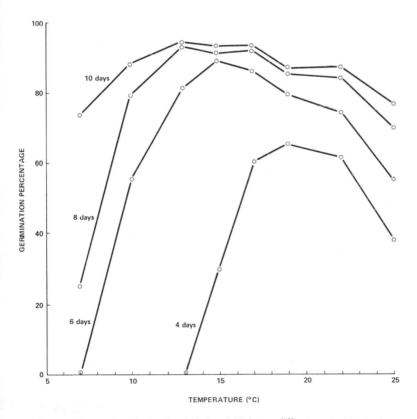

Figure 23.1   Germination of oats after 4, 6, 8 and 10 days at different constant temperatures. (From Edwards (1932) after Atterberg (1928))

each showing the extent of germination at different temperatures for a different time after the start of the experiment. More recently Thompson (1970) has used thermogradient bars to construct curves of this type more accurately and to use them for characterising the temperature-response pattern of various species and ecotypes in relation to their natural habitats (see Chapter 3).

Kotowski (1926) investigated both the level and rate of emergence of a number of vegetable seeds sown in sand held at a range of constant temperatures from 4°C to 30°C, and J. F. Harrington (1962) germinated the seeds of several vegetable species on blotters at a range of constant temperatures from 0°C to 45°C. Harrington noted that the extreme temperatures of germination could be influenced by factors such as genetic variability, age of seed and the criterion used for germination. He used seeds of high germination percentage and equated this with high seed vigour. G. T. Harrington (1923), using a number of constant and alternating temperatures within the range 15°C to 35°C, showed that for 2 carrot seed lots the final level of germination was reduced only in temperature regimes which included a period at 35°C.

The work of Kotowski (1926) and J. F. Harrington (1962) was carried out using one seed lot only to represent each vegetable type. Harrington speculated, however, that if germination conditions had been less favourable, for instance if seeds of low vigour had been used or the tests had been run in soil, the temperature extremes of germination would not have been so wide.

Torfason and Nonnecke (1961) studied the effect of a number of low-temperature stress treatments on the laboratory germination of three varieties each of beetroot, carrot, onion and swede. The seeds were held at one of 4 temperatures in the range of approximately 2°C to 10°C (35–50°F) for between 1 and 22 days before being transferred to room temperature for germination to be completed. The germination of onions was the same in all the treatments but in the other crop types final germination was reduced by one or more of the treatments. There were significant differences of response between the cultivars of each crop type used and, for beetroot and swede at least, there appeared to be a cultivar × temperature interaction. An interaction between seed stock and temperature was reported by Heydecker (1962) quoting the work of Hepton (1957). Two stocks of cauliflower seed were germinated at a range of temperatures (2–35°C) and differences in their tolerance to high and low temperatures were found.

The published data on carrots suggest that the minimum temperature for germination is between 0°C and 5°C, with germination below the highest attainable percentage between 5°C and approximately 10°C. Some damage may be sustained by exposure to temperatures just above 0°C. Since temperatures such as these are often recorded at seed depth during the typical period when sowing is carried out in the field (*Table 23.1*), low temperature could be expected to have an effect on germination and emergence. In addition, differences between seed lots in the temperature response patterns of germination such as those shown by other crop types could be important in determining both the final level and the rate of germination and emergence.

PERCENTAGE GERMINATION AND EMERGENCE

In the experiments to be described germination tests were carried out either in plastic germinators with the seeds placed either on blotting paper which was kept wet by wicks carrying water from a reservoir below, or in petri dishes (10 cm^2) with the seeds placed on four thicknesses of germination paper, wetted with 12·5 ml water. The seeds were kept at constant or alternating temperatures in the dark. Field emergence studies were carried out with seeds dusted with Orthocide Seed Dressing (75 per cent a.i. captan) at a rate of 0·3 per cent w/w. Full details have been given previously (Hegarty, 1971; Perry and Hegarty, 1971).

Table 23.2 PERCENTAGE GERMINATION AT A RANGE OF TEMPERATURES FROM 5°C TO 25°C OF 4 CARROT SEED LOTS IN 1969

Seed lot	Temperature (°C)				
	5	10	15	20	25
1	86	91	92	91	94
2	75	82	88	89	89
3	47	68	75	80	84
4	39	78	88	90	93

In 1969, laboratory germination tests were carried out on 11 seed lots at a range of temperatures from 5°C to 25°C. The seed lots were also sown in the field on three occasions: 1 April, 29 April and 27 May. The germination percentages (radicle emergence) of four of these seed lots are shown in *Table 23.2*. Whilst on the whole there was a tendency for the seed lots to show lower germination at lower temperatures, there were obvious differences in the response patterns of the different seed lots. The highest and the lowest germination percentages of seed lot 1 differed by only 8 per cent, whereas for seed lot 4 the difference was 54 per cent. A sharp drop in germination was apparent between 10°C and 5°C in seed lots 3 and 4. Seed lot 2 showed an intermediate pattern of response. These data are wholly in agreement with the low temperature end of the 'inverted-u' temperature response curve but show that the position of the downward-sloping portion of the curve can differ with the seed lot. This displacement is evident over precisely the range of temperatures that occur in the field during germination.

For the 11 seed lots correlation coefficients were calculated between the emergence percentages at the field sowings and germination percentages at the five temperatures (*Table 23.3*). The best correlations were between emergence at sowings 1 and 2 and germination at 10°C. The mean soil temperatures between sowing

Table 23.3   CORRELATION COEFFICIENTS FOR THE RELATIONSHIP BETWEEN PERCENTAGE FIELD EMERGENCE AND PERCENTAGE LABORATORY GERMINATION AT A RANGE OF CONSTANT TEMPERATURES FOR 11 SEED LOTS SOWN ON 3 OCCASIONS IN 1969 (1 APRIL, 29 APRIL, 27 MAY)

Germination temperature (°*C*)	*Sequential sowing number*		
	1	*2*	*3*
5	0·69*	0·60	0·54
10	0·85†	0·85†	0·67*
15	0·62*	0·58	0·50
20	0·50	0·54	0·51
25	0·43	0·46	0·71*

* P = 0·05
† P = 0·001

and half emergence at the three field sowings were 10·6, 10·2 and 15·0°C, respectively. It seems likely, therefore, that the 10°C germination test could be a useful indicator of emergence in the field for early sowings when the mean soil temperature is close to this value.

In order to test this, 20 seed lots were sown in 1970 on 2 April and, using the regression equation calculated between field emergence at sowings 1 and 2 in 1969 and germination at 10°C (G_{10}), the field emergence was predicted from the G_{10}s of the 20 seed lots. The agreement between the actual and predicted emergence was good (*Figure 23.2*). For seed lots of the cultivar Chantenay the equation obtained for the relationship between field emergence and laboratory germination was:

$$\text{Emergence } (\%) = 0.8 \times \text{Germination at } 10°C \ (\%)$$

It appears that the temperature sensitivity of a seed lot in respect of its germination is relevant to its ability to emerge in the field under certain field conditions. An arbitrary definition of temperature sensitivity is that the G_{10} is significantly less ($P < 0.05$) than the

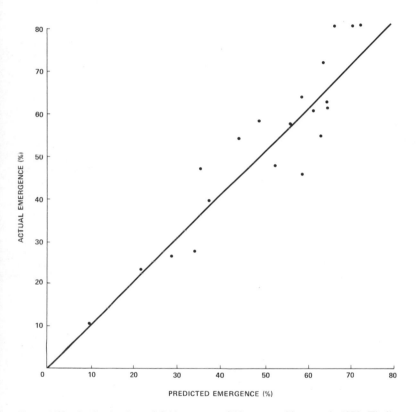

Figure 23.2   Predicted and actual field emergence of 20 carrot seed lots sown in 1970. The line y = x is shown. (From Hegarty, 1971)

G_{20}. In 1971 an investigation showed that temperature sensitivity occurred in 19 out of 50 commercially available Chantenay carrot seed lots and if this applies generally almost 40 per cent of seed lots will show temperature sensitivity of germination. Data obtained over 3 years have shown that temperature sensitivity is in no way associated with the ability of the seed lot to germinate under optimum conditions. The data in *Table 23.4*, for five of the seed lots used in these studies, show this point clearly.

In a further study of temperature effects on germination, two seed lots were germinated in a number of temperature regimes in the laboratory (*Table 23.5*). These included constant temperatures between 8°C and 26°C at 3°C intervals and, using a cam-controlled

Table 23.4 PERCENTAGE GERMINATION OF 5 SEED LOTS AT 10°C AND 20°C

| Seed lot | Temperature (°C) | |
	10	20
1	90	91
2	67	90
3	18	74
4	49	68
5	67	65

Table 23.5 GERMINATION IN PERCENTAGES AND ANGLES (ARC SIN TRANSFORMATION) OF 2 SEED LOTS (A AND B) AT A RANGE OF CONSTANT TEMPERATURES AND AT ONE CYCLING TEMPERATURE WITHOUT AND WITH TEMPERATURE STRESS TREATMENTS; THESE TREATMENTS WERE APPLIED FOR 12 h EACH AND THE SEEDS THEN TRANSFERRED BACK TO THE CYCLING REGIME

| Temperature regime (°C) | Seed lot | | | |
| | A | | B | |
	percentage	angles	percentage	angles
26	88·9	70·6	86·1	68·2
23	89·5	71·3	88·3	70·3
20	88·5	70·3	83·9	66·4
17	86·6	68·4	78·9	62·8
14	88·7	70·8	76·8	61·3
11	89·2	70·9	72·8	59·0
8	85·2	67·5	46·3	42·8
2/14	88·0	69·8	72·4	58·4
2/14 *with temperature stress treatments*				
after 1 day— −3	88·3	70·1	72·5	58·5
after 1½ day—26	88·0	69·9	75·5	60·5
after 1 day— − 3 then 26	85·4	67·8	76·9	61·5
after 7 days— −3	79·8	64·2	67·9	55·6
after 7½ days—26	84·8	67·1	78·2	62·2
after 7 days— − 3 then 26	71·4	58·1	71·7	57·9
S.E.M. (39 df)		2·20		2·03

incubator, a regime cycling diurnally between 2°C and 14°C, the mean minimum and mean maximum daily temperatures recorded in April over 3 years (*Table 23.1*). The cam was cut to give a sine-curve fluctuation and the agreement between the desired and the actual temperature (measured with thermistors on the germination paper inside the petri dishes) was good (*Figure 23.3*). In addition a

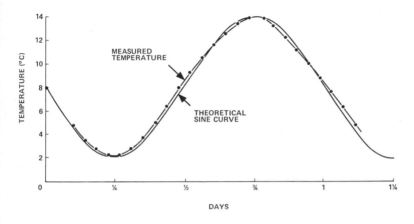

Figure 23.3   The approximation of the 2/14°C cycling temperature regime to the sine curve relationship, temperature (°C) = 8–6 sin 2πt, where t is the number of days from the start of the experiment

number of temperature stress treatments were imposed on this alternating cycle (*Figure 23.4*). These were based on the extreme soil temperatures of −3°C and 26°C recorded in April, 1969. For each treatment the seed lots were transferred from the cycling regime to a constant temperature of −3°C or 26°C for 12 h, with the −3°C and 26°C stresses applied in the first and second halves, respectively, of a stress day in order to keep the treatments 'in phase' with the temperature cycle. A combination of −3°C followed by 26°C for 12 h each was also included. After the stress treatments the seed lots were transferred back to the cycling regime. The stresses were applied on one of two occasions, either after 1 or 1½ days (for the low and high temperature stresses, respectively), during the late stages of imbibition, or after 7 or 7½ days, when the first seeds were seen to be germinating. As one seed lot (A) germinated more rapidly than the other, the treatment, applied simultaneously to the two seed lots, may not have acted in quite the same way on each seed lot. The results are presented in *Table 23.5*.

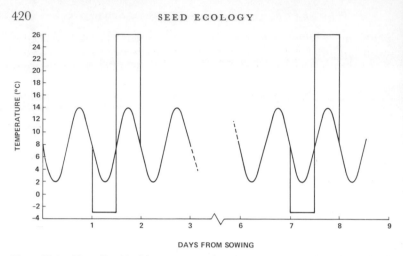

DAYS FROM SOWING

Figure 23.4   The cycling 2/14°C temperature regime with the temperature stresses imposed on it

At constant temperatures the germination percentages of seed lot A did not differ significantly ($P < 0.05$) although there was a slight drop in germination at 8°C, the lowest constant temperature. The cycling temperature did not reduce germination, nor was there any effect of any of the stress treatments applied at $+1$ or $+1\frac{1}{2}$ days. But when applied at 7 days the cold treatment, particularly in combination with the 26°C treatment, did result in a reduction of germination. Unless some form of dormancy was induced which inhibited germination, it would appear that here seeds were killed by the stress treatment. Since the treatments applied in this experiment were more extreme in both intensity and duration than the conditions normally prevailing in the field the implication is that the level of germination of seed lot A would be little affected within the average range of field temperatures.

In contrast seed lot B showed temperature sensitivity of germination. The effect of the fluctuating temperature was to reduce germination below the maximum level reached at higher temperatures, although in this case the additional temperature stress treatments applied caused no further reduction in germination. In seed lot A none of the treatments resulted in an inhibition of germination; only when the temperature stress was experienced as radicle emergence was commencing were a number of seeds killed. In seed lot B many of the treatments resulted in inhibition of germination and possible death, but there is no evidence that the -3°C stress treatments caused additional death of seeds. This may have been because the seeds were not 'at risk' when stressed.

Evidence from other experiments suggests that in temperature

sensitive seed lots, seeds can be either inhibited or killed by comparatively mild stress treatments. For instance, of the 19 seed lots found in
1971 to exhibit temperature sensitivity of germination, i.e. reduced
germination at 10°C compared to 20°C, full 'recovery' of germination did not take place after transfer from 10°C to 20°C. The mean
germination percentages at constant temperatures of 20°C and 10°C
were 79·3 per cent at 20°C and 65·9 per cent at 10°C rising to only
71·9 per cent after transfer to 20°C for a further 14 days.

The results indicate that under the average soil temperature
fluctuations of April the germination of the temperature insensitive
seed lot would not be inhibited, although exceptional frost conditions
occurring at or near the time of radicle emergence might result in

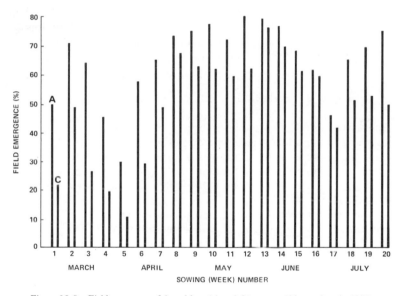

Figure 23.5   Field emergence of 2 seed lots (A and C) sown on 20 occasions in 1971

some deaths. In the temperature sensitive seed lot inhibition of
germination would be expected with resultant killing of a proportion
of the inhibited seeds, presumably due to the extended period that
seeds were at risk to pathogen attack at lower temperatures.

This may be illustrated by data obtained in 1971 for carrot seed
lots sown in the field on 20 weekly occasions from 9 March. The final
percentage emergence was recorded when the majority of the
seedlings had reached the 3–4 leaf stage. The results for two of the
seed lots (A and C) are shown in *Figure 23.5*. Emergence of both

P

was sharply reduced in April and June, on the former occasion apparently due to the formation of a soil cap and on the latter occasion due to a combination of high soil temperature and exceptionally low soil moisture. Differences between the seed lots were large on some occasions but small on others, representing differences in the ability of the seed lots to withstand adverse conditions—i.e. differences in seed quality. Finally, the emergence of seed lot A reached values exceeding 65 per cent in March, April, May, June and July. This potential consistency of emergence occurred over a period when the mean daily soil temperatures during germination and emergence rose from approximately 7°C in March to 22°C in July. Seed lot A was the same as the seed lot A used in the germination experiment described above, and this potential consistency of emergence over a wide range of soil temperature conditions indicates that in a temperature insensitive seed lot temperature *per se* (within the range normally found in the field in spring) need have little or no effect on the final level of emergence.

RATE OF GERMINATION AND EMERGENCE

Apart from its effect on the final level of emergence, temperature has an effect on the rate ('speed') of emergence in the field. In the experiment just described the time to half emergence $(T_{\frac{1}{2}})$ (i.e. the time when half the seedlings ultimately emerging had done so) varied from approximately 6 weeks for the first sowing in March to 10 days in July. Dr. P. A. Thompson (private communication) has shown that there is a linear relationship in carrots between rate of germination and temperature over a restricted temperature range. Using data obtained in the laboratory for seed lots A and B the relationship is demonstrated in *Figure 23.6*. The rate value used here is the mean of 4 replications of the reciprocal of $T_{\frac{1}{2}}$. In seed lot A the percentage value of half germination did not vary greatly over the range of constant temperatures used, but in seed lot B it was lower at lower temperatures. For seed lot B the rate of germination at 26°C was lower than would have been expected with a linear relationship and it seems that in this seed lot the effect of superoptimal temperature was being shown; in seed lot A there was no apparent deviation from linearity over the range 8°C to 26°C. Over this range for seed lot A, and 8°C to 23°C for seed lot B, the correlation coefficients between rate of half germination $(R_{\frac{1}{2}})$ and temperature (θ) were 0·998 and 0·999, respectively. The differences between the slopes of the regression equations were significant (P < 0·05).

Because temperatures in the field are normally fluctuating rather than constant the effect of the cycling temperature regime 2/14°C on rate of germination was investigated in the laboratory and compared with the rate of germination at constant temperatures. In theory, because rate of germination is linearly related to temperature, rates of germination over a range of temperatures should be

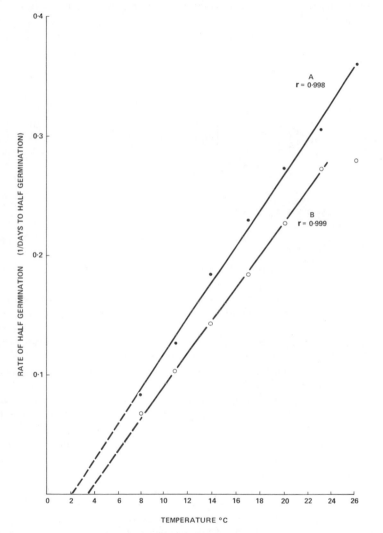

Figure 23.6   Rate of half germination plotted against temperature for two carrot seed lots (A and B). The point for seed lot B at 26°C was not included in the calculation of the correlation coefficient or regression equation

additive and, for any temperature cycle fluctuating symmetrically about the mean of its extreme values, the progress made towards germination during one complete daily cycle should be the same as that made in one day at the constant temperature mean.

For precise prediction of the time to half germination ($T_{\frac{1}{2}}$) in a cycling temperature it is necessary to use a model of the situation. It has been established that there is a linear relationship between rate of half germination ($R_{\frac{1}{2}}$) and temperature (θ):

$$R_{\frac{1}{2}} = P + Q\theta \tag{1}$$

where P and Q are constants for any particular seed lot over a certain temperature range (*Figure 23.6*). Over any small period of time, Δt, when the temperature is θ, the seed lot makes some progress towards half germination ($\Delta G_{\frac{1}{2}}$), which is complete when, by definition, $G_{\frac{1}{2}} = 1$. Hence:

$$\Delta G_{\frac{1}{2}} = R_{\frac{1}{2}} \times \Delta t \tag{2}$$

and, incorporating equation (1):

$$\Delta G_{\frac{1}{2}} = (P + Q\theta) \times \Delta t \tag{3}$$

In the situation where the relationship between θ and t can be described mathematically, as in the case described earlier (*Figure 23.3*), an equation may be established between $G_{\frac{1}{2}}$ and t alone and the value of $T_{\frac{1}{2}}$ may be calculated from the integral of this equation. The use of a model such as this would be particularly helpful in situations where the temperature fluctuation is not symmetrical about the mean of the two extreme temperatures. In the present situation it is useful both for predicting precisely values of $T_{\frac{1}{2}}$ and for demonstrating the importance of two assumptions made. The first is that the rate–temperature relationship is linear down to the value of θ at which $R_{\frac{1}{2}} = 0$ (the limiting low temperature of germination); the second is that if temperatures lower than this minimum temperature are encountered in the temperature cycle, negative values of $\Delta G_{\frac{1}{2}}$ should be summed in the integral. This may or may not be a real situation.

For seed lot A there is evidence from a subsequent experiment that the rate–temperature equation was linear at least down to 5 °C. The estimated minimum temperature of germination was 2 °C, the extreme low temperature of the cycle used. The model described above should therefore have been valid for seed lot A, at least as far as the temperature limits are concerned. The actual $T_{\frac{1}{2}}$ for the

2/14°C cycle was 10·3 days, which compares with the value at 8°C constant of 11·9 days (significantly different for logarithmically transformed data) and the predicted value of 11·7 days. Germination was thus significantly faster in the fluctuating regime.

The process of germination involves physical and chemical processes, both of which are likely to be temperature-dependent. At the low-temperature limit germination ceases although some at least of the physical processes do not. The cessation of the germination process is presumably due to the effect of temperature on one or more chemical processes. However, it is possible that accumulation or loss of various substances into, from or within the seed can continue, thus allowing a faster rate of germination at a subsequent higher temperature when this process would otherwise be limiting. In this way the rate of germination at a fluctuating temperature with its low temperature extreme at or close to the minimum temperature of germination could be faster than expected.

The results from a separate experiment support this hypothesis. Seed lot A was germinated at a cycling temperature of 4/24°C (linear change of temperature) and the observed and predicted values of $T_{\frac{1}{2}}$ agreed closely. One difference between the 2/14°C and the 4/24°C regimes was the period of time spent at low temperature— e.g. in the former approximately $7\frac{1}{2}$ h and in the latter only 1 h below 5°C in each daily cycle. This is further evidence that events occurring at low temperature may have been the cause of the difference in $T_{\frac{1}{2}}$ between the 2/14°C regime and the constant 8°C regime.

This suggests that except at low temperature, rate of germination should give a linear plot with $\bar{\theta}$, where $\bar{\theta}$ is the mean daily temperature of a symmetrical cycling temperature regime. Even at low temperature this is approximately true, since rates of germination are low in any case and the mean rate is therefore relatively little affected by the deviations that would be expected to occur.

The application of this treatment of results to data obtained from the field requires caution. The processes affecting rates of emergence in the field after germination need not be the same as those affecting germination itself; the daily cycle of temperature in the soil at seed depth is rarely symmetrical; and there is temperature drift during the period of germination and emergence. Nevertheless, using the reciprocal of time to half ultimate emergence as a measure of rate of emergence, and mean soil temperature between sowing and half emergence as a measure of temperature, a reasonably good linear relationship between $R_{\frac{1}{2}}$ and $\bar{\theta}$ has been shown to exist. *Figure 23.7* gives the results for seed lot A sown on 20 consecutive weekly occasions in 1971. Point X is clearly anomalous. This sowing was made into soil at 3–4 per cent moisture content which did not increase until rainfall occurred 8 days later. On the simple assumption that the start of germination was delayed by 8 days the rate and mean temperature values have been recalculated and the point (X′) fits well in the overall relationship. Point Y, the sowing made one week after sowing X, represents seeds held in dry soil for one day before rainfall occurred and has been adjusted accordingly (Y′). No

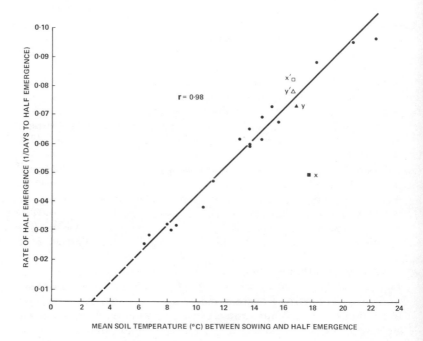

Figure 23.7 Rate of half emergence plotted against mean soil temperature between sowing and half emergence for carrot seed lot A. Point X (■) adjusted to point X′ (□) and point Y (▲) to Y′ (△). The regression line includes the adjusted values

other sowing date was affected in this way by soil conditions such as these. For the 18 uncorrected and 2 corrected points the correlation coefficient was 0·98. The estimated minimum temperature for germination from the graph was 1·7 °C which was in good agreement with the estimated value of 2·2 °C obtained from the rate–temperature relationship for this seed lot in the laboratory. Similar relationships were found between rate of emergence and mean soil temperature for three other seed lots. Thus the same sort of relationship between rate of germination and temperature that holds in the laboratory appears to be valid for rate of emergence and mean soil temperature in the field. It is interesting that this relationship holds over a relatively wide temperature range (approximately 2°C to 26°C under constant temperature conditions for one carrot seed lot at least) and there is a strong implication that except under conditions of extreme moisture stress, the effects of soil moisture on rate of emergence may be small compared to the effect of temperature.

Kotowski (1926), using his 'coefficient of velocity of germination' as a measure of rate of germination, showed that Q_{10} (temperature coefficient) values were high at low temperatures and decreased with an increase in temperature. For example, the values for carrot were 4·69 at 8–11°C and 1·38 at 25–30°C. He suggested that the speed of growth of seedlings (speed of emergence) may depend on the rate of chemical reactions that are responsible for the characteristic metabolism of each kind of germinating seed.

Calculations using data on carrot emergence in the field have shown that the coefficient of velocity used by Kotowski and the rate of half emergence used in the experiments described above differ little. Indeed, a number of different methods of calculating rate of germination give values that are essentially similar (Nichols and Heydecker, 1968). Accepting that the rate measures used are reasonably comparable, Q_{10} values were calculated for the carrot germination and emergence data in the above experiments. The values (calculated from the regression equations) for the intervals 8–11°C and 25–30°C were 4·09 and 1·49, respectively, for the germination data and 3·63 and 1·48 for the field emergence data. It is a property of linear relationships with temperature such as the one discussed here that Q_{10} values must be higher at low than at high temperatures. The similarity of pattern of response with Kotowski's values suggested that his own data might be reinterpreted on a simple rate–temperature basis, and the results for carrots, beetroot, cabbage and onion are shown in *Figure 23.8*. Over the range 5°C to 30°C there is evidence to suggest that linear relationships hold between rate of emergence and temperature for all these crop types. For other species that Kotowski used the evidence is less strong, or the relationship may not be linear.

Germination, as defined by radicle emergence, is the outward biological expression that a large number of chemical and physical processes have occurred in the seed. At any particular instant the rate of progress towards germination is governed by the slowest process involved. According to Mayer and Poljakoff-Mayber (1963), the fact that germination over the *whole* temperature range in which it occurs is not characterised by a simple temperature coefficient may

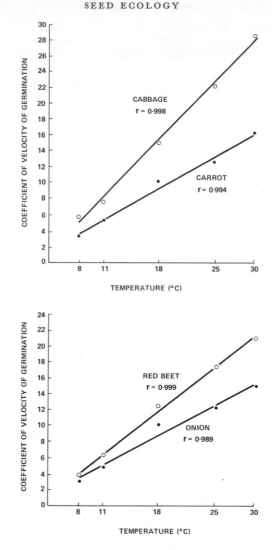

Figure 23.8   Coefficient of velocity of germination plotted against temperature for four crop types. Data of Kotowski (1926)

be understood if it is appreciated that germination is a complex process and a change in temperature will affect each constituent step individually, so that the effect of temperature which is observed will merely reflect the overall resultant effect.

There is, therefore, no *a priori* reason to suppose that there should be a single, dominant rate limiting process during germination and emergence, even over the restricted temperature range studied (approximately 5°C to 25°C), although the striking linear relationship between rate of germination or emergence and temperature suggests that this might be the case.

For many enzyme-activated reactions there is a linear relationship between the logarithm of the rate of reaction and the reciprocal of the absolute temperature (the Arrhenius relationship). The data from the carrot experiments described above are not linearly related when plotted in this way and a single rate-limiting reaction of this type is clearly not operating.

Processes with a high Q_{10} at temperatures below a critical value and a low Q_{10} at temperatures above this are common in biological systems (Johnson, Eyring and Polissar, 1954; Kuiper, 1964). Again, the data obtained in the present experiments do not fit the relationships used to demonstrate this as well as they fit the linear rate-versus-temperature relationship.

Kramer (1940) has shown that uptake of water into the roots of sunflower plants, whether held in water or soil, is approximately linearly related to temperature over the range 0°C to 40°C. However, he suggested that this may be the fortuitous resultant effect of two processes, both of which affect uptake rate, acting against each other. This sort of possibility must be borne in mind.

In the present circumstances there is no evidence to suggest what the rate limiting process or processes might be. However, it may be relevant to consider germination and emergence in relation to water uptake, using this as an example of one of a number of physical process that could be rate limiting. In highly endospermic or perispermic seeds (such as carrot, onion or beetroot) sown in the soil, germination (radicle protrusion from the seed) is a more or less arbitrary stage in the growth of the embryo leading to field emergence. In view of this continuity of growth, the factors affecting the process of water uptake by the embryo within the seed (rather than by the seed as a whole), could well be similar to those affecting water uptake by the protruding radicle and the developing root in the soil. If such a process were rate limiting its effect and response to temperature could be continuous throughout the stages of both germination and emergence. This would help to explain why the same sort of rate–temperature relationship governs both germination and emergence.

It may, therefore, be more logical to look as closely for a physical explanation of this relationship than to attempt to explain, as Kotowski (1926) did, the difference in Q_{10} values at different temperatures purely on a chemical basis.

CONCLUSIONS

The range of temperatures found in the field during spring sowings might be expected to affect both the final level and the rate of emergence. In carrots it is apparent that, even within the wide range of temperatures which can occur in carrot seed beds, temperature itself is not likely to be a factor limiting the final level of emergence if a temperature-insensitive seed lot is used. The use of a temperature-sensitive seed lot, however, can lead to a reduction in emergence. Differences in seed quality associated with low-temperature sensitivity do not appear to be confined to carrots; the results of Hepton (1957) suggest that this is also true for cauliflower. In addition, the cultivar × temperature interaction of beetroot and swede indicated in the results of Torfason and Nonnecke (1961) could well be due to differences in seed stocks rather than cultivars.

Rate of germination is closely correlated with temperature in the laboratory, as is rate of emergence with mean soil temperature in the field. Under the field conditions found in 1971, temperature was the dominating factor affecting rate of emergence in the field. The linear relationship between rate of germination (emergence) and temperature over a wide temperature range (5°C to 25°C approximately) for a number of unrelated crop species suggests that the rate of germination and emergence may be limited as much by physical as by chemical processes.

The differences in germination characteristics found within one carrot cultivar could be due to a number of factors, e.g. maturity of the seed at harvest, seed age, and physiological status (the extent to which the seed has deteriorated in storage). For many agricultural and other species these same factors could affect seed quality. It therefore seems dangerous to generalise about genetic (varietal or ecotype) differences in germination behaviour between seed lots unless the environmental factors affecting seed quality prior to sowing have been eliminated.

REFERENCES

ATTERBERG, A. (1928). 'Die Nachreife des Getreides', *Landw. Versuchsst.*, **67**, 129–143

BLEASDALE, J. K. A. (1963). 'The bed system of carrot growing', *Short Term Leafl. Minist. Agric. Fish Fd.* No. 27, 19

DALE, H. M. (1970). 'Germination patterns in *Daucus carota* ssp. *carota*. I. Variations in the 1967 collections', *Can. J. Bot.*, **48**, 413–418

EDWARDS, T. I. (1932). 'Temperature relations of seed germination', *Q. Rev. Biol.*, **7**, 428–443

HABERLANDT, F. (1874). 'Die oberen und unteren Temperaturgrenzen für die Keimung der wichtigeren landwirtschaftlichen Sämereien', *Landw. Versuchsst.*, **17**, 104–116

HARRINGTON, G. T. (1923). 'Use of alternating temperatures in the germination of seeds', *J. agric. Res.*, **23**, 295–332

HARRINGTON, J. F. (1962). 'The effect of temperature on the germination of several kinds of vegetable seeds', *16th Int. hort. Cong.*, **2**, 435–441

HEGARTY, T. W. (1971). 'A relation between field emergence and laboratory germination in carrots', *J. hort. Sci.*, **46**, 299–305

HEPTON, A. (1957). 'Studies on the germination of *Brassica oleracea* var. *botrytis* (Linn) with special reference to temperature relationships', *B.Sc. (Hons.) Dissertation*, Univ. of Nottingham

HEYDECKER, W. (1962). 'From seed to seedling: factors affecting the establishment of vegetable crops', *Ann. appl. Biol.*, **50**, 622–627

JOHNSON, F. H., EYRING, H. and POLISSAR, M. J. (1954). *The kinetic basis of molecular biology*, Wiley, New York

KOTOWSKI, F. (1926). 'Temperature relations to germination of vegetable seed', *Proc. Am. Soc. hort. Sci.*, **23**, 176–184

KRAMER, P. J. (1940). 'Root resistance as a cause of decreased water absorption by plants at low temperatures', *Pl. Physiol.* **15**, 63–79

KUIPER, P. J. C. (1964). 'Water uptake of higher plants as affected by root temperature', *Meded. Landb. Hoogesch. Wageningen* **64**, 1–11

MAYER, A. M. and POLJAKOFF-MAYBER, A. (1963). *The Germination of Seeds*, Pergamon Press, Oxford

NICHOLS, M. A. and HEYDECKER, W. (1968). 'Two approaches to the study of germination data', *Proc. int. seed Test. Ass.*, **33**, 531–540

PERRY, D. A. and HEGARTY, T. W. (1971). 'Effect of captan seed dressing on carrot emergence', *Pl. Path.*, **20**, 29–32

REYNOLDS, J. D. (1968). 'Maincrop carrot variety trials, 1959–65', *J. natn. Inst. agric. Bot.*, **11**, 307–321

SACHS, J. (1860). 'Physiologische Untersuchungen über die Abhangigkeit der Keimung von der Temperatur', *Jahrb. wiss. Bot.*, **2**, 338–377

THOMPSON, P. A. (1970). 'Characterization of germination response to temperature of species and ecotypes', *Nature*, **225**, 827–831

TORFASON, W. E. and NONNECKE, I. L. (1961). 'A study of the effects of temperature and other factors upon the germination of vegetable crops. IV. Root crops', *Can. J. Pl. Sci.*, **41**, 360–370

DISCUSSIONS

Edwards: Where the seed coat does not present an appreciable resistance to oxygen diffusion, as it does in intact charlock seeds, the rate of oxygen supply is likely to be considerably influenced by temperature, because oxygen diffuses to the embryo in solution and its solubility is higher at lower temperatures. The rate of growth of excised charlock embryos is directly proportional to the oxygen concentration up to 10 per cent O_2.

Monteith: The relationship between rate of germination and temperature has a sharp cut-off at a certain low temperature; this is not a property of a straightforward physical relationship. Nor does the rate of physical processes increase so extraordinarily rapidly over a narrow temperature range as the rates of germination shown by Dr. Hegarty. It is interesting that this linear relationship applies to many other developmental processes in plants, for instance the appearance and extension of leaves as shown by J. N. Gallagher.

Gallagher: Irwin in 1931 published data on the emergence of cereals throughout the U.K. These show a linear correlation of 0·95 between soil temperature and emergence rate. One also finds a linear relation between rate of cell division and temperature and the same kind of relationship in some developmental processes of animals. We are at present at a loss to explain the mechanism governing a process whose Q_{10} appears to be infinitely high at low temperatures and tends towards zero at high temperatures.

Thompson: Caution is needed with these linear relationships. They certainly do apply to most vegetables, even those with a minimum germination temperature as low as $2°C$ or tomatoes with a minimum close to $10°C$. But with lettuce seeds, it is the logarithm of the reciprocal of the time to 50 per cent maximum germination that is a linear function of temperature and it is questionable whether this complicated approach is fruitful. With many species, a linear relationship holds only over a limited part of the temperature range and it cannot be extrapolated with confidence. Therefore, even though many vegetable seeds show an apparently simple linear relationship, there are many obvious exceptions.

Hegarty: Neither do all Kotowski's and Harrington's data on vegetable seeds fit the straight line relationship.

Harrington: A linear relationship can be fitted to the data of Kotowski who terminated his germination records at a specified time, but not to mine; I recorded germination until no more seeds germinated.

Harrington: Is it known why some seed lots of carrot do, but others do not, germinate at low temperatures?

Hegarty: We are not sure at present why some seed lots fail to germinate at sub-optimal temperatures. It could be due to ageing but it could also be a characteristic of immature embryos which according to Borthwick germinate slowly even at optimal temperatures. But doubtless the environment during seed production is important: seeds from crops sown in different parts of a country from the same seed stock have shown different temperature sensitivities.

Storey: Are carrot seeds harvested successively as they mature? If not, are some seeds inferior owing to premature harvest?

Harrington: In California, since carrot seeds do not shatter, the whole crop is allowed to mature before harvest. This results in greater weathering of the seeds of the first order than of later order umbels.

Miles: Grains and vegetable seeds may form one ecological group because they have been selected, consciously or unconsciously, for thousands of years for certain qualities, for instance, for storage under dry conditions.

Hegarty: Harvest conditions are very important, probably more so than genetic ones.

24

ESTABLISHMENT OF SEEDLINGS IN A CHANGEABLE ENVIRONMENT

R. L. GULLIVER and W. HEYDECKER

*Department of Agriculture and Horticulture,
University of Nottingham School of Agriculture*

Παντα ρει
'The only thing that is unchanging is the phenomenon of change'.
Heraclitus 6th Century B.C.

INTRODUCTION

The environment is changing, from season to season, from day to day and from second to second. It therefore seems reasonable to study the influence of fluctuating—though controlled—conditions on germination and we set out to do this comparing the effects of fluctuating and constant conditions. The work has been largely confined to two major environmental factors which affect seeds, namely temperature and water, and has included detailed laboratory studies, experiments with soil in seed trays placed in growth cabinets, and field experiments. In this paper the findings will be dealt with in that order. Within this framework, effects on imbibition, on time characteristics of the germination process of a population of seeds and on final establishment will be considered.

WATER UPTAKE

TEMPERATURE DEPENDENCE

Once imbibition starts, which is not always immediately the seed comes into contact with water, the rate of water uptake is initially

433

rapid and then slows down progressively (see Chapter 25): this is the curvilinear phase. There then follows a more active, more or less linear phase once the axis has begun to grow in length. Depending on species and circumstances, the two phases may merge imperceptibly or may be quite distinct. The classic work of Brown and Worley (1912) clearly demonstrates both the exponential nature of the rate of water uptake and the effect of temperature on the process. The fact that the rate of water uptake increases with temperature shows that it depends on biological processes (e.g., conversion of reserves) in addition to the simple physical gradient of moisture potential.

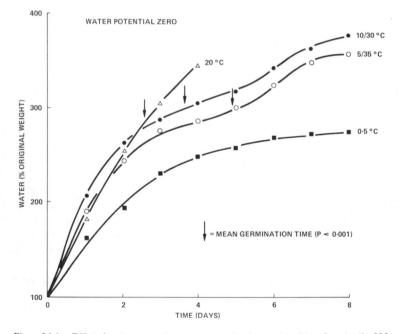

Figure 24.1   *Effect of temperature regime on water uptake of pea seeds, cultivar Surprise (lot 393)*

When peas were set to germinate on moist blotting paper, alternating temperature regimes of 10/30°C and 5/35°C (12 h each) increased the time taken until the radicles emerged, compared with, 20°C constant (*Figure 24.1*), showing that widely fluctuating temperatures are here less beneficial than an equable regime.

AVAILABILITY OF WATER

Ease of water availability allows a higher rate of water uptake by the seed. Radicle emergence then occurs not only at an earlier time

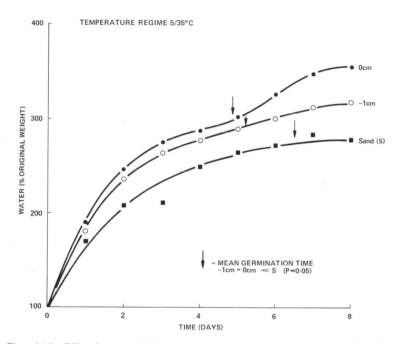

Figure 24.2   Effect of water availability on water uptake of pea seeds, cultivar Surprise (lot 393)

but also at a higher water content than when water is less readily available, provided that conditions are aerobic. This is shown by *Figure 24.2* where water content on day 5 is significantly greater at moisture potentials of 0 and −1 cm than on sand (S). In this experiment the seeds were placed on blotting paper with water up to the surface of the paper, on blotting paper with water up to 1 cm below the surface of the paper and on sand at 'field capacity', all in tightly sealed germination boxes (Orphanos and Heydecker, 1968). The difference between the two blotting paper treatments was almost certainly one of contact area, the water contents of the paper being 450 per cent and 380 per cent, respectively. The effects of a reduced contact area were even more marked with seeds on top of sand. Differences in mean germination time between the same water regimes in a further more elaborate experiment were very highly significant ($P < 0.001$).

These and other results show that there is an inverse relationship between time to radicle emergence and seed water content at radicle emergence in response to different rates of water supply, i.e. slow radicle emergence is associated with low seed water content at radicle emergence. This relationship is subject to two limiting

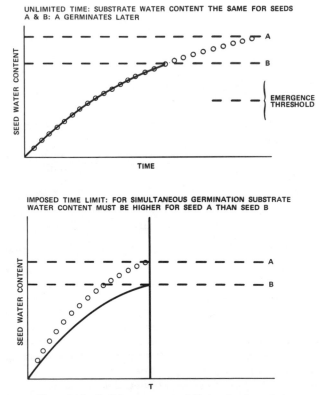

Figure 24.3   Radicle emergence at minimal seed water content

factors: the time required for the radicle to emerge under conditions of ideal water supply and the seed water content at which the radicle can just grow sufficiently to emerge. The *minimum* time to radicle emergence is determined by the highest possible rate of hydration, activation and extension of the radicle. The *maximum* time for radicle emergence is determined by the lowest soil water content at which the radicle can emerge at all; this in turn depends on for how long the seed can survive in the soil, i.e. the length of time available for water uptake before the seed dies. These relationships are shown schematically in *Figure 24.3*. In practice, the situation is more complicated, owing to fluctuation of the soil water content around the minimum for water uptake, and owing to the soil pathogen load. Under such near-drought conditions, the time to radicle emergence also depends on how well the radicle can compete with the cotyledons for water in the early stages of imbibition. The work of Stiles (1948) has shown that even at this stage the radicle has a greater water absorbing capacity than the cotyledons. Brown (1946), working with barley, attributed this solely to the process of cell vacuolation. He observed two phases of water uptake, between 0 and 48 h and between 60 and 96 h. The second but not the first phase coincided with a period of increase of dry weight of the embryo. Throughout the whole 96 h there were no indications of cell division having taken place.

LABORATORY EXPERIMENTS ON GERMINATION TEMPERATURE

EFFECTS OF CONSTANT TEMPERATURE

At constant temperatures, and taking radicle emergence as the index of germination, the proportion of seeds which germinate commonly rises up to a certain temperature, but falls again at still higher temperatures. The rate of germination also shows a rise with temperature to a peak and then falls. Like Kotowski (1926) we have expressed this rate as 100 × the reciprocal of the arithmetic mean germination time in days (see Appendix 3, 'time to germination'). Examination of the data of Kotowski (1926) and Harrington (1963) shows that the rise in the rate of germination up to the maximum rate is frequently linear with temperature. This relationship is discussed by Hegarty (Chapter 23), and is exemplified by the results with peas (*Figure 24.4*). The data of Kotowski (1926) and Harrington (1963) can also be used to show that the optimum temperature for the rate of germination is higher than that for peak numbers, and this point is well illustrated by the findings of Hepton (1957) on cauliflower, presented as *Figure 24.5*. A likely explanation of this relationship is that, as temperature rises, physiological processes are speeded up until, at super-optimal temperatures, weak

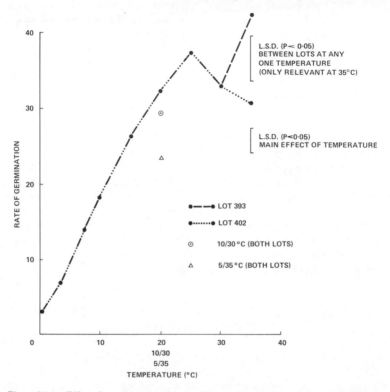

Figure 24.4   Effect of temperature on the rate of germination of 2 lots of peas, cultivar Surprise, sown on blotting paper

seeds are unable to synchronise their biochemical processes correctly and are eliminated. Those seeds which can cope with these faster reaction rates do of course germinate faster. At even higher temperatures (above the optimum for germination rate), however, the metabolic imbalance becomes so severe that the rate of germination is also reduced, usually showing a progressive decline.

In one lot of pea seeds, cultivar Surprise (lot 393), which was investigated the rate of germination rose as the temperature increased from 30°C to 35°C (*Figure 24.4*). This anomalous rise was due to a very few seeds germinating rapidly. The same phenomenon has been demonstrated with wallflower seeds (*Cheiranthus cheiri*) (O'Connor,

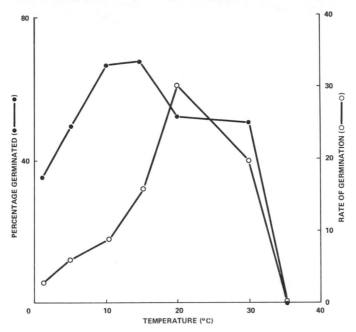

Figure 24.5   Response of cauliflower cultivar All the Year Round to temperature

1972), but it was not observed in a second lot of peas (lot 402) (*Figure 24.4,* superimposed) or in the species used by Harrington (1963), presumably because the critical high temperature for the respective seed lots was not included.

The behaviour of seeds at high temperatures is likely to be important from an agronomic as well as a physiological standpoint, for example where efforts are made to produce large numbers of seedlings quickly using high temperatures, as might be done in a growing room. Here one would employ the temperature which produces the highest rate that can be obtained without loss of numbers. The possibility that the optimal temperature for subsequent seedling growth may differ from that for radicle emergence should, however, be borne in mind (Edwards, Pearl and Gould, 1934). One might, however, use temperatures near the optimum for germination rate in order to select lines which would germinate rapidly and grow well at high temperatures, hence further shortening the early stages of seedling life. Should individuals selected out by high temperature also show an increase in 'vigour' over the basic population, then this process would provide a useful basis for a

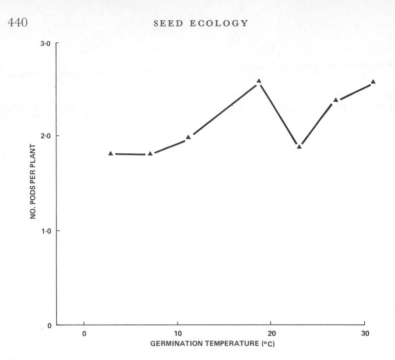

Figure 24.6   Effect of germination temperature on performance of peas. N.B. germination per-centage not stated by the authors. (From: Highkin and Lang, 1966)

breeding programme. Highkin and Lang (1966) may have un-consciously carried out such a selection. In *Figure 24.6* the number of pods per plant is plotted against germination temperature, but a similar response pattern was obtained with other measures of performance, e.g. number of seeds per plant and plant height.

Temperature, as well as affecting the number germinated and the rate of germination, can also affect the uniformity of germination. At low temperatures the scatter of days over which germination occurs is large and is reduced progressively as the temperature rises (*Figure 24.7*). As the mean of a distribution changes in value, the variance tends to change concomitantly. At low temperatures the rate of germination is very low, hence the potential for spread of the distribution of germination in time is great. One can test whether the change in uniformity is solely due to the mathematical relation-ship between the variance (1/uniformity) and the mean (i.e. mean germination time or 100/rate) over a range of conditions by cal-culating the respective coefficients of variation (variance/mean). This has been done for the data presented in *Figure 24.7* and there were no significant differences of coefficients of variation between temperatures, i.e. changes in uniformity could be attributed to variation in the rate of germination at different temperatures. By contrast, analysis of the data of Nichols and Heydecker (1968) on lettuce germination reveals a significant increase in the coefficient of variation at higher temperatures (20–25 °C) but this may be the first manifestation of thermodormancy.

This mathematical linkage between rate and uniformity does not however imply an agronomic linkage. Lateness of emergence may not be a disadvantage in itself, but the poor uniformity usually associated with low germination rates may cause the crowding out of later emerging individuals, hence resulting in a low effective final stand.

EFFECT OF ALTERNATING TEMPERATURE

Changing temperatures can have a range of effects. When two lots of peas were germinated on blotting paper (*Figure 24.4*), the germination percentage of both was vastly superior at an alternation between 5 °C and 35 °C than at a constant temperature of 35 °C which was virtually the thermal death point, a relationship also found by Hepton (1957) for cauliflower. However, both the numbers germinating and the germination rate were lower at 5/35 °C than at 20 °C constant, the mean temperature of the fluctuation. Even at 10/30 °C germination was slightly, but significantly, slower than at 20 °C, although numbers germinating were the same. The rate of germination at alternating temperatures has often been found to be higher than at the constant mean temperature (see Chapters 23 and 25). The present findings do not agree with this experience but as a constant temperature of 30 °C, compared with 25 °C, reduced the rate of germination (which indicates some physiological stress) this result is not unexpected.

INTERACTIONS BETWEEN SEED LOTS AND TEMPERATURE REGIMES

Work with peas also provides evidence that responses to temperature fluctuations can vary with seed lot. The effects of temperatures of 10 °C, 14 °C and 10/14 °C were compared using three lots of peas placed on moist blotting paper in germination boxes. In contrast to previous experiments, none of these temperatures was sub- or super-optimal. Two of the three lots performed well in the field (cultivar Jade, lots 104 and 108); the third performed poorly (cultivar Surprise, lot 393). The *mean number* germinating was

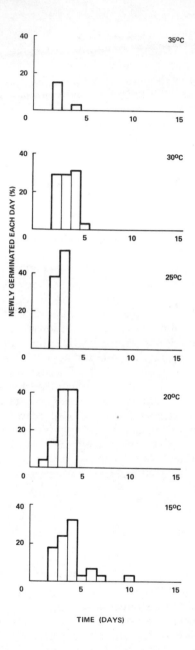

Figure 24.7 Germination of peas at different temperatures, cultivar Surprise (lot 393)

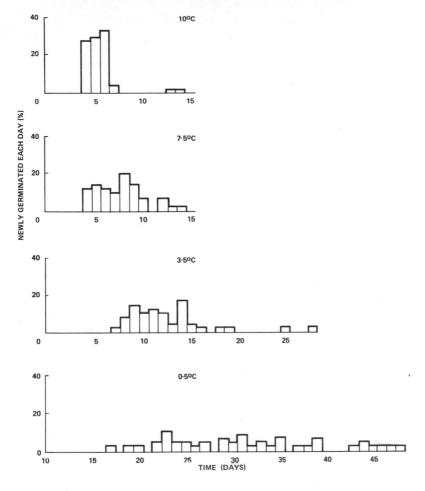

significantly reduced at 10/14°C compared with 10°C or 14°C, but this was almost entirely due to the poor performance at 10/14°C of the lot (No. 393) which showed poor emergence in the field. On the other hand, the *rate of germination* for all lots was worse at 10°C than at 14°C constant, but at 10/14°C it was almost the same as at 14°C (*Figure 24.8*). In other words, the rate of germination was virtually unaffected by the lowering of the temperature from 14°C to 10°C

444

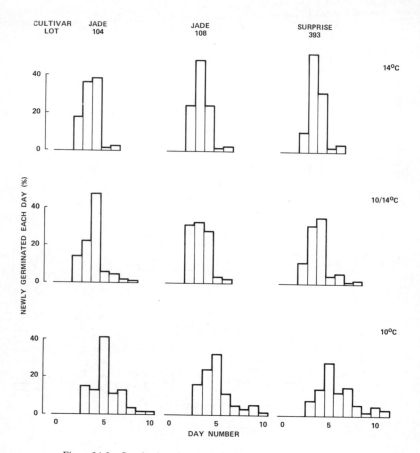

Figure 24.8   Germination of peas at constant and alternating temperatures

for half of each day, and in contrast to numbers germinating, all lots behaved similarly (i.e. there was no lot × temperature interaction for rate). These findings agree with those of Hegarty (Chapter 23). The effect of temperature on *uniformity* was proportional to that on the rate of germination (i.e. the coefficient of variation did not differ significantly). Regarding the three seed lots, although two of these had good field emergence, only one of them (lot 108) had a significantly higher rate of germination. The uniformity of each lot varied with its germination rate, though the differences in uniformity of the three lots were not quite significant. To summarise, the general trend, which was most marked in the lot with poor field emergence (lot 393), was as follows: at 10/14°C the numbers germinated were lower than at 14°C or 10°C, although the rate and uniformity were as high at 14°C and higher than at 10°C constant.

The pattern with carrots germinating at these same three temperature regimes was totally different. Rate was at least as low at 10/14°C as at 10°C, but there was no difference in numbers at any temperature. Went (1957) and Lang (1965) have suggested that the low rates of germination that occur at very low and very high temperatures are due to a relatively rapid production of inhibitors under these conditions. It is impossible to substantiate or reject these hypotheses on the evidence so far published, but such a mechanism could explain the adverse germination rate observed with carrots at 10/14°C because 10°C is a hazardously low temperature for carrot seed germination (Chapter 23).

GERMINATION AND WATER SUPPLY

EFFECT OF CONSTANT LEVELS OF WATER SUPPLY

Seeds show the same response pattern to an increase in water supply as they do to an increase in temperature, despite the fundamental dissimilarity of these two environmental factors. There is an optimal substrate water status for percentage germination, with lower germination on either side of this optimum status, and an optimum for rate at a higher substrate water status. Here a super-optimal water supply appears to act in the same way on germination percentages as do super-optimal temperatures: it 'stresses' the seeds and eliminates some, but allows faster germination of those which survive. 'Stress' here may not only result from too rapid water uptake, but also from reduced oxygen availability, as well as from increased microfloral competition for oxygen. *Figure 24.9* shows the response of red beet seeds to differences in water supply produced by increasing the number of filter papers in petri dishes which contained identical amounts of water, a technique also used by Negbi, Rushkin and Koller (1966). Rate of germination rises to a peak, which occurs at a higher substrate water content than does the peak of percentage germination; and there is also a secondary rise in rate at very high water availability, directly comparable with that which occurred at high temperature with one lot of peas (*Figure*

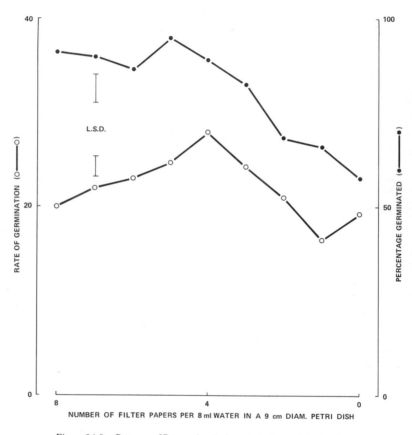

Figure 24.9   Response of Beta vulgaris *L. seeds to the availability of water*

24.4). Potential germination of beet may be further reduced at low substrate moisture contents because only small amounts of the endogenous inhibitors are leached out of the fruits. This problem was partly overcome in this experiment by using pre-washed material.

The complex relationship between soil moisture and effective inhibitor concentration is discussed in a later section.

The results obtained by Doneen and MacGillivray (1943) for red beet (*Figure 24.10*) show that with progressively lower soil water content the percentage of seeds which emerge declines more rapidly on a per seed than on a per cluster basis. This suggests that one of the first effects of progressively drier conditions is to reduce the number of seedlings emerging per cluster. Doneen and MacGillivray's results do not show the decline in beet germination at high soil water

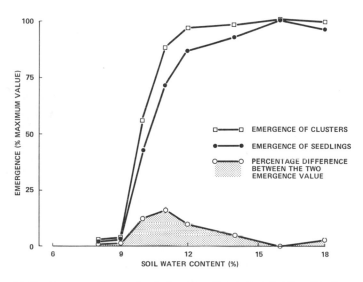

Figure 24.10   Emergence of Beta vulgaris *L. cultivar Detroit Dark Red from Yolo fine sand at different soil water contents (After: Doneen and MacGillivray, 1943)*

content demonstrated by Longden (1971) for sugar beet (*Figure 24.11*). This is almost certainly because the fine sand they used drained sufficiently easily to allow good aeration even at field capacity, in contrast to many cultivated soils including the soils used by Longden and the Sutton Bonington sandy loam used by Gulliver and Heydecker. An experiment with peas (cultivar Jade, lot 108) sown in this soil at different soil water contents (*Figure 24.12*) showed that here the percentage germination was substantially reduced at high soil water contents. This is again in contrast to the results of Doneen and MacGillivray who found with peas, as with beet, no decline in percentage germination at high soil water contents.

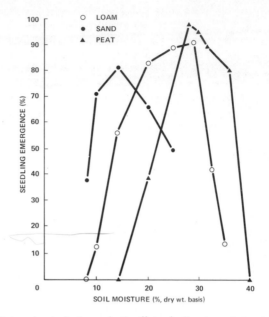

Figure 24.11 Beta vulgaris *L. (sugar beet), effects of soil water content and soil type on seedling emergence (After: Longden (J. agric. Sci., in press.))*

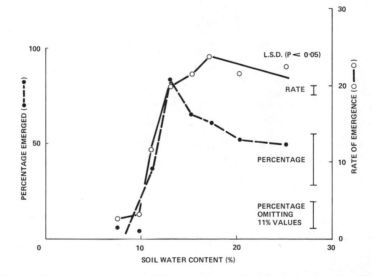

Figure 24.12 Emergence of peas at different soil water contents (dry wt. basis), cultivar Jade (lot 108)

Presumably, once more, the difference is due to the nature of the soil types. *Figure 24.12* shows that for peas, as for beet, the optimal soil water content for the rate of germination is well above the one for the highest percentage emergence. Judging by the widespread occurrence of a higher optimum for rate than for percentage in respect of temperature, it may well hold for crops other than beet and peas in their response to water. In addition it appears that the distribution of water in soil over very small distances can affect the rate of germination (*Figure 24.15*).

FLUCTUATING WATER AVAILABILITY

Soil moisture fluctuations have been simulated using a rain-like spray to apply measured amounts of water to soil (here Sutton Bonington sandy loam) at fixed intervals. A further treatment involved periodically reinstating 'field capacity' by sub-irrigation and

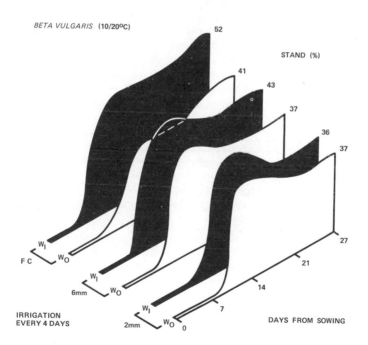

Figure 24.13   Response to three water regimes. W_O fruits not washed, W_I fruits washed before sowing, FC soil brought to field (tray) capacity by capillary rise of water

capillary rise. The method has many advantages; it is however difficult to set up a constant soil moisture regime (the mean of the fluctuations) as a 'control'.

In one experiment with three cultivars of beet—one red beet and two monogerm sugar beet—given water at 4-day intervals, the effect of washing out the water soluble inhibiting chemicals from the fruits before sowing (MacKay, 1961) was investigated, and on this treatment was superimposed the replenishment of the soil to three different soil water contents. In all cases seedlings from washed fruits emerged sooner and reached a higher number; but with the two partial (surface) irrigations the numbers then fell again quite steeply, due to a post-emergence attack of *Phoma betae*.

By contrast, the numbers from unwashed fruits rose to a lower peak but fell to a lesser extent, to result in the same final stand as that obtained from the washed fruits (*Figure 24.13*). (Interactions between lot and treatment disappeared early in the experiment, and hence for the sake of clarity, the results have been pooled over cultivars.) Though initially discouraging, the results show that if pathogens could be adequately controlled at germination and during the earliest stages of growth, washing would be advantageous. [In other experiments it has been found that treatment with the fungicide thiram, not necessarily the thiram soak (Maude, Vizor and Shuring, 1969), was beneficial although occasionally the loss of seedlings was only delayed.] In the wettest treatment, i.e. the sub-irrigation 'field capacity', seedlings from washed fruits performed much better than from unwashed fruits, and here the benefit persisted: the number of seedlings emerging rose to a higher peak than in other treatments and this rise was not followed by a decline. We take this to indicate that at high soil moisture contents either the pathogenic *Phoma betae* suffered a reduction in infective capacity, or the more consistently turgid cells of the well-watered plants were better able to resist the attack; whichever is true, the result was a higher ultimate stand. With the unwashed beet fruits, however, this stand was not obtained because of the greater effect of the inhibitors in wet, as well as in dry, conditions. This apparent paradox is further discussed by Heydecker, Chetram and Heydecker (1971). That fluctuating water levels at their best ensure an adequate but non-excessive water supply—which prevents flooding of the tissues, leakage of metabolites and oxygen starvation (see Lang, 1965)—is shown by the results of Heydecker and Kohistani (1969) who found successive small applications of water to peas in petri dishes superior in effect on rate and percentage germination to the same quantity given in fewer, larger doses. *Phaseolus vulgaris* beans also show the same response pattern (Russell, 1966).

SOIL WATER AND FIELD EMERGENCE

Work has also been carried out in the field where environmental fluctuations are tolerated, or at best modified, rather than controlled. Here, one very important secondary effect of soil moisture fluctuation has been tackled, namely crusting (or capping) which often occurs when flooded soil surfaces dry rapidly. Although none of the attempts with a range of soil conditioners has proved markedly successful, it should be pointed out that the control of crusting may well be a more relevant problem for the reliable production of stands on many soils than an excess or deficiency of water *per se*. Pending the arrival of a reliable soil conditioner at an economic price, it is worth recollecting that Sale and Harrison (1964) showed that one of the most certain preventive measures against crusting is to keep the soil continuously moist, preferably by very fine spray irrigation, until the seedlings have emerged, which resolves the problem into a question of economics.

Our own contribution to a non-chemical approach to the crusting problem is illustrated by an experiment in which the interaction of

Table 24.1 PERCENTAGE EMERGENCE (AVERAGED OVER DEPTHS) FOR RED BEET CULTIVAR BOLTARDY

	mm irrigation		
	0	2	16
Not firmed	66	76	54
Firmed	60	64	70

P < 0·01 for Chi-square on raw data

depth of sowing, post-sowing firming of the soil and irrigation (simulated rain), applied immediately afterwards, on the establishment of red beet were studied. In all cases sowing deeper than 25 mm (1 in) resulted in inferior stands, and the decline in numbers was linearly related to depth.

At the more favourable shallow depths the situation was as follows (*Table 24.1*): when the soil remained dry or received only light irrigation (2 mm), stands tended to be better in the absence of firming. After heavy irrigation (16 mm), however, the position was reversed, and here the stand was worse on unfirmed soil. This can be tentatively attributed to a greater water stability of the soil after firming: unfirmed soil has large enough pores to permit unstable water-borne particles to re-sort themselves, eventually to form a crust when the water evaporates. By contrast, firmed soil presents a greater resistance to a rearrangement of its particles by infiltrating water, and is therefore less liable to crust on drying than is unfirmed soil.

PARAMETERS OF SUCCESSFUL SEEDLING ESTABLISHMENT

Under realistic conditions, and especially where fluctuations are liable to add to the risk, efforts to assess 'good establishment' are dogged by the problem of what criterion of success to employ. For instance, *the number of seedlings* emerged in the field may appear to be a reasonable measure, but this number often rises to a peak and then falls *(Figure 24.14)*. (The emergence values are means of four lots of

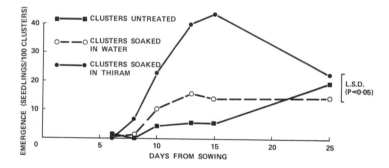

Figure 24.14   Effect of treatment on emergence and survival of Beta vulgaris *L. seeds (Sutton Bonington, 1971)*

low, but similar, viability.) Furthermore, the time when a stand rises to its maximum can vary with seed lot, seed treatment, and soil conditions. Thus a single count can be very misleading. One can, however, be sure that the longer a seed spends in the ground, the longer it is at risk, and hence the *mean emergence time* is a useful parameter. But, as will be shown, *uniformity in time to emergence* is an equally important parameter, especially in view of present sophisticated systems of crop production, where uniformity in harvesting dates and in the size of the final produce is often essential. As has been pointed out, average time and uniformity are frequently linked.

The rate of growth of seedlings may or may not interact with the uniformity in time to emergence in the following ways. Where two populations differ in mean germination time, the exponential growth of the earlier emerging population will 'increase' its initial advantage. Populations with a long mean germination time will also have low uniformity, and therefore there will be a greater tendency for late emergers to be lost than in a population with a short mean germination time.

Similarly, within a stand early emergers will always be ahead of late ones. Furthermore, in dense stands the growth of earlier emerging seedlings will progressively restrict the supply of light, water and nutrients to the later emerging ones; thus late emergers in such a population will be at a double disadvantage. The differences in size between plants within one population resulting from these two processes (lateness *per se* and competition) have often been attributed to an inherent linkage between ability to germinate rapidly and to grow rapidly due to a combination of properties vaguely termed 'vigour'. Whilst the possibility cannot be ruled out, superior performance by virtue of earlier emergence alone must not be mistaken for evidence of such a linkage.

A high degree of uniformity of a population is an advantage, whichever explanation of inter-plant differences holds. If the vigour concept holds, then high uniformity tends to indicate many vigorous plants in the population. If superior performance of early emergers is purely due to earlier utilisation of environmental resources, then high uniformity minimises the tendency for the late emergers to be lost.

One might argue that the most far-sighted assessment of success—either of seedbed conditions or of the quality of seed lots relative to each other—is the mean individual seedling weight on a specified day after sowing. However, mean weight—like mean time to germination—is meaningful only when the population is reasonably uniform. The many late emerging seedlings characteristic of non-uniform emergence will greatly affect the mean time to emergence, and also the mean seedling weight; and yet, by remaining so under-sized, will make an insignificant contribution to the ultimate population.

Only a measurement of rate which discards the useless tail of the germination curve is truly valid. One convenient way of obtaining this parameter is to plot the cumulative number germinated against time (or log time) on probability graph paper (see, e.g., Roberts, 1972, who used the technique to study the loss of viability of seeds under different storage conditions). The estimate of mean germination time (the reciprocal of rate) thereby obtained is not affected by the late germinators, unlike the estimate of rate obtained using the Kotowski formula (see p. 437). The same technique could be applied

454

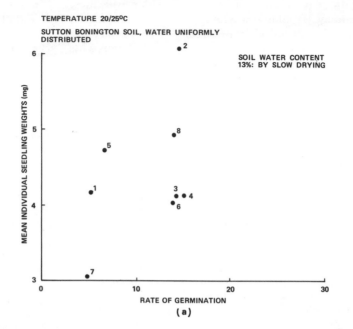

(a)

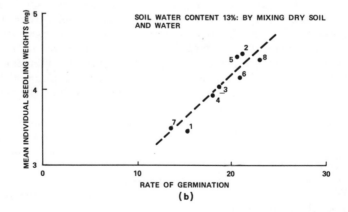

(b)

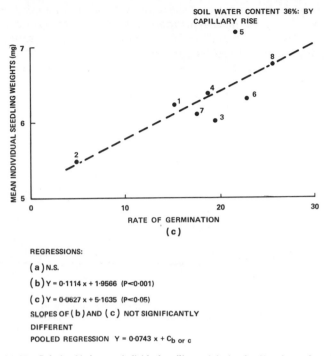

Figure 24.15   Relationship between individual seedling weight (on day 8) and rate of germination for 8 cultivars (1–8) of cabbage (rate = 100/days)

to mean seedling weight, but as it would involve weighing each seedling individually, its use would rarely be justified.

The results show that the rate of emergence of an individual cultivar (or even seed lot) may not be well related to the resulting mean seedling weight (*Figure 24.15(a)*). This may be broadly for two reasons. First, lots with the same mean time but different uniformities may be expected to have different mean weights, due to the increasing suppression of the late germinators. Secondly, two seed lots may be similar in their relationship of growth rate to germination rate under one, but differ under another set of conditions, as is implied in *Figure 24.15(b)* and (*c*). The whole problem of rate and uniformity of emergence and subsequent seedling growth merits much further attention.

ALL-FACTOR INTERACTIONS

So far the effect of environmental factors (temperature, water) and seed factors (cultivar, seed lot) have been considered independently, but these factors do, not surprisingly, interact. An experiment with peas provides a simple but useful example of seed lot × temperature interaction (*Figure 24.16*). Here one lot of cultivar Jade (lot 81) showed quite acceptable values of germination rate, compared with other seed lots, at low constant temperatures; but at higher, alternating temperatures it lagged behind the others.

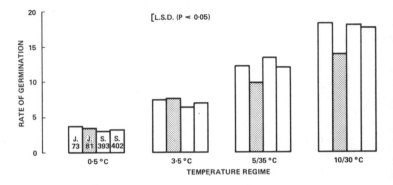

Figure 24.16   Rate of germination of pea seeds at different temperature regimes, cultivars Jade (lots 73, 81) and Surprise (lots 383, 402) (rate = 100/days)

Interactions between temperature, seed lot and water supply became apparent in an experiment on *Phaseolus vulgaris* (French bean) cultivar Sprite sown in seed trays containing Sutton Bonington sandy loam. Two seed lots, harvested in 1966 and 1969, were sown (in 1971) at 15°C, a reputedly marginal, and 20°C, a favourable, temperature. They were watered by spraying from above, as in the beet experiment already described, so as to replenish the soil at four day intervals to the equivalent of 2, 4, or 6 mm of rain above the original soil water content of 12·4 per cent. Seed lot, temperature, amount of water applied and frequency of application significantly affected rate of emergence, final numbers and individual seedling weight (*Figure 24.17*). The results show that the most limited water supply, the higher temperature and, surprisingly, the younger seed lot were inferior. However, these effects were entirely due to the one combination of the two environmental factors which brought out the inferiority of the younger seed lot under these testing conditions.

With the present increasing emphasis on preparatory treatments of seeds, the position is further complicated. This is shown by a field experiment which was carried out to determine the respective and combined effects on the establishment of two lots of sugar beet and

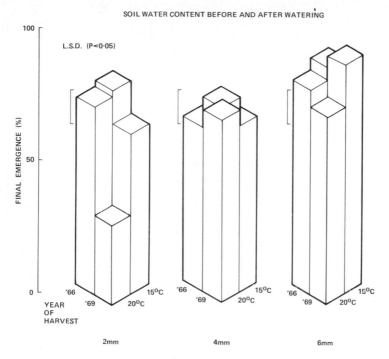

Figure 24.17   Emergence of French beans in Sutton Bonington soil in relation to soil water, temperature and age of seeds (purchased in 1966 and 1969)

one of red beet of the following: (*a*) washing the fruits for 24 h in running water at 10°C, (*b*) soaking them for 24 h at 30°C, (*c*) treating them with thiram fungicide and applying fairly heavy irrigation (18 mm) to the seed bed after sowing. The treatment which had the most clear-cut effect was the application of thiram which significantly improved the emergence of two lots though not of the third (monogerm B). The third lot was somewhat atypical in that it was depressed in its performance by being washed or soaked in water, irrespective of thiram treatment. An equally, if not more, important finding was that the effects of the other treatments varied greatly with individual seed lots, as shown in *Figure 24.18*.

The many interactions which occur in the response of a seed to treatments such as those demonstrated, can be handled using one of

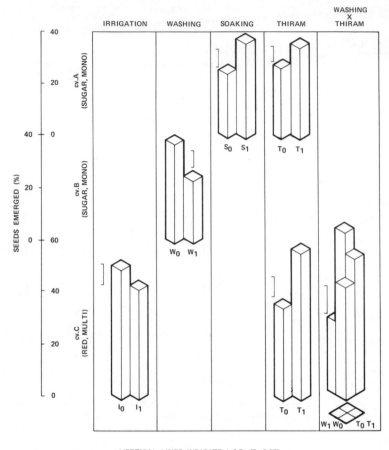

VERTICAL LINES INDICATE L.S.D. (P<0·05)
SUFFIX 0 = NOT TREATED, 1 = TREATED
TREATMENTS EFFECTS NOT SHOWN ARE NOT SIGNIFICANTLY DIFFERENT WITHIN CULTIVARS

Figure 24.18 Beta vulgaris *L. response to seed treatments and irrigation*

two possible research strategies: either testing each seed lot in-
tensively in order to specify its strengths or weaknesses, or relying on
the known main effects of a treatment, which has proved beneficial
in many cases, and taking the calculated risk that this treatment
may be wasted on any particular lot, and might occasionally even
be detrimental.

CONCLUSION

Thus, however much is learned about the establishment of seedlings in a changeable environment, we shall always be hampered by the fact that what applies to one seed lot need not necessarily apply to another. Moreover, differences in response between seed lots of one cultivar can be more important than differences between cultivars and possibly even species.

Is there anything that can be done about the situation? At least one possibility presents itself. By increasing the seed water content prior to sowing, one can reduce the time from sowing to field emergence, and hence curtail the period in which the sown seeds could be affected by an adverse change in the seed bed from the presumably favourable conditions in which they were sown. Such an increase in seed water content can be brought about either by equilibration of the seeds with a humid atmosphere (*Table 24.2*), or by contact with a solution of strongly negative osmotic potential, which results in enough water being taken up to allow the preliminary processes of germination to proceed but not enough to permit radicle emergence. Both techniques are described by Levitt and Hamm (1943). Furthermore, certain seed treatments involving the imbibition of physiologically active substances may be of special value in promoting germination and early seedling growth (Joshua and Heydecker, Appendix 2), especially as the beneficial effects persist for some time even when the treated seeds are dried back to facilitate sowing.

Table 24.2 EFFECT OF EQUILIBRATION ON RATE OF GERMINATION (RATE EXPRESSED AS $100 \times$ RECIPROCAL OF MEAN GERMINATION TIME)

	Non-equilibrated	Equilibrated
Peas cv. Surprise: Lot 393	25·5	33·6
French Beans cv. Processor	21·5	27·3
Onions cv. Rijnsburger	17·6	41·0

L.S.D. (P < 0·05) 4·8

A fuller, more rapidly established and more uniform stand should result from such practices. The problems which could arise from increased fungal attack of the moist pretreated seeds might be countered by a suspension of a non-pathogenic fungus which would release antibiotics and thereby protect the seeds in the soil (Wright, 1956); this would be ecologically more attractive than the use of fungicides. More intimate contact between the seeds and soil water might be achieved by coating them with a mucilaginous material with an inherently low affinity for water but pronounced hydraulic conductivity (Elliott, 1967).

Such ideas for shortening the period during which fluctuating conditions can play havoc with the seeds before and during germination may sound as fantastic now as the idea of precision drilling did only a few years ago. But we dare hazard a guess that it is, more or less, in this direction that the future lies.

ACKNOWLEDGEMENTS

The work on which this paper is based was largely financed by a grant from the Agricultural Research Council. We thank Mrs. C. K. Parsons, Mr. J. L. Churchman, Mrs. F. V. Murray and Miss V. Attenburrow for technical help.

REFERENCES

BROWN, R. (1946). 'Studies on germination and seedling growth. III. Early growth in relation to certain aspects of nitrogen metabolism in the seedling of barley', *Ann. Bot.*, **10**, 73–76

BROWN, A. J. and WORLEY, F. P. (1912). 'The influence of temperature on the absorption of water by seeds of *Hordeum vulgare* in relation to the temperature coefficient of chemical change', *Proc. Royal Soc. London B.*, **85**, 546–553

DONEEN, J. D. and MACGILLIVRAY, J. H. (1943). 'Germination (emergence) of vegetable seed as affected by different soil moisture conditions', *Pl. Physiol.*, **18**, 524–529

EDWARDS, T. I., PEARL, R. and GOULD, S. A. (1934). 'Influence of temperature and nutrition on the growth and duration of life of *Cucumis melo* seedlings', *Bot. Gaz.*, **96**, 118–135

ELLIOTT, J. G. (1967). 'The sowing of seeds in aqueous fluid', part of 'Work in progress on grassland' by the Agronomy Section of the Department of Weed Control, *Weed Research Organisation, Second Report 1965–66*, 31–32

HARRINGTON, J. F. (1963). 'The effect of temperature on the germination of several kinds of vegetable seeds', *16th Int. Hort. Cong.*, **2**, 435–441

HEPTON, A. (1957). 'Studies on the germination of *Brassica oleracea var. botrytis* (Linn.) with special reference to temperature relationships', *B.Sc. (Hons.) dissertation*, Univ. Nottingham

HEYDECKER, W., CHETRAM, R. and HEYDECKER, J. C. (1971). 'Water relations of beetroot seed germination. II. Effects of the ovary cap and of the endogeneous inhibitors', *Ann. Bot.*, **35**, 31–42

HEYDECKER, W. and KOHISTANI, M. R. (1969). 'Hollow heart and poor stands of peas (*Pisum sativum* L.)', *Ann. appl. Biol.*, **64**, 153–160

HIGHKIN, H. R. and LANG, A. (1966). 'Residual effect of germination temperature on the growth of peas', *Planta*, **68**, 94–98

KOTOWSKI, F. (1926). 'Temperature relations to germination of vegetable seeds', *Proc. Am. Soc. hort. Sci.*, **23**, 176–184

LANG, A. (1965). 'Effects of some internal and external conditions on seed germination', in the *Encyclopedia of Plant Physiology*, (ed. Ruhland, W.) Vol. 15, No. 2, 850–993, Springer-Verlag, Berlin

LEVITT, J. and HAMM, P. C. (1943). 'A method of increasing the rate of seed germination of *Taraxacum kok-saghyz*', *Pl. Physiol.*, **18**, 288–293

LONGDEN, P. C. (1971). (Summary of findings). '[Sugar beet] seed production', 'Factors affecting yield'; 'Soil type and moisture content', *Rep. Rothamsted exp. Stn. for 1970*, Pt. 1, 262–263

MACKAY, D. B. (1961). 'The effect of pre-washing on the germination of sugar beet', *J. nat. Inst. agric. Bot.*, **9**, 99–103

MAUDE, R. B., VIZOR, A. S. and SCHURING, C. G. (1969). 'The control of fungal seed-borne diseases by means of a thiram seed soak', *Ann. appl. Biol.*, **64**, 245–257

NEGBI, M., RUSHKIN, E. and KOLLER, D. (1966). 'Dynamic aspects of water relations in germination of *Hirschfeldia incana* seeds', *Plant Cell Physiol.*, **7**, 363–376

NICHOLS, M. A. and HEYDECKER, W. (1968). 'Two approaches to the study of germination data', *Proc. int. Seed Test. Ass.*, **33**, 531–540

O'CONNOR, D. E. (1972). 'An investigation into some aspects of wallflower seed production', *B.Sc. (Hons.) dissertation*. Univ. Nottingham

ORPHANOS, P. I. and HEYDECKER, W. (1968). 'On the nature of the soaking injury of *Phaseolus vulgaris* seeds', *J. exp. Bot.*, **19**, 770–784

ROBERTS, E. H. (1972). 'Storage environment and the control of viability', in *Viability of Seeds*, (ed. Roberts, E. H.), 14–18, Chapman and Hall, London

RUSSELL, A. (1966). 'Investigations into the soaking injury of *Phaseolus vulgaris* L. seeds', *B.Sc. (Hons.) dissertation*, Univ. Nottingham

SALE, P. J. M. and HARRISON, D. J. (1964). 'Seedling emergence as affected by soil capping', *J. hort. Sci.*, **39**, 147–161

STILES, I. E. (1948). 'Relation of water to the germination of corn and cotton seeds', *Pl. Physiol.*, **23**, 201–222

WENT, F. W. (1957). 'The experimental control of plant growth', *Chronica bot.*, Waltham Mass.

WRIGHT, J. M. (1956). 'The production of antibiotics in soil. VI. Production of antibiotics in coats of seeds sown in soil', *Ann. appl. Biol.*, **44**, 561–566

DISCUSSIONS

Skinner: Field capacity is a very imprecise term. It would be better to control the soil moisture by means of a tension apparatus such as a Buchner funnel.

Heydecker: The bottom centimetre of our seed trays was placed in water which then rose by capillarity, and the trays were then allowed to drain. Soil at this 'tray capacity' has a slightly higher water content (a lower water tension) than the same soil at field capacity in the field, but the procedure is quite reproduceable.

Perry: Satisfactory emergence of *Beta vulgaris* seedlings from soil at field capacity is contrary to our experience. What kind of soil was it? Were the seeds treated with a fungicide?

Gulliver: The seeds were not treated with fungicide. It must be remembered that whilst the soil, a sandy loam, was brought to 'tray' capacity every fourth day, considerable evaporation occurred in the intervening period.

Heydecker: Dr. Côme's paper (Chapter 9) sheds light on this situation: if the phenolic inhibitors contained in the beetroot fruit structures operate by cutting down the oxygen supply to the seeds, then a wet regime is likely to interfere with germination. This was also our experience. However, when we washed the clusters before sowing and thereby largely eliminated the inhibitors we found that ultimately a larger stand was established, especially in wet soil, though at a slower rate under these conditions than in a drier regime.

Maude: The higher incidence of post-emergence damping off in the drier regimes reported in this communication is not uncommon. Perhaps seedlings which are infected are capable of growing away from the infection as long as the water supply is adequate.

Heydecker: This may be so, but one must bear in mind that in our case the rate of emergence in the wetter soil was slower and yet high numbers of seedlings were obtained and they did not diminish later.

Storey: It is important to know the pathogen involved.

Heydecker: We think in our case it was *Phoma betae* but are not certain.

Matthews: Did the continued germination in the wetter soil perhaps cover up losses from disease?

Heydecker: No, at 'tray capacity' (in contrast to drier regimes) there were virtually no post-emergence losses either while further seedlings emerged or during one month after emergence had ceased. As Mr. Maude has pointed out, the seedlings may have grown away faster from the attack than seedlings in drier soil.

25

THE SEED–SOIL SYSTEM

J. A. CURRIE

Rothamsted Experimental Station, Harpenden, Hertfordshire

INTRODUCTION

A germinating seed has few requirements. They can be met simply
in an ideal environment and have been discussed in detail at this
Easter School. The soil, however, is often far from being an ideal
environment. Indeed, at times it can be quite hostile. Much has been
written about soils, and can be read in such standard texts as, for
example, Keen (1931), Russell (1961), Baver (1956) and Rose (1966).
Here, following a brief description of the fabric of the soil, those
properties which influence the seed during germination are discussed.

THE SOIL

Soil is essentially a three-phase system consisting of solids, liquids
and gases in varying proportions. In most soils the solids are predominantly mineral, derived from rock materials, but may include
some organic matter. The amount of organic matter is determined
partly by the rate at which fresh plant residues are added, and partly
by the rate at which these are decomposed by the microflora and fauna.
The primary soil particles are classified by size. On the international
scale, for example, the groups are clay (diameter < 2 μm), silt
(2–20 μm), fine sand (20–200 μm), coarse sand (0·2–2·0 mm), gravel
and stones (> 2 mm). Other scales use other divisions, often determined by experience of the field behaviour of soils. For example, the
Soil Survey of England and Wales currently sets the division between
silt and fine sand at 50 μm, and further subdivides the sand fraction
at 100, 200 and 500 μm. The clay particles are usually finely divided

463

crystalline alumino-silicates formed by weathering of the minerals in the parent rock. The coarser particles are usually either quartz or unweathered rock fragments. These particles are found mixed in various proportions to give recognisable textural classes to soil, e.g. clays, silty clays, clay loams, silts, sandy loams, sands, etc. The matrix formed by such a mixture has a fundamental pore size which reflects the proportions of the ingredients. Thus clay soils have many small pores, sands have fewer but larger pores.

The clay particles have electrically charged surfaces which enable them to cohere, and to adhere to the less active silt and sand, often with the help of intermediate films of organic matter (Emerson, 1959). In addition, many clays swell when wet and shrink when dry. The basic matrix becomes moulded during successive cycles of wetting and drying into a series of discrete coherent micro-porous units called peds, separated by a continuous system of fissures. These fissures may be little more than planes of weakness when the soil is wet and fully swollen, but may become large open crevices in summer when the soil dries. This heterogeneous pattern is termed structure and, on the basis of the size and shape of the peds, is another criterion for classifying soils.

But structure has a somewhat wider meaning. The ploughed layer of an arable soil has artificial structure caused by cultivations and other work in growing a crop. The natural peds are distributed to become the clods of the newly ploughed field and these in turn, during subsequently cultivations, hopefully are broken down to become the crumbs of the seed bed. This tilth, or crumb structure, is traditionally considered desirable, though recent work with minimum cultivation may show that this is not always true. The larger inter-crumb pores permit surplus water to drain quickly from the soil, and admit air. The water required by the plant is retained within the finer pores of the crumb. Plant roots can move freely through the inter-crumb pores to all parts of the soil, in search of nutrients and water. Such a soil is also easily cultivated. However, the strength of the forces holding the crumb together determine its stability and hence the durability of the crumb structure. Weather and the operations inevitable in some farming systems are mainly responsible for degrading structure. Organic matter enhances structure and increases stability either because coarser fractions hold the crumbs apart or because decomposed material cements the smaller particles together.

Soil structure is important to germinating seeds. When sown the seeds fall initially into the larger inter-crumb pores where many germinate. Those which do not germinate at once may, in time, as the structure collapses and is rebuilt, be incorporated within the micro-porous matrix of the crumb. Similarly, if after the seed is sown the structure of an unstable soil collapses under, say, rain impact or a passing wheel, the seed will again be effectively 'incarcerated' in the fine-pored matrix of a crumb or clod. The distribution of water and air in the inter-crumb pores and within the crumb differs and the effect of this on the seed must be considered.

The liquid phase is water which may contain small amounts of solids and gases in solution in equilibrium with the solid and gaseous phases. When a soil is saturated, for example in winter, most of its pores are filled with water. As water is lost by drainage or evaporation, the larger more continuous inter-crumb pores drain first. The water is replaced by air drawn down from the soil surface. At what is known as field capacity, the inter-crumb pores tend to be air-filled while the crumbs remain saturated, and further loss of water is from within the crumb. The tenacity with which water is held (its 'minus potential') is inversely proportional to the size of the pore it occupies. This can be measured and is most conveniently displayed as the moisture characteristic, effectively a desorption curve showing the suction, ψ, which must be applied in order to withdraw further water from the soil at water content, θ.

The gaseous phase, the soil air, occupies those pores not already filled with water. In practice, the amount of air space can be controlled only by managing the structure and water content. The composition of the soil air also varies in a way that influences germination.

WATER UPTAKE

During imbibition, the seed takes up water from the soil. This is essentially a transport problem governed by a transport equation, which, for water, is most conveniently written

$$F = - K \frac{d\psi}{dx}$$

where the flux, F, the quantity of water moving through unit cross-section, in unit time along an axis x, is proportional to the hydraulic conductivity K, and to the potential gradient $d\psi/dx$, i.e. the change of potential, ψ, with distance, x. Alternatively, the flux may be written

$$F = - D_l \frac{d\theta}{dx}$$

where D_l, the so-called diffusivity, is equal to the product of the hydraulic conductivity, K, and the slope, $d\psi/d\theta$, of the moisture characteristic. Both these equations may be encountered. While the first is preferable to understand the mechanism involved, the second is useful because it is easier to calculate D_l having measured the flux and moisture gradient $d\theta/dx$, than to calculate K using $d\psi/d\theta$ from a separate experiment. As written here, both equations refer to steady state flow, and are the basis of the second-order differential equations used to describe transient flow.

During imbibition, the effect of potential difference and hydraulic conductivity are best considered separately. The initial rate of transfer is set by the initial difference in soil water potential between the soil and the seed. As imbibition proceeds, the seed gets wetter, the soil gets drier, the potential difference decreases and the rate of water transfer to the seed decreases. *Figure 25.1* (after Hadas, 1970)

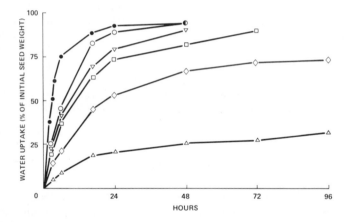

Figure 25.1   Water uptake of chickpea seeds (Cicer arietinum *L.*) *as a function of time in distilled water* ● *and in a loessial silty loam at* ○ *17%,* ▽ *14%,* □ *11%,* ◇ *8% and* △ *5% soil moisture, dry weight basis (after Hadas, 1970)*

shows how the initial water content of the soil affects imbibition. *Figure 25.2* (after Williams and Shaykewich, 1971) shows the effect of water content and water potential on germination.

In considering the effects of hydraulic conductivity, the path followed by the water may conveniently be divided into three sections —within the soil, across the seed/soil interface, and within the seed. Within both the soil and the seed, the hydraulic conductivities are functions of water content and may change by orders of magnitude over the range of water contents found in the field. In saturated soil,

467

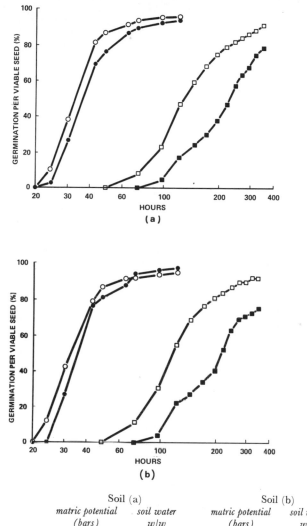

	Soil (a)		Soil (b)	
	matric potential	*soil water*	*matric potential*	*soil water*
	(bars)	*w/w*	*(bars)*	*w/w*
○	−0·6	34·2	−0·6	24·9
●	−2·8	23·7	−2·8	17·6
□	−7·8	16·5	−7·8	13·2
■	−10·8	—	−10·8	—
No germination	−15·1	13·8	−15·1	11·5

Figure 25.2   Germination of rape seeds in (a) a silty loam and (b) a loam (after Williams and Shaykewich, 1971)

468

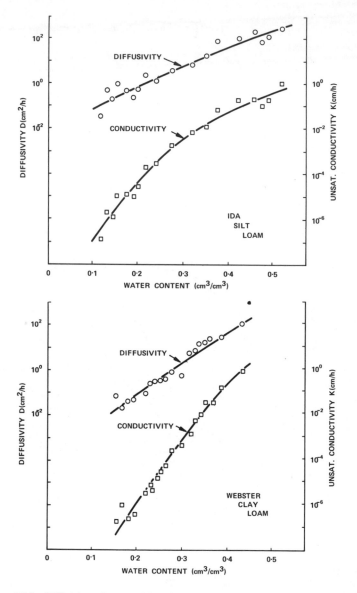

Figure 25.3   Diffusivity and unsaturated conductivity as a function of soil water contents in Ida silt loam and Webster clay loam (after Kunze and Kirkham, 1962)

the water movement is entirely within the liquid phase which is continuous: hydraulic conductivity is then at a maximum and is determined largely by pore diameter. As the soil dries, the liquid phase decreases in volume, the liquid-filled paths become more tortuous with a corresponding decrease in cross-section, and hydraulic conductivity decreases. In dry soils, the water films are discontinuous and water must now move by molecular diffusion in the vapour phase (Rose, 1963). *Figure 25.3* (after Kunze and Kirkham, 1962) shows a typical relationship for two soils. As water moves from the soil into the seed, the soil around the seed becomes drier (*Figure 25.4,*

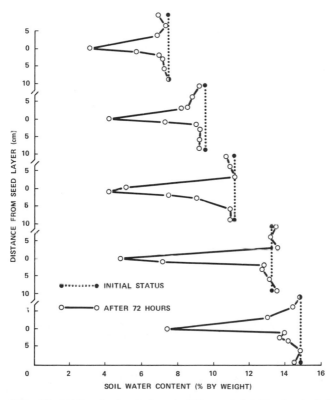

Figure 25.4   The depletion of soil water by seeds of Oryzopsis holciformis *after 3 days from different initial soil water contents (after Dasberg, 1971)*

after Dasberg, 1971) and restricts transfer further. By contrast, as the seed gets wetter, hydraulic conductivity within the seed increases and tends, therefore, to favour quicker distribution of water within

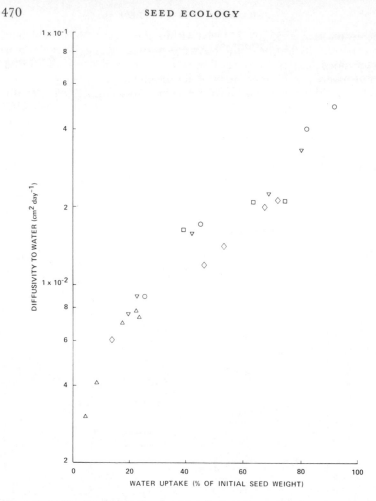

Figure 25.5   The dependence of the water diffusivity of chickpea seeds on the water uptake. Soil moisture (dry wt. basis) : ○ *17%,* ▽ *14%,* □ *11%,* ◇ *8%,* △ *5% (after Hades, 1970)*

the seed. *Figure 25.5* (after Hadas, 1970) shows the change in water diffusivity in a seed during wetting.

When a seed is newly sown it will, as already discussed, come to rest within an air-filled pore space. Though technically surrounded by wet soil, it will have few points of contact with the soil at which water, moving in the liquid phase, might pass into the seed. Furthermore, the rough seed coat of many seeds decreases the efficiency of physical contact between soil and seed.

The extent of interfacial resistance to water transfer is simple to demonstrate. Batches of 20 peas were weighed out. One batch was placed in a petri dish having a water film 1 mm thick in the bottom; this provided perfect interfacial contact at the base of the seed over about one-tenth of its total surface. Others were placed on saturated sintered glass plates subjected in advance to a suction from below of 10 cm water. A suction of 10 cm makes little difference to the potential difference between soil and seed, or to the hydraulic conductivity of the sintered plate, but ensures that excess water is removed from the interfacial region. A final batch was placed on a coarse screen of nylon gauze fixed above a saturated pad of filter paper so that a 1 mm air gap separated water and seed. Water uptake by these seeds is plotted in *Figure 25.6*. The peas in direct contact with water

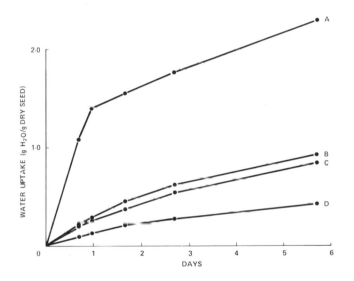

Figure 25.6   The effect of interfacial resistance at the soil/seed surface on water uptake of peas (Pisum sativum L.): *(A) Seeds resting with their bases in water; (B) seeds on a grade 5 (fine) sintered glass plate at 10 cm suction; (C) seeds on a grade 1 (coarse) sintered glass plate at 10 cm suction and (D) with 1 mm air gap between the seeds and pad of saturated filter paper*

imbibed their own weight of water in 16 hours and germinated within 3 days. By contrast, those on the sintered plates neither had taken up their own weight of water nor had germinated after 6 days. Those on the gauze, not in direct contact with water, imbibed at half the rate of those on the sintered plates. Allowing for the shorter distance between the sintered plate and the seed, the rate of water uptake by the seeds is not inconsistent with transfer by vapour diffusion. Indeed, in soil the seed may often have to rely on this mechanism during imbibition. In this experiment sinters with decreasing pore size from grade 1 to grade 5 were used. Decrease in pore size is associated with an increase in surface smoothness which, it was thought, might have improved contact between the plate and seed. The results show little evidence of this. In all the sinters, the water menisci probably lay, as indeed they would in soil, well down within the pores and out of reach of the seed. Collis-George and Hector (1966) demonstrated that if the seeds were placed on a sintered plate with the water initially at zero suction, and if a small suction were then applied, a significant decrease in germination time was observed. They supposed that in this, their pre-wetting treatment, an improved contact between seed and sinter allowed more rapid uptake of water compared with the predrained treatments used above.

In practice, it is usual after sowing to compact seed beds slightly. Whatever reason may be claimed for this, the seed/soil contact is almost certainly improved to the benefit of imbibition. It should be recorded, however, that whereas compaction may improve imbibition and hasten germination, it may restrict subsequent root and shoot elongation and decrease emergence.

The seed that has lain long dormant in the soil and which has become intimately incorporated into the crumb matrix will not have the same contact problem, but presumably has failed to germinate from the lack of some other stimulus to germination.

RESPIRATION

The seed respires during and after imbibition: oxygen is absorbed and carbon dioxide evolved. In response to the concentration differences that develop between the interior and exterior of the seed, oxygen moves through the surrounding soil into the seed. At the same time, carbon dioxide moves out through the seed, and through the surrounding soil. Movement is primarily by molecular diffusion and the transport equation in its simplest form is

$$\frac{dq}{dt} = - D \frac{dc}{dx}$$

where the flux, dq/dt (the quantity q moving in time t), moves in response to a concentration gradient, dc/dx (the change in concentration c over distance x), through a medium in which D is the effective coefficient of diffusion of the gas. In the dry seed, diffusion will occur through the air-filled intercellular spaces, but as the seed wets and these spaces fill with water, diffusion will be in solution at a much restricted rate. Diffusion of oxygen within the seed has been discussed by Edwards (Chapter 10). In the soil it is convenient to consider the effective diffusion coefficient, D, as a fraction of D_0, its value in the absence of all constricting soil particles. The ratio D/D_0 is then a property of the soil and is independent of the gas. The diffusion pattern may be simple or complex depending on where the seed lies within the soil structure. For the newly sown seed lying in the intercrumb pores (as do seeds of most crop plants in a stable structure) it is necessary to consider only diffusion through the macro-pores. D/D_0 has a maximum value when the soil is dry and decreases in a manner determined by the soil type as it wets. In a pure sand in which all the pores are about the same size (unimodal pore size distribution) there is a large initial decrease in D/D_0 as the sand is wetted. By contrast, in a cultivated soil with a well-developed crumb structure, D/D_0 decreases little as the soil is wetted until all the crumb pores are filled with water. Not until further water is added to the inter-crumb pores is there a large decrease in D/D_0 when the pattern becomes the same as for sand (Currie, 1961a). Thus, the water held in a cultivated soil at field capacity does little to restrict gas diffusion through the soil in bulk. It is of interest to compare the diffusion pattern for horticultural (heat expanded) vermiculite with that for a good soil and a sand (*Figure 25.7*). In its dry state, vermiculite has a porosity greater than 0·90. Half this pore space can be filled with

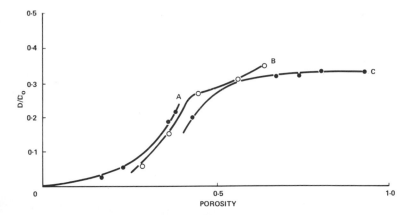

Figure 25.7   The relationship between diffusion and air filled porosity as packings of (A) sand, (B) clay loam (1–2 mm crumbs) and (C) vermiculite are wetted

water without increasing gaseous diffusion by more than 10 per cent, giving adequate aeration and generous water-holding capacity, two ideals for seed germination and for seedling growth.

If the seed lies inside the crumb matrix, it is necessary to consider diffusion within the crumb. Orientation and bonding between particles in the crumb make the pores more complex so that the diffusion coefficient is correspondingly smaller. Also, the crumb pores contain some water for much of the year so that diffusion is restricted even further. At field capacity, when the crumb is saturated, diffusion of gases within the crumb is entirely in solution in the soil water, giving a drastic increase in diffusion resistance of four orders of magnitude. Thus, while it is most improbable that aeration within the macropores would be inadequate at field capacity, it is possible that crumbs greater than a few millimetres in diameter could be anaerobic at their centres (Currie, 1961b; 1962).

But if the seed were the sole sink for oxygen in the soil, it is unlikely that there would be an aeration problem. In most soils, however, the organisms feeding on the organic residues from previous crops respire in competition with the seed, at many times the rate of the seed, increasing the need for adequate ventilation. A number of factors affect the ambient soil respiration rate. The quantity and

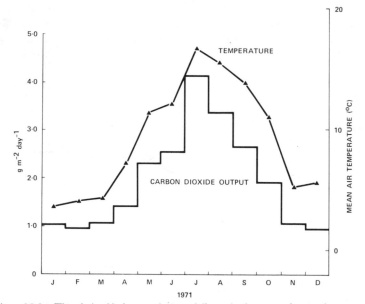

Figure 25.8   The relationship between the mean daily respiration rate and mean air temperature for an uncropped soil during 1971

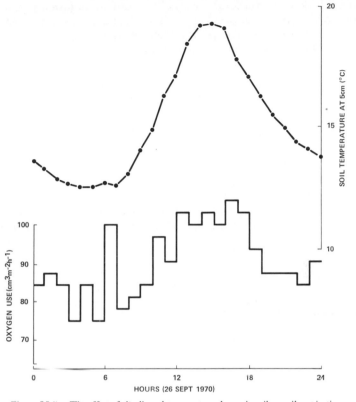

Figure 25.9 The effect of the diurnal temperature change in soil on soil respiration

freshness of the organic matter set a basic level, the water content of the soil determines the extent to which that level is attained, and the prevailing soil temperature not only determines the instantaneous rate of respiration reactions, but also affects the number of organisms at work. The greatest temperature effect is seasonal. Weather causes day-to-day variations and diurnal temperature changes cause hourly variations. Some of these changes, measured in respirometers in the field, are shown in *Figures 25.8* and *25.9* and these effects are to be discussed in detail elsewhere (Currie, 1973).

It is perhaps fortunate that in summer, when the soil is warm and competitive respiration is likely to be most keen, the soil is most likely to be dry and gaseous diffusion least restricted. Seed beds should not be over-watered in summer. This decreases the air filled pore space and restricts diffusion. It may cause structural break-down, decreasing the number of larger pores, and still further restricting diffusion. If structural collapse is complete, diffusion within the seed bed may be entirely in the liquid phase, and any seedlings may subsequently have great difficulty in emerging through the resultant cap or crust when the soil surface dries.

CHEMICALS

Nutrients move by diffusion in the soil water and therefore require continuous water films. It is not necessary to consider the movement of nutrients during germination, but one aspect of diffusion in the liquid phase deserves mention. Some seeds or dispersal units produce substances which inhibit their own germination (Audus, 1959; Heydecker, Chetram and Heydecker, 1971). If these substances are to diffuse away from the seeds, there must be adequate contact between the seed and the water in the soil, and within the surrounding matrix.

LIGHT

Some seeds germinate in response to a light stimulus. It is improbable that light penetrates soil, except down drying cracks, to depths of more than about 1 mm. The finely divided nature of the mineral particles causes light to be scattered and absorbed within this distance.

TEMPERATURE

The temperature of the soil is determined by the balance between incoming and outgoing radiation at the soil surface, and by the way that this energy is redistributed throughout the soil as heat. By day heat is gained, by night it is lost. In winter the net daily gain is negative, in summer positive. Thus two cyclic temperature patterns may be observed, one diurnal, the other annual. The surface temperature at time t may, as a convenient approximation, be written

$$T(0, t) = T_a + T_0 \sin 2\pi t/P$$

where, whether for a day or a year, T_a is the average temperature, T_0 the amplitude of the temperature variation, P is the period time (24 h or 365 days) and t is in hours or days. The temperature at depth z is then given by

$$T(z, t) = T_a + T_0 \exp\left(-2\pi z/l\right) \sin 2\pi \left(\frac{t}{P} - \frac{z}{l}\right)$$

This equation is discussed in a different form but in greater detail by Van Wijk and De Vries (1963).

Two things emerge in theory and practice. The amplitude of the temperature variation decreases with depth by the factor exp $(-2\pi z/l)$ and maximum and minimum temperatures lag behind those at the surface by a time $(z/l)P$. The parameter l may be regarded as a wavelength and is the vertical distance between points at which successive maximum or minimum temperatures occur simultaneously (Keen, 1931). Its value is given by $l = (4\pi k P/c)^{\frac{1}{2}}$ where k is the thermal conductivity of the soil, and c is the volumetric heat capacity of the soil. At a depth of 35 cm the daily temperature lag is of the order of 12 h and the amplitude of the variation only 0·04 times the surface value. By contrast, at the same depth, the seasonal temperature lag will be 9·5 days, but the amplitude will be 0·84 times the surface value. This may be of importance to seeds which require a large change in temperature to trigger germination. Seeds buried at plough depth will experience an almost equable daily temperature with only a slow seasonal change. When brought to the surface at a subsequent ploughing, they will experience a large daily change which may break dormancy.

Both the thermal conductivity of soil and the volumetric heat capacity increase with water content. Thus, although a wet soil may transmit heat more readily, more heat is required to raise its temperature. This is demonstrated clearly, for example, in the above equations where the ratio k/c (the thermal diffusivity) determines both the degree of attenuation and the phase lag. The effect of water content on soil temperature is thus often less than sometimes imagined.

The relationship between the rate of germination and temperature is linear over a restricted range, but Hegarty (Chapter 23) demonstrated that seeds subjected to a harmonic temperature variation germinated more quickly than those held steady at the mean temperature. As the temperature at the soil surface is seldom steady, this effect is important and merits a speculative explanation. Many biological processes respond to temperature according to the relationship:

$$R_2 = R_1 \cdot Q_{10}^{(T_2 - T_1)/10}$$

where the rate of the reaction, R_2, at temperature T_2 is given in terms of that, R_1, at temperature T_1. The value of Q_{10}, the so-called Q-ten, is a constant equal to between 2 and 3. This means, in effect, that for an increase in temperature of 10°C, the reaction rate increases by a factor of from 2 to 3. *Figure 25.10* illustrates, for example, what might happen to a seed having some metabolic process which is

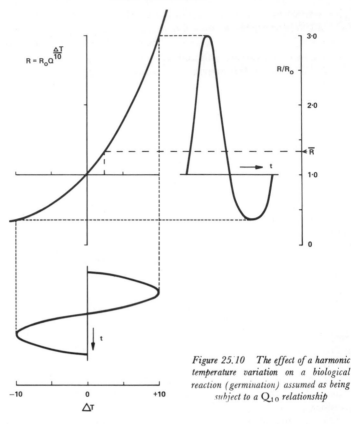

$$R = R_o Q^{\frac{\Delta T}{10}}$$

Figure 25.10   The effect of a harmonic temperature variation on a biological reaction (germination) assumed as being subject to a Q_{10} relationship

necessary for germination and which is subject to $Q_{10} = 3$ when exposed to a temperature variation of amplitude 10°C. At the bottom of the figure is the temperature sine wave. At the top left is the Q_{10} relationship. This is used as a transfer characteristic to give, by 'reflection', the rate of the reaction in the seed as illustrated top right. If the reaction rate at the mean temperature is set at 1·0, the mean reaction rate, $\overline{R}$, integrated over the complete cycle is 1·325. The same rate would have been achieved had the temperature been steady throughout the period at 2·5°C above the cyclic mean.

CONCLUSIONS

The number of soil types is large, the variety of physical conditions in which each may be encountered is wide, and so the possible fates of seeds in the soil may seem limitless. This paper has shown the way in which those processes affecting germination operate, and gives some indication how this operation varies in different soils. These processes have necessarily been considered singly, with only a hint as to the interactions which occur between, for example, water and air, temperature and air. The processes which affect seeds also affect other organisms in the soil, for example, the pathogens. This may mean that the interpretation of the direct effects of soil condition on the seed is often obscure. But what is most important is that those who are concerned with the germination of seeds in the soil should be increasingly careful to observe, for their colleagues and successors, the relevant soil parameters outlined here.

REFERENCES

AUDUS, L. J. (1959). *Plant Growth Substances,* Leonard Hill, London

BAVER, L. D. (1956). *Soil Physics,* John Wiley & Sons, New York

COLLIS-GEORGE, N. and HECTOR, J. B. (1966). 'Germination of seeds as influenced by matric potential and by area of contact between seed and soil water', *Aust. J. Soil Res.,* **4**, 145

CURRIE, J. A. (1961a). 'Gaseous diffusion in porous media. Part 3—Wet granular materials', *Br. J. appl. Phys.,* **12**, 275

CURRIE, J. A. (1961b). 'Gaseous diffusion in the aeration of aggregated soils', *Soil Sci.,* **92**, 40

CURRIE, J. A. (1962). 'The importance of aeration in providing the right conditions for plant growth', *J. Sci. Fd. Agric.,* **13**, 380

CURRIE, J. A. (1973). 'Seasonal changes in respiration in cropped and uncropped soil', (in preparation)

DASBERG, S. (1971). 'Soil water movement to germinating seeds', *J. exp. Bot.,* **22**, 999

EMERSON, W. W. (1959). 'The structure of soil crumbs', *J. Soil Sci.,* **10**, 235

HADAS, A., (1970). 'Factors affecting seed germination under soil moisture stress', *Israel J. agric. Res.,* **20**, 3

HEYDECKER, W., CHETRAM, R. S. and HEYDECKER, J. C. (1971). 'Water relations of beetroot seed germination. II Effects of the ovary cap and of the endogenous inhibitors', *J. exp. Bot.,* **35**, 31–41

KEEN, B. A. (1931). *The Physical Properties of the Soil,* Longmans, Green, London

KUNZE, R. J. and KIRKHAM, D. (1962). 'Simplified accounting for membrane impedance in capillary conductivity measurements', *Soil Sci. Soc. Am. Proc.,* **26**, 421

ROSE, C. W. (1966). *Agricultural Physics,* Pergamon Press, London

ROSE, D. A. (1963). 'Water movement in porous materials. Part 2—The separation of the components of water movement', *Br. J. appl. Phys.,* **14**, 491

RUSSELL, E. W. (1961). *Soil Conditions and Plant Growth,* Longmans, Green, London

VAN WIJK, W. R. and DE VRIES, D. A. (1963). 'Periodic temperature variations', in *Physics of Plant Environment* (ed. Van Wijk, W. R.) North Holland Publishing Company, Amsterdam

WILLIAMS, J. and SHAYKEWICH, C. F. (1971). 'Influence of soil water matric potential and hydraulic conductivity on the germination of Rape (*Brassica napus* L.)', *J. exp. Bot.,* **22**, 586

DISCUSSIONS

Fearn: Is there an ecological explanation for the need of many kinds of seed for fluctuating temperatures to enable them to germinate?

Currie: This requirement may be of ecological importance because such fluctuations are most marked near the soil surface as a result of solar radiation.

Monteith: There may not be a mathematical conflict between Currie's assumption of a curvilinear relationship and Hegarty's, of a linear relationship between germination rate and temperature (Chapter 23), although the latter suggests a cut-off at a certain base temperature whilst the former does not. But there may be a biological conflict: a linear response of germination rate to temperature also suggests a linear response of respiration to temperature in view of the close relation between respiration rate and growth rate. This requires scrutiny.

Gordon: Seed beds in forestry nurseries are prepared by throwing up the soil until late spring, then rolling to compact it, sowing on top of the compacted soil and covering with very light grit particles, presumably in order to reflect sunlight and keep the temperature down. This appears to be a highly unnatural system. Is there a physical explanation?

Newton: It possibly mainly serves to adjust the water relationship by permitting drainage during the winter, and later a supply of water to the seeds from beneath as well as adequate aeration by means of the covering grit.

Heydecker: Very little critical work seems to have been done on the preparation of seed beds.

Heydecker: The high RQ which results from an excess of water in contact with seeds has sometimes been interpreted as resulting in a dangerous excess of CO_2. Is the explanation not more simply sought in a lack of oxygen (see Chapter 10)?

Currie: This may be so; but apart from any possible narcotic effect of an excess of CO_2, every ten-fold increase in the CO_2 concentration may decrease the pH of the soil by as much as one half of a unit and the pH in the immediate vicinity of a seed in a less well buffered soil may therefore be lowered by 2–3 pH units.

E. H. Roberts: Very high concentrations of CO_2 (about 10 per cent) are required to depress germination, whereas lower but still high concentrations (about 2·5 per cent) may have a stimulating effect on the germination of some seeds. Furthermore, according to R. Brown (1940, *Ann. Bot.*, **4**, 379–395) the resistance of covering structures of seeds to the inward diffusion of oxygen is far greater than their resistance to the outward diffusion of CO_2.

Edwards: Does the diffusivity of seeds alter in the course of imbibition?

Currie: The diffusivity of any dry porous system is small but increases as the system becomes wetter. Usually a vapour front precedes the entering water front.

Edwards: By what process does oxygen enter the soil?

Currie: Over 90 per cent of oxygen movement into the soil is by diffusion.

26

THE MECHANISATION OF SEED SOWING

J. S. ROBERTSON

Agricultural Development and Advisory Service, Cambridge

'. . . they used to have a chylde to go in the forowe before the horses with a bagge or hopper of corne, and by lyttel and lyttel he casteth it in the sayd forowe. . . .' (From FitzHerbert's *Boke of Husbandry*, 1523 and C. Culpin's *Farm Machinery*, 1957.)

The apparently simple task of making a machine to place a specified quantity of seeds in the soil in the optimum position for germination and growth has occupied the ingenuity of men for many years. Jethro Tull described his new mechanical seed drill in 1762 and this is usually regarded as the first successful attempt at mechanising seed sowing.

Tull's drill had what we now call a cup-feed mechanism. It is remarkable that this type of mechanism has remained in use in principle, though possibly with considerably altered detail, in many of the seed drills employing a metered feed right up to the present day.

It is important to differentiate between drills employing a metered feed and those which do not. Where a drill uses a metered feed, the quantity of seed delivered is constant (within the limitations of a given drill) in relation to the ground covered. That is, the seed rate per acre should remain the same regardless of the forward speed of the drill within reasonable practical limits.

Another widely used class of drill employs gravity only to deliver the seed from the seed box (hopper) to the coulter. The seed flow is restricted by placing a metal disc with a hole punched in it at the outlet to the box (the size of hole can be varied for different types of seed), and seed delivery is maintained by means of a wavy disc agitator within the box. Here the seed flow is constant as regards time rather than distance. Thus, a constant forward speed is essential when using this type of drill, since to alter forward speed will automatically alter the seed rate in inverse ratio.

Any examination of mechanised seed sowing must look critically at the material to be sown as well as at the mechanism which sows it. Traditionally, vegetables were grown by sowing ample seed to be sure of a stand of sufficient young plants to produce the crop (either direct in the field or as transplants). When the seed germinated and the seedlings grew, excess plants were scrapped by hand-hoeing or simply by transplanting the strongest from the plant bed to the field and destroying the rest. Crudely put, the underlying theory is 'one for the rook, one for the crow, one to rot and one to grow' and for most crops this was good enough when coupled with the grower's own experience and judgement.

With the increasing demand for crops grown to as rigid a specification and time scale as possible, it has become important to try to be more precise in terms of final plant population in the field and this has meant that more accurate assessment must be made of the number of seeds 'to rot and to grow' in setting the drill mechanism. Additionally, seeds of some newer hybrid (F_1) varieties tend to be much more expensive so that it is also desirable not to use a higher seed rate than necessary. This assessment of seed rate has been expressed by Bleasdale (1963):

The seed required in lb/acre =

$$\frac{272 \times \text{No. plants/ft}^2 \text{ (field area)}}{\text{No. seeds per oz.} \times \% \text{ Lab. germ.} \times \text{Field factor}}$$

The main purpose of repeating this formula is not to act as a reminder of the calculation: it is to emphasise that provision of a mechanism capable of planting an exact number of seeds in a specific area of ground is only one part of the overall problem of sowing seeds to produce a given plant population.

It will be useful to consider the various factors which affect the operation of drilling mechanisms and to see over which, if any, the drill manufacturer, designer or operator have any control.

SEED CHARACTERISTICS

1. Different crops have different shaped and sized seeds. Unfortunately, not only do these characteristics vary between species of crop, but to a large extent variation, particularly of size, also occurs within a given species and even within a given variety. This size variation has a double effect. The purely practical and obvious one is that a drill at a given setting will sow more individual seeds when they are small than when they are large. One might claim that, as most batches of seeds are well mixed, the drill performance would correspond to the average seed

count, but it is likely that separation of different sizes of seed can take place within the seed box due to the combined effects of vibration and flow of seed out of the box to the sowing mechanism.

2. It has been shown that the germination figure is linked to the specific gravity of the seed (Austin and Longden, 1967). Thus, to get an even crop it is highly desirable that seed graded by specific gravity is used. It is as well to note that in theory it does not matter what level of specific gravity a particular grade may have: the important point is that all seeds should fall within known limits and thus have *predictable* 'germination vigour'. Once the value of this factor is known, due allowance can be made for it.

3. Seed shape is mainly a function of species. Some seeds, notably brassicas, are ideal for use, particularly in precision drills, in that they are regular in shape and of a suitable size. Others, like lettuce, carrot, onion or beet are more difficult to deal with, being either too irregular in shape or too small, or both, to be easily handled by a seeding mechanism on an individual seed basis. It is possible, of course, to do something about this, and the pelleting of small seeds in an inert clay or other material, and the smoothing ('rubbing') of awkwardly shaped seeds like those of beet, are well-known processes. It is important, however, with these processes to be sure that germination is not impaired.

4. Laboratory germination of a sample of seed is simple to determine. There are, currently, arguments for changing the basis to include a factor for vigour of germination as well as total germination; but these will not be considered in this chapter (see Chapter 17).

FIELD FACTOR

This is *the* 'one for the rook, one for the crow' factor, and assessment of its value depends entirely on the skill of the farmer or his advisers. No matter how great this skill, the number of variables in the shape of climate, soil, pests and diseases, weeds, etc., can completely upset the calculation; and although irrigation, herbicides, pesticides and fungicides and other measures can do much, they cannot ensure the completely uniform, predictable conditions which are necessary to grow crops exactly to a specification.

Some attempts to improve the field factor by controlling the microclimate in which the seed initially germinates will now be described.

TAPED SEED

With this system, seed is sealed at the appropriate intervals in a patent water-soluble plastic tape. The tape is wound on to reels holding between 1500 and 6000 m (5000–20000 ft). It is then laid out in a furrow by means of a specially developed drill. In the United States, where the system has been used since about 1967, claims are made for greater uniformity of emergence, plant stand and maturity and higher yields.

Rae (1969) reported on trials carried out in England in 1968 and 1969 on a total of 14 sites located in Lancashire, Warwickshire, Worcestershire, Cambridgeshire and Norfolk, using sugar beet, onion, cabbage and lettuce seed. With the exception of onions, the taped seed performed unpredictably and was on average less satisfactory than seed sown by conventional methods. It was suggested that the depth of planting, soil consolidation and water requirements after drilling all needed further study and the 'shelf life' of the tape was also doubtful.

An apparently more successful paper-based tape is produced in England. This is being used commercially to a limited extent but suffers from the disadvantage of using a bulkier taping medium.

FLUID DRILLS

One of the problems in establishing an even stand of crop from direct drilling is that the micro-environment directly around the seed varies considerably and is only controllable to a relatively small degree by normal methods. A particular problem is lack of moisture at drilling and this has been a factor leading to the development of a so-called 'fluid drill'.

The National Institute of Agricultural Engineering (N.I.A.E.) and the Agricultural Research Council Weed Research Organisation (W.R.O.) together developed a fluid drill which sows seeds in a thin stream of a thixotropic aqueous gel (Elliott and Parker, 1965; N.I.A.E., 1970). This is not a precision drill, as the even distribution of seeds along the row relies upon their having been evenly distributed in the gel before this is put in the machine. In its present form it is more suited to trial and plot work for which it was designed in the first instance.

In the United States a prototype machine, claimed to be a precision drill, injects dry seeds into a water stream which is, in turn, pumped into the seed furrow. But at the time of writing this equipment appears to be in the early stages of development.

Fluid drilling has many apparent advantages in that the right moisture content for germination can be ensured and, in addition, fertiliser, insecticide, fungicide and herbicide could also be incorporated in the fluid. A disadvantage can be that whilst the fluid induces prompt germination, it may soon evaporate if dry conditions follow sowing. A means of overcoming this is the use of some form of mulch placed in strip form over the seed furrow. The American machine referred to above also places, if desired, an 'anti-crustant' material, e.g. vermiculite, over the line of seeds. Where soil crusting (capping) is a problem, existing drills can be obtained with various types of skeleton or cage press wheels to try to minimise this effect.

VACUUM SEEDERS

N.I.A.E. (1971) have developed an experimental vacuum-operated seed-metering mechanism, designed to operate more readily with irregularly shaped seeds and with those which vary in size. Commercial vacuum-operated seeders of French and German origin are now available.

DRILL TYPES

Vegetable crops grown from seed can be divided into two types. There are those which require individual plants to be widely spaced in the field and in which the plant population is measured in terms of, for example, square metres of field per plant. Crops of this type include lettuce and most brassicas. The other type of crop is one in which population can be measured in terms of plants per square metre of field area and these include canning and ware carrots, onions, parsnips and round red beet. This is relevant to the following discussion.

Drilling mechanisms can be categorised into precision, semi-precision and non-precision drills.

Precision drills are capable, depending on the shape and size of the seed, of placing single seeds or groups of seeds accurately at regular intervals down the row. Current versions are limited in forward speed to about 3·25–4·75 km/h (2–3 mph) and a number of types of mechanism are in current use. Amongst the most common is the punched belt, in which an interchangeable endless belt with holes punched in it takes single seeds or groups of seeds and carries them to an ejection point of the mechanism, where they can be dropped to the bottom of the furrow.

R

Another common precision-drill mechanism uses a cell or selector wheel with cups or indentations machined into the perimeter. The cells or cups have the same purpose as the holes in the belt-type drill. In some types, selector wheels are interchangeable; in others a single selector wheel in each drill unit has a series of different sized or spaced cells in parallel rings running around the perimeter of the wheel, and the ring of cups required for use is selected by means of a sliding gate.

The vast majority of precision drills in use today make use of one or other of the mechanisms described above, or slight variations on the theme. Other mechanisms are being, or have been, used. One such is the 'helical' drill, having a combination of a cylinder with a helical screw thread machined in it and an adjustable cut-off plate. The seeds are metered by being screwed along the thread in the cylinder. Very accurate seeding should theoretically be obtained with this mechanism but it is not widely used at present and the author has had no practical experience of its use.

The drills so far described have individual, independent, units at ground level and a length of drop of probably not more than 13 mm ($\frac{1}{2}$ in).

Another seeder uses a 'pure' cup-feed principle, with 30 plastic cups (colour coded and interchangeable in sets for a wide range of different seeds) to meter the seeds. The cups are tipped by cam action to discharge the seeds at about 20° before top dead centre of the revolution of the disc holding the cups. This disc is about 170 mm (7 in) in diameter.

Several types of *non-precision drilling* mechanisms are currently in use for vegetable crops. Brief descriptions of the main types are as follows:

Rotary Disc drill

High-level seed units draw seeds from a common seed box and meter them by means of a rotating horizontal disc drilled with holes of a varying number and size according to seed type and rate. The seeds drop down coulter tubes to the seed furrow. A variety of seeds may be sown, up to the size of natural red beet.

External Force Feed (Cereal) drill

This type of seeding mechanism dispenses seeds at high level from a common seed box by means of a fluted or studded roller and delivers them to the ground through the coulter tubes and coulters. It will handle most vegetable seeds including the smaller beans and is widely used for cereals. Drive is via a multi-speed gearbox from land wheels.

Pneumatic or Centrifugal drill

A single central-adjustable dispenser, land-wheel or power take-off driven, delivers seeds pneumatically or centrifugally to a distribution head. From this they are delivered in fractions to each coulter. The machine handles a wide range of seeds and sowing rates can vary from 2·5–40 kg/ha (2–36 lb/acre).

Brush Feed drill

A traditional pattern of drill which discharges seeds from a common high-level seed hopper by means of rotating brushes. It will handle most small seeds and can be obtained in machine widths and row widths to suit the requirements of the purchaser.

Disc and Agitator drill

This seeder, already referred to, relies on a hole in a disc at the bottom of the seed hopper for metering of the seeds. A wavy disc agitator keeps the seeds falling through the hole whilst the drill is in motion, but has little effect on sowing rate which is governed almost wholly by the forward speed. The disc can be rotated to present different-sized holes for different types of seed.

As pointed out earlier, one of several points dividing precision from non-precision seeders is the length of drop from the metering unit to the furrow bottom. For example, by calculation

$$V^2 = U^2 + 2\,AS$$

where V = terminal velocity, U = initial velocity, A = acceleration and S = elapsed time, a free-falling seed (ignoring air resistance) will take about 0·18 sec to fall 150 mm (6 in) or 0·36 sec to fall 600 mm (24 in). Even small variations in the falling time of each seed (caused by vibration or roughness of the coulter tube, static electricity, etc.) will be sufficient to upset the regularity of the seeding pattern considerably. So for true precision sowing, the seed must be ejected as close to the ground as possible.

The type of work which the main types of seed drill are capable of doing is given in *Table 26.1*. Note that there may be several makes of drill of each type, but not all makes are necessarily capable of sowing all the seeding patterns stated for a particular type.

Table 26.1. TYPES OF DRILL AND THEIR POSSIBLE SEEDING PATTERNS

Drill type	Seeding pattern possible
Belt	Single seeds, spaced from 25 mm (1 in)
	Groups of seeds spaced from 25 mm (1 in) upwards, across or along the drill
	Thin lines
	Grouped rows or beds from 25 mm (1 in) inter-row spacing (units in tandem)
Cell wheel	Single seeds spaced from 19 mm ($\frac{3}{4}$ in) upwards
	Groups of seeds spaced from 19 mm ($\frac{3}{4}$ in) upwards
	Thin lines
	Groups of rows or beds from 38 mm ($1\frac{1}{2}$ in) inter-row spacing (units in tandem)
Cup feed	Semi-accurate placing of single seeds
	Groups of seeds
	Thin lines
	Bands
	Rows spaced at 205 mm (8 in) or upwards unless 'splitting back'
Rotating disc	Thin lines
	Bands, with special coulter
	Groups of rows or beds from 75 mm (3 in) inter-row spacing
External force feed (fluted or studded roller)	Thin lines
	Bands
	Groups of rows or beds from about 100 mm (4 in) inter-row spacing
Pneumatic/centifugal	Thin lines
	Bands
	Groups of rows or beds from about 100 mm (4 in) inter-row spacing
Brush feed	Bands, row spacing to order
Disc and agitator	Thin lines
	Bands
	Groups of rows or beds from about 75 mm (3 in) inter-row spacing

It has been necessary to go into the types of drill design and factors affecting drill performance and operation at some length in order to give the background to some of the present work which is being carried out on drills.

Whilst many crops can only be drilled 'to a stand' (so as to have plants exactly, and only, when required) by a particular type of drill, certain other crops can be grown in a number of different patterns for the same purpose. It has been shown (Salter and Currah, 1970; 1971) that with carrots, for example, provided the plant population in terms of numbers of plants per unit of field area is correct, the actual spatial arrangement is not so critical as was formerly thought. Thus, one seeding pattern for canning carrots, for example, is thin lines at 380 mm (15 in) centres. *Table 26.1* shows that any one of seven different types of drill can be used to sow carrots in this arrangement. The question has been put by growers: If I now have, for example, an external force-feed multi-purpose drill, do I necessarily have to buy a special-purpose drill to achieve the same result? To begin to answer this sort of question, mechanisation and horticultural advisers of the Agricultural Development and Advisory Service (A.D.A.S.) conducted a trial to compare the performance of six different vegetable drill mechanisms on just such a drilling pattern.

The trial was carried out in 1971 in Norfolk on a medium sand typical of many of the mineral soil carrot-growing areas of East Anglia. The six machines were:

1. Rotating disc non-precision drill.
2. Punched-belt precision drill.
3. Disc and agitator non-precision drill.
4. Cell-wheel precision drill.
5. Multi-purpose force-feed non-precision drill.
6. Semi-precision cup-feed drill.

The drills were all borrowed from local farmers and were pre-calibrated to deliver 9·5 kg carrot seed per ha (8$\frac{1}{2}$ lb/acre). The Bleasdale formula indicated this seed rate to provide a final plant population of 430 plants/m^2 of field area (40 plants/ft^2). The drills were calibrated both static and in motion over a measured length of the level floor of a barn.

Drilling depth was set at 13 mm ($\frac{1}{2}$ in). The drills comprised at least 5 units each and each drill made two passes separated into two blocks across the field. Four rows from each pass, chosen at random for sampling, constituted one plot. There were therefore eight sample rows per treatment, each of which was treated as a sub-plot. The sample rows were 486 mm (19·2 in) in length. This length was chosen as representing 2 ft^2 of field area with rows spaced at 15 in centre to centre (2 ft^2 = 0·186 m^2 and 15 in = 0·357 m). Plant counts were made at intervals from first emergence throughout the growing season. Finally, all the plots were lifted and the roots graded and weighed. Though, at the time of writing, computer analysis of

Table 26.2　MEAN DRILLING DEPTHS (TARGET DEPTH 13 mm (0·5 in))

	Drill no.*											
	1		2		3		4		5		6	
	inch	mm	inch	mm	inch	mm	inch	mm	inch	mm	inch	mm
Mean depth	0·8	20	0·75	19	0·7	18	0·6	15	0·9	23	0·6	15

* The drill types, details of which are in *Table 26.1*, are listed on page 489

the figures has not been completed, some results are available and are given below.

Although all drills were calibrated with the intention of giving the same seed rate, field populations varied considerably from drill to drill. Since for various reasons the seed was not weighed into and out of the drills it is unfortunately not possible to state categorically that the differences were due to different seeding rates. This is, however, a reasonable inference since all other conditions were more or less constant.

Mean depths of drilling were measured as the length of white hypocotyl on the newly emerged seedlings, from the junction of root and hypocotyl to the start of the green colour below the base of the cotyledons. These depths varied from between 15 and 23 mm (0·6 and 0·9 in) (*Table 26.2*); nevertheless, emergence from five of the drills peaked on the 12th day after emergence was first observed and the sixth peaked on the 13th day (*Table 26.3*).

Figure 26.1 shows the emergence pattern from each drill. It also

Table 26.3　RATE OF EMERGENCE AS PERCENTAGE OF MAXIMUM

Days after drilling	Drill no.*					
	1	2	3	4	5	6
12	27·6	35·5	30·7	28·2	19·3	19·6
14	63·3	71·4	64·7	66·9	65·9	56·3
16	80·2	94·9	76·0	82·5	83·8	78·0
19	92·4	93·07	89·7	95·4	93·4	92·7
21	100	100	100	100	99·1	100
23	96·8	99·5	99·7	97·3	100	99·7
28	94·4	96·6	96·1	96·5	98·0	96·1

* For details see *Table 26.1* and list on page 489.

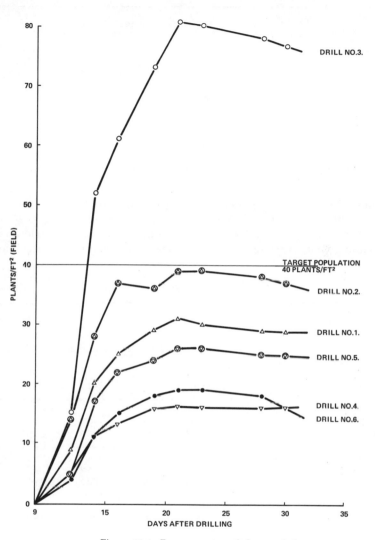

Figure 26.1 Emergence rates and plant populations

Table 26.4 ORDINATES FOR *Figure 26.1* (CORRECTED TO NEAREST WHOLE NUMBER)

Days after drilling	Drill no.*					
	1	*2*	*3*	*4*	*5*	*6*
12	9	14	25	5	5	4
14	20	28	52	11	17	11
16	25	37	61	13	22	15
19	29	36	73	16	24	18
21	31	39	81	16	26	19
23	30	39	80	16	26	19
28	29	38	78	16	25	18
30	29	37	77	16	25	16

* For details see *Table 26.1* and list on page 489

shows how near each final emergence came to the target of 40 plants/ft^2 of field area. *Table 26.4* gives the ordinates for *Figure 26.1*.

Table 26.5 gives the mean density, total yield, canning yield and canning yield expressed as a percentage of the total for each drill at

Table 26.5

	Drill no.*					
	1	*2*	*3*	*4*	*5*	*6*
Mean density						
Plants/ft^2	28	35·7	65·4	16·0	22·6	17·5
Plants/m^2	300	385	700	172	243	188
Canning yield ($\frac{3}{4}$-1$\frac{3}{4}$ in dia)						
Tons/acre	11·7	13·6	13·9	9·1	10·7	10·2
Tonnes/ha	29·4	34·0	34·8	22·8	26·8	25·6
Total yield						
Tons/acre	22·2	24·5	23·4	28·0	24·1	29·7
Tonnes/ha	55·5	61·5	58·5	70·0	60·3	74·5
Canning as % of total	52·6	55·4	59·4	32·5	44·4	34·5

* For details see *Table 26.1* and list on page 489

harvest time. It is interesting to note that the precision drill No. 2, which came nearest the target density, produced 34 tonnes/ha (13·6 tons/acre), whilst the non-precision drill No. 3, giving around 160 per cent of the target density, produced 35 tonnes/ha (13·9 tons/acre) of canning grade—virtually the same quantity.

CONCLUSIONS

The results from each drill are valid as they stand. It would be unwise, however, to draw any more than a very general conclusion from a trial of this nature, for various reasons. These are chiefly its rather limited scope, the different ages of the different drills and the fact that the trial has so far been carried out only in one year. Nevertheless, the results do allow one to conclude tentatively that under the conditions under which the trial was conducted, provided the plant population is at the right level, there does not appear to be very much difference in final yield of the desired grade (i.e. in this case canning carrots $\frac{3}{4}$–$1\frac{3}{4}$ in diameter), irrespective of whether a precision or non-precision drill is used.

Perhaps the most valuable lessons learned from this trial are the non-quantitative ones. The first is that accurate calibration of present-day drills is far from easy, as field performance can obviously vary considerably from the original calibration performance. With the small quantities of seed used for vegetables per unit area, it is difficult for a farmer to judge whether his seed rate is correct until a reasonable acreage has been sown, and even then only estimates can be made. This raises the question whether a population in the field is different from the calculated target because of a wrong estimate of the field factor or because the drill has been at fault through not putting on the right quantity of seed in the first place.

The tables show that the field performance of some drills can deviate from the calibrated setting by a considerable margin. Detailed population figures from individual row counts (not included in this paper) also show large deviations from coulter to coulter within the same drill. Deviation of even a single coulter can result in yet another kind of error, since it is not only the size grade of carrots in one row that is affected, but this in turn, by varying the competition effect, could alter the grade in adjacent rows.

Because the performance of a particular drill was good or bad on carrots, for example, it does not necessarily mean that its performance on a different kind of seed will also be bad. Quite the opposite is likely to be the case, since all the drills in the trial are widely used for a large number of crops with satisfactory results. There is clearly a need for much more detailed investigation of drills used for vegetables, and if it has done nothing else the A.D.A.S. investigation described here has illustrated this need. It has also highlighted the difficulty of making such an apparently simple investigation in terms of time and labour required and in deciding upon the particular aspects which most require detailed examination.

SUMMING UP

Although many developments have occurred, both in seeds and in the mechanisms for sowing them, there are still many biological and engineering problems to be overcome. Notably, it seems that we still have a great deal to learn about our existing equipment and the ways in which it can be improved before we pass on to new or more highly sophisticated mechanisms.

REFERENCES

AUSTIN, R. B. and LONGDEN, P. C. (1967). 'Some effects of seed size and maturity on the yield of carrot crops', *J. hort. Sci.*, **42**, 339–53

BLEASDALE, J. K. A. (1963). *The Bed System of Carrot Growing*, Min. of Ag. Fish. and Food, Short Term Leaflet No. 27

ELLIOTT, J. G. and PARKER, J. D. (1965). *The Sowing of Seeds in an Aqueous Fluid*, Agricultural Research Council Weed Research Organisation (unpublished report)

N.I.A.E. (1970). *Seed Metering, Delivery and Covering*, National Institute of Agricultural Engineering, Report 1 April 1967–31 March 1970, p. 20

N.I.A.E. (1971). *Seed Metering, Delivery and Covering*, National Institute of Agricultural Engineering, Report 1 April 1970–31 March 1971, p. 20

RAE, R. (1969). *Summary of 'Polyox' Tape Planting Experiments in the United Kingdom.* N.A.A.S. Eastern Region Experiments Committee (unpublished report)

SALTER, P. J. and CURRAH, I. E. (1970). *Rep. natn. Veg. Res. Stn.* for 1969, 72

SALTER, P. J. and CURRAH, I. E. (1971). *Rep. natn. Veg. Res. Stn.* for 1970, 74

DISCUSSIONS

Harrington: Neither hydro-seeders nor seed tapes are as yet used commercially in California.

Richardson: 1. *Seed placement.* An extensive project on accurate seed placement is in progress at the Rowcrop Department, National Institute of Agricultural Engineering (N.I.A.E.). The programme looks at the problems in dynamic terms, for example the velocity of the released seed, its angle of impact on the soil and the consequent effect upon seed displacement. Studies are also being made, or proposed, on seed metering problems, seed release from the metering mechanisms, seed delivery to the soil, seed covering and soil pressing.
2. *Seedling emergence.* The drill mechanism may affect the capability of the seed to grow by causing damage. It affects the seed environment by opening a furrow, filling the furrow and then pressing the soil. Studies on drill coulter, coverer and press wheel design are envisaged, but more knowledge of what seeds require from the environment is needed to guide this work.
3. *Future possibilities.* The present stage of the work looks mainly at the

optimisation of conventional seed drills and drilling methods with which the operation is conducted at present. But limitations imposed by seeds and soil conditions have encouraged the search for other ways of achieving crop establishment. Among these the following possibilities may be listed:

(a) The use of seeds which have commenced to germinate under controlled conditions. It may be possible to use some sorting process in order to select a batch of seeds all of which have commenced germination within some specified time range.

(b) The use of specially prepared seed-covering materials designed to provide an optimal environment for the seeds. Protection against damage to the environment by subsequent weather conditions, notably for the prevention of soil crust formation, would be incorporated.

(c) The protection from pests and diseases, the provision of water for germination and initial seedling development (perhaps in the form of a gel which will remain with the seed and may also contain nutrients for the seedling) are other possibilities.

There are complex interactions between the biological requirements and any attempted engineering solutions and when the biological requirements are better understood the engineer will be in a better position to study them and eventually to provide the ways and means of meeting these requirements in an economic manner.

27

SEED ECOLOGY—PRESENT AND FUTURE

I. H. RORISON

Nature Conservancy Grassland Research Unit,
Department of Botany, University of Sheffield

INTRODUCTION

It was hoped that, although they are not all ecologists, the contributors of this Easter School Symposium on seed ecology would approach the subject from an ecological standpoint. Nevertheless, the papers have largely been concerned with seed physiology seen through the eyes of crop physiologists, biochemists and pathologists.

Regarding the ecology of agronomic situations, attention has been concentrated more on conditions in which optimum yield and productivity, rather than bare survival, are achieved. With increasing mechanisation of crop husbandry it is not only optimum yield but also uniformity of plant development from sowing to canning which is required (see Chapters 24 and 26). However, with increasing concern for extreme environments, such as are encountered during the reclamation of derelict and disturbed land, the importance of studying responses at both the upper and lower limits of a plant's tolerance becomes apparent. Such studies allow some prediction of a plant's potential adaptability in both competitive and non-competitive situations (Rorison, 1969).

For both approaches a knowledge of the basic mechanisms which regulate the response of the seed to its environment is needed. It is important to think of the seed, or seedling, not only as a complex of mechanisms, each possibly of some adaptive significance, but also as a whole organism in relation to its environment. Desirably, we are not only concerned with the 'average' plant in relation to the environment but with a comparative study of species in relation to their habitat. Any adaptational difference could be of vital interest to the ecologist concerned with plant distribution and also to the experimenter at the cellular level concerned with revealing the controlling mechanisms involved. But not all differences are necessarily of adaptational significance.

As in so many approaches to scientific problems, most of the papers at the Symposium have been arbitrary in the sense that areas of progress are strongly influenced by suitability of techniques and by the availability and responses of certain plant species. This applies particularly to studies at the cellular level. The techniques concerned or the species investigated are not always those which are immediately relevant to key ecological factors in the field and as with other studies, such as mineral nutrition, one of our major tasks now is to ensure maximum interdisciplinary co-ordination between field and laboratory studies.

This summary of the Symposium follows the progress of the seed from the time that it is dependent upon the parent to the time that it becomes an independent autotrophic plant. Points of particular interest to the practising ecologist are commented upon and an attempt is made to bridge the gap between laboratory and field-based experimentation.

PARENTAL REPRODUCTIVE 'STRATEGY'

Parental reproductive strategy is a suitable starting point and yet has received little direct attention at this conference, despite its ecological significance.

Stebbins (1971) states that the complexity and diversity of adaptations for seed reproduction are partly a result of the separate and sometimes conflicting demands made during seed development, dispersal and seedling establishment.

According to Harper, Lovell and Moore (1970), the proportion of effort that is put into seed production varies roughly according to the habitat which the plant occupies. *Table 27.1* shows the broadest categories, from specialised grain crops at 40 per cent of net annual assimilation down to 15–0 per cent by herbaceous perennials which also reproduce vegetatively.

The question arises: is the proportion allocated in this way fixed and characteristic of a species or group of species or is it plastic, being subject to change in response to environmental stress (Harper, (1967)? Extreme environmental stresses may result in a variation in these figures but, for example, using a fairly wide range of pot volumes resulting in 85-fold differences in plant productivity, Harper and Ogden (1970) found that *Senecio vulgaris* and related annual

Table 27.1 (COURTESY: HARPER, LOVELL AND MOORE, 1970)

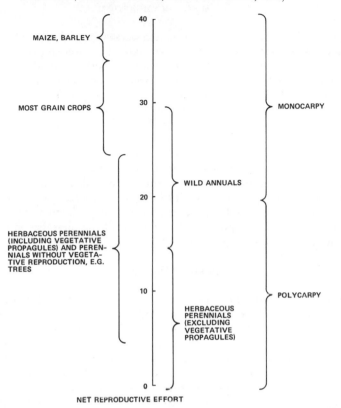

species maintained a very steady seed production at between 20 and 30 per cent of total dry weight.

It is practicable to refer to only a few of the contributions of Harper and his co-workers. They are making an incisive contribution to the important topic of reproductive strategy and their results provide a firm basis for further ecological thinking (Harper and White, 1970) allied to specific physiological experimentation.

It was particularly interesting in this regard to hear from Dr. Gutterman (Chapter 4) that photoperiodic treatments of the parent, even before flowering, can affect both the rate and extent of germination of subsequently formed seeds of *Lactuca sativa*. L. Interspecific differences in parental response to photoperiod and their relation to hormonal activity may go far to explain the mechanisms behind different reproductive strategies, but much remains to be done.

One of Dr. Gutterman's points, the occurrence of seeds of different degrees of maturity on the same plant, has been noted in several native British species and its importance to their survival, by ensuring discontinuous germination, has been appreciated. For example, Cavers and Harper (1966) reported such polymorphism in seeds of *Rumex crispus*. Whether seed polymorphism occurs in many plant species has yet to be confirmed. What is essential is to remember that bulk samples of seeds may, as Salisbury (1965) found with *Plantago major*, mask features of biological significance, while reliance on seeds from a single plant or inflorescence may lead to the recording of a typical responses. These findings are well known but, it is suspected, widely ignored and since they are important in the interpretation of both physiological and ecological experiments, the potential importance of seed polymorphism is one of the messages that should emerge from this conference.

PREDATION AND DISEASE

Localised polymorphism may be due to injuries caused by predation and disease. The importance of seed predation in plant distribution was noted by Darwin and has been reviewed most recently by Janzen (1971). Not only are insect, bird and mammal predators active against some species of all plant families but their attacks can occur at various stages. Defoliation of parent plants, eating of flowers and attacks on seeds, before and after dispersal, all add to pre-germination depletion (Maun and Cavers, 1971a; 1971b). Conversely, attempted predation can end, for some seeds, in improved dispersal and even in germination. Structures serving to resist or avoid predation or to aid dispersal have to be accounted for in the plant's reproductive budget. This is another subject little touched upon in published research.

Seed-borne diseases (Christensen, 1972) again offer a fascinating topic ripe for large-scale comparative study. Maude (Chapter 18) points out that pathogens may be carried superficially or deep in the tissues and the latter are the more difficult to eradicate. Systemic fungicides offer great promise as a means of control in crop husbandry. When a wide enough range of them is available they might also be used as tools to assess the effect of seed-borne fungi on the survival rate of naturally occurring species. The diagnostic tetrazolium tests described by Professor Moore (Chapter 20) could also be employed in assessing the causes of damage done to seeds of naturally occurring species in a range of habitats (some of the limitations of the tetra-zolium method are given by MacKay, 1972).

All these influences must be borne in mind when considering reproductive strategy in terms of seed numbers and seed size.

SEED NUMBERS AND SEED SIZE

Salisbury (1942) suggested that in general, in temperate Britain, plants of closed (stable) communities produce a few large seeds and plants of open (pioneer) communities produce many small seeds. *Figure 27.1* (Harper, Lovell and Moore, 1970) shows seed weights (mean and range) of groups of species from various habitats in Britain: open sites, short grass, and meadows (iv); woodland margins (v); woodland ground flora (vi); woodland shrubs (vii); and woodland trees (viii). The mean weights of the seeds vary over three orders of magnitude and lie well within the ten orders of magnitude which cover the flowering plant kingdom. Salisbury suggested that for autotrophic plants the capacity to colonise in the face of competition appears to be associated with the amount of food reserve which the seed contains and that large seeds have an initial advantage under competitive conditions. This simple idea has been modified in the light of further research into interspecific differences in relative growth rate (Hunt and Grime, in preparation), the ecological significance of mycorrhizal associations (Harley, 1971) and other factors which may influence the establishment of a species in a community.

Stebbins (1971) reports that seed number is subject to very great phenotypic modification, depending on the environment. In contrast, Harper, Lovell and Moore (1970) stress that mean weight per seed produced by a species can be constant and is the least plastic of all the components of reproductive yield. It is also their main thesis that variation in seed numbers and seed size represent alternative strategies in the disposition of reproductive resources.

Although mean seed weight may be constant, the range of weight around the mean may vary considerably from species to species, and the proportions and absolute sizes achieved in any one year may be a reflection of environmental factors.

If the problem of size is examined more closely, intraspecific seed size may imply differences in weight and/or in degree of differentiation. If differences in seed size involve differences only in weight and volume but not in stage of development, then relative growth rates are often equal, although absolute rates vary with differences in initial capital (Black, 1957; Rorison, 1961). In optimum competitive

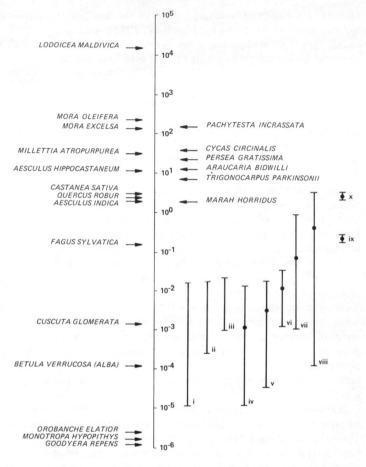

SEED WEIGHT (g)

Figure 27.1   Ranges of variation in seed weights (from Harper, Lovell and Moore, 1970)

situations, individuals from large seeds have therefore a better chance
of survival than those from small seeds—micro-topography permitting (Currie, Chapter 25; Harper, Williams and Sagar, 1965;
Harper and Benton, 1966). If differences in seed size involve varying
degrees of differentiation and maturity this may lead not only to
different initial growth rates, but also to differences in rate and time
of germination, and these may blur experimental responses to a
prescribed range of conditions (e.g. nutrients) unless taken into

account. In the field it may be that, because of discontinuous germination, the few stragglers which remain dormant during unusual climatic or other disturbances in the community are, in some seasons, the only effective propagules. It can therefore be important to determine the responses of seedlings derived from the whole spectrum of seeds produced by a species and not only from the largest and most rapidly developed.

It is agricultural practice to harvest seeds at one time when the majority are mature but of various weights. The smallest seeds whose embryos are least likely to be completely differentiated are discarded on cleaning (Scott and Longden, Chapter 5). Under natural conditions all seeds do not necessarily leave the parent plant at once. Any process which ensures departure only at maturity has advantages. There is, however, virtually no information in the literature on how long nutrient supply from the parent continues and to what degree competition between seeds in an inflorescence occurs.

SEED PRODUCTION

There is a great deal of variation between species, both crop and native, in the timing and extent of their seed production. Having heard about several crop species, let us briefly examine a field and laboratory study of a native species *Scabiosa columbaria* because it exposes sources of variations which may arise unnoticed in isolated studies. *Scabiosa columbaria* reproduces exclusively by seed and is a perennial of uncertain age (perhaps up to 50 years). The seeds have no dormancy, their germination is discontinuous and the species has an edaphic range limited in Great Britain to dry calcareous soils. It depends for survival upon the longevity and overall seed production of the parent plant. For it must be remembered that *time* is a vital factor and that in terms of survival it is as important to know the rate of effective seed production over several years of varying climate as it is to know the effect of climate and site on the establishment of seedlings in any one year. To maintain its number in a population it needs to produce one successor; that is to say, perhaps one in 30 years from 10 000 to 30 000 seeds. It requires the right gap under the right conditions.

Flowers are arranged in inflorescences which produce symmetrical seed heads with no apparent spiral sequence of seed size (see *Figures 27.2* and *27.3*). When a very large seed occurs it is usually adjacent to a very small one (this is unlike several species which exhibit distinct size patterns). The number of seeds in heads from any one site in any one year is remarkably uniform (*Table 27.2*). The table

Figure 27.2 Scabiosa columbaria. *Flowering head in the early stages of expansion. Outer florets open first followed by some at the centre*

Figure 27.3   A mature seed head

506

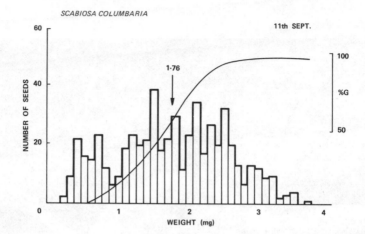

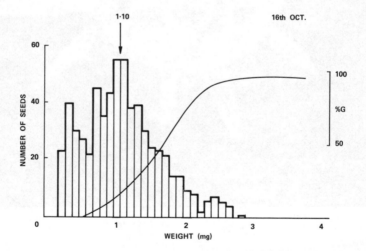

Figure 27.4 Scabiosa columbaria. *Distribution of seed weights into 0·1 mg categories from 10 ripe heads collected at random from the same site in September, 1962 and in October, 1962 (histograms); percentage germination determined after 20 days at 20°C in the laboratory (superimposed curves)*

Table 27.2 SEED HEADS OF *SCABIOSA COLUMBARIA*

Date collected	No./head ± SEM	Wt./seed (mg)	% effective germination at 20°C
11. 9.62	56·7 ± 8·7	1·76	55
27. 9.62	59·4 ± 15·2	1·56	46
16.10.62	61·9 ± 17·7	1·10	23
16.11.62	55·1 ± 13·3	1·16	29

shows mean numbers of seeds per head, and mean weight per seed, of samples collected at four different times in the autumn of 1962. Each sample comprised ten ripe heads collected at random at the same site. Percentage effective germination of each sample was calculated from data given in *Figure 27.4*.

The mean seed weight drops significantly as the season progresses and the proportion of lateral to apical heads increases. Seed heads collected at different times of the year have a different range of seed weights with a preponderance of heavy seeds in the earlier months (*Figure 27.4*). The heaviest seeds have the highest percentage germination. Therefore, because of the higher percentages of large seeds ripening in September, effective germination is approximately double that obtained from seed collected in October and November (*Table 27.2*).

GERMINATION IN RELATION TO TEMPERATURE

Not only does effective germination decrease as winter approaches, but so also does the mean soil temperature. Do the seeds germinate immediately on ripening or do they overwinter first? What is the temperature and moisture regime before and during germination?

To find out, five-weekly sowings were made in the field throughout the year at sites at which soil and air temperature 2 cm above and below ground level were measured continuously.

When seeds were sown during the period from October to March no germination occurred either immediately or subsequently. Seeds sown in March and April, when the soil was at field capacity, germinated as soon as suitable soil temperatures were reached (*Figure 27.5*). Between April and September, soil temperatures were adequate but soil moisture stress at times delayed germination and caused the death of young seedlings, particularly on a south-facing slope (Rorison, in preparation). Seeds sown both early and late in the year took longest to germinate and had the lowest germination.

SCABIOSA COLUMBARIA

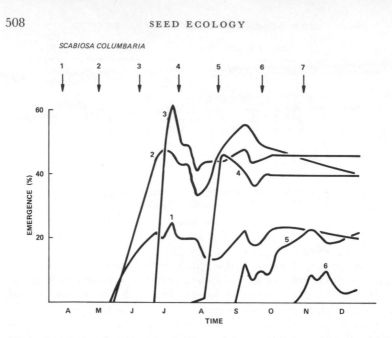

Figure 27.5 Scabiosa columbaria. *Seedling emergence in the field. Downward pointing arrows indicate seven sowing dates. The six numbered curves indicate seedling emergence and subsequent survival throughout the season from May until December—the seventh sowing failed to emerge*

The optimum conditions for germination in 1971 were in May and June, two months before the current year's seeds began to fall.

Records of daily max.–mean–min. soil temperatures throughout the period suggested that a mean of 8–10°C was required to initiate germination (*Figure 27.6*).

Laboratory responses were checked using a thermal gradient bar of the type mentioned by Dr. Thompson (Chapter 3) and this gave results which could be reconciled with field conditions (*Figure 27.7*). The lowest temperature at which 50 per cent germination was achieved was about 9°C and the highest 34°C.

To determine the effect of diurnal fluctuations in temperature two other sets of results were obtained, one with a 5°C and the other with a 10°C range. As Thompson (1970) has reported for other species, the uppermost temperature at which 50 per cent germination was reached remained constant whatever the diurnal fluctuation (*Figure 27.7*). Thus, at steady temperatures the maximum for 50 per cent germination was about 34°C. With a five degree fluctuation the maximum range was 29–34°C (mean 31·5°C) and with a 10°C fluctuation the maximum range was 24–34°C (mean 29°C).

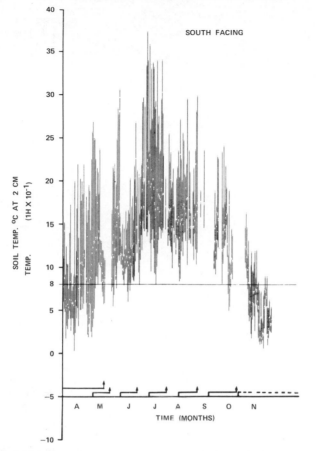

SOUTH FACING

SOIL TEMP. °C AT 2 CM
(1H × 10⁻¹)
TEMP.

TIME (MONTHS)

Figure 27.6   A computer drawing of soil temperature at 2 cm depth showing, by means of individual vertical lines, daily maximum, minimum and mean temperatures. A horizontal line is drawn at the 8°C level (see text) and the arrowed lines at the base indicate the time between sowing and emergence after each sequential sowing of Scabiosa columbaria *(see Figure 27.5)*

At the lower end of the tolerance range, 50 per cent germination was achieved at a minimum mean temperature of about 9°C, whether the regime was steady or with a diurnal fluctuation of either 5°C or 10°C.

In short, with diurnal fluctuations, *Scabiosa columbaria* was able to germinate while spending at least 12 h per day at temperatures at least 4°C lower than tests at steady temperatures had indicated.

Its failure to overwinter in soil or to shed its seeds during the time when conditions are optimum for its germination suggest one more reason why it is restricted to habitats with low competitive pressures.

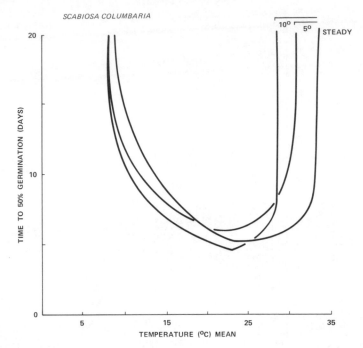

Figure 27.7 Germination curves for Scabiosa columbaria, *showing time taken to reach 50 per cent germination over a temperature range of 5–40°C at steady temperatures; at mean temperatures derived from a 5°C diurnal variation; and at mean temperatures derived from a 10°C diurnal variation. The two horizontal bars at the upper right-hand side of the curves indicate the upper limits of temperature variation at which 50 per cent germination in the two ranges was achieved (in a time well in excess of 20 days)*

In contrast, a highly competitive ruderal species, *Rumex acetosa,* is able to germinate when sown through the year (Vincent, 1970) and shows tolerance of mean temperatures 5°C lower than *Scabiosa columbaria* (Rorison, unpublished).

These results have been given to emphasise that: in some species seed size is not only variable but also has a strong influence on effective germination; to reinforce Dr. Hegarty's findings (Chapter 23) that it is important to consider the effects of diurnal fluctuations in temperature when looking for early planting times and resistant species or strains; and to give a simple example of the value of combining field with laboratory experimentation.

Plant physiologists often work at steady laboratory temperatures of around 25°C because they find metabolic systems are then most active. Such temperatures are of fleeting relevance in the ecology of native British species. Diurnal fluctuations may not influence optimum responses but they may be vital in extending the limits of tolerance of species adapted to extreme environments.

SOIL CHEMICAL FACTORS

SOIL REACTION

Apart from soil temperatures, there are other edaphic factors which may have a drastic effect on germination and survival.

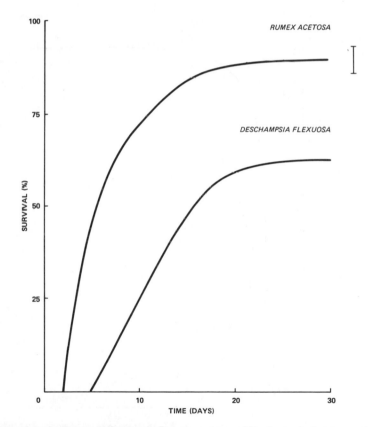

Figure 27.8   Seedling survival curves for Rumex acetosa *and* Deschampsia flexuosa, *each grown on 9 different soils, ranging in pH from 3·5 to 7·4. The vertical bar indicates the 95 per cent confidence limit within which the results for all the soils lie*

Soil chemical factors, particularly those of acidity, alkalinity and salinity, have a critical influence on seedling survival (Rorison, 1960) and, in some cases, even on germination (Rorison, 1967).

A brief example concerns three species of different ecological distribution, *Rumex acetosa*, *Erigeron acer* and *Deschampsia flexuosa*. When their seeds were sown in a bioassay under controlled laboratory conditions where soil type was the only variable, the widespread, fast-growing ruderal species *Rumex acetosa* germinated equally well on nine soils ranging in pH from 3·5 to 7·4. So did the strict calcifuge *Deschampsia flexuosa* (*Figure 27.8*). Subsequent seedling survival of the two species reflected their natural distribution, with *R. acetosa* widely indifferent and *D. flexuosa* failing in soils of high pH according

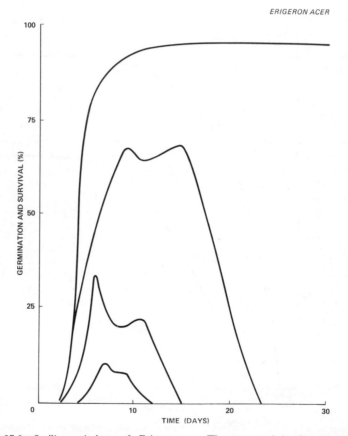

Figure 27.9   Seedling survival curves for Erigeron acer. *The upper curve derives from calcareous soils and the lower three from soils of increasing acidity (for details see Rorison, 1967).*

to their level of calcium carbonate. In contrast, both the germination and survival figures of the strict calcicole *Erigeron acer* fell with increasing soil acidity (*Figure 27.9*). These patterns have been repeated in various degrees with other species.

The different chemical factors involved—toxicities of polyvalent cations, nitrogen source and nutrient deficiencies—have been studied extensively, but the influence of seed reserves in combating external nutrient disorders has received less attention. It would be particularly interesting to determine the relative influence of available reserves whose proportions might vary within a species and according to seed size.

NITRATE IONS

There are widespread reports of stimulation of germination in nitrate solutions, albeit at about 10^{-2}M, a somewhat higher concentration than in most soil solutions. What happens in soils whose sole or major source of nitrogen is ammonium nitrogen, i.e. acidic mineral soils? Certainly the seedling growth of some species is related to the nitrogen source in the soil (Gigon and Rorison, 1972). The questions arise: is their germination stimulated by nitrate ions and, if not, is there an alternative? A comparative experiment in which such calcifuge species as *Spergula arvensis* and *Deschampsia flexuosa* are included might well extend our appreciation of the problem.

ETHYLENE

Soil physical conditions may be no less severe for some species grown under extreme conditions than chemical ones and the two may interact. Soil moisture is an obvious limiting factor and associated atmospheric conditions may be equally so. Processes limited by oxygen availability have already been mentioned by several contributors and the influence of ethylene under conditions of waterlogging and compaction (Smith and Restall, 1971) is potentially of ecological interest.

It is not possible to attribute distinct ecological distributions to the three groups of weed species listed by Olatoye and Hall in *Table 13.1*, and none of the five species whose germination is promoted by ethylene has any particular affinity for waterlogged soils where ethylene concentration might be relatively high. Forms of *Hypochoeris radicata* and *Rumex crispus* occur even on sand dunes, which are not usually anaerobic, but it would be of interest to work with samples from dune slack areas known to be waterlogged for part of the year. Conversely, the three species whose germination is inhibited by ethylene, *Chenopodium rubrum*, *Plantago major* and *P. maritima*, may occur in soils which are damp and occasionally even waterlogged. Theirs may be an adaptive mechanism to prevent germination when conditions are too wet.

One would suggest that the inclusion of species with known ranges of tolerance to anaerobic conditions is necessary before any ecological significance can be attributed to ethylene concentrations in the soil.

DORMANCY AND GERMINATION

Discontinuous germination has ecological advantages in that even if transitory adverse environmental conditions are lethal to early germinating seeds, more seedlings may appear after conditions have again become favourable.

As contributors have shown, there is an impressive amount of information identifying mechanisms which control dormancy and show that they can reside innately in the embryo or be induced by the parent or enforced by the environment (Harper, 1957, quoted by Roberts, 1972). The factors most important in breaking dormancy are frequently listed as light, temperature and nitrate ions. Professor Roberts (Chapter 11) has stressed the importance of studying the interactions of these factors, since they are unlikely to operate in isolation, and has related their effects to the single purpose of stimulating the pentose phosphate pathway which he suggests leads in many species to loss of dormancy.

LIGHT EFFECTS

Temperature and nitrate have already been mentioned but light has been the factor most discussed. Interest centres round phytochrome and the absorption of red and far-red light. In particular, the reversible promotion and inhibition of the germination of light-sensitive seeds by alternate exposures to red and far-red light has been shown by Cumming (1963) to have ecological significance. He suggested that the germination of several species of *Chenopodium* may be restricted in areas shaded by green leaves. He found that there was more germination in light with red/far-red ratios, similar to that of sunlight (1·3), than in sunlight transmitted through green foliage (0·70–0·12). Roberts (1972) quotes unpublished work of Pigott showing that this factor also suppresses the germination of *Urtica dioica* under a woodland canopy. The same, no doubt, might be found to be true of several other herbaceous species, e.g. *Digitalis purpurea* and *Chamaenerion angustifolium,* which normally do not occur under dense woodland canopies but are common at woodland margins or in clearings (see also Stoutjesdijk, 1972).

The germination rates of seed of light-sensitive species exposed to different red/far-red light regimes could possibly provide a rapid method of testing for shade tolerance in plants. An equally interesting possibility has been raised by Professor Smith (Chapter 12), who has suggested that 'phytochrome acts to "inform" the plant whether or not it is growing in the shade of another and elicits developmental responses appropriate to the conditions.' As he says, this would be important not only for seed germination but for many other developmental processes in shaded seedlings. The measurement of responses of a range of light-sensitive species from open and from closed communities would be most interesting in this context.

SEED STORAGE

Seed storage investigations may have seemed remote from the general ecology of seeds in so far as the ideal conditions for storage of seed of many species in the laboratory—low moisture content, O_2 and temperature—are often violated by fully imbibed seeds which may lie, viable but dormant, for years in the field or even in the laboratory (Villiers, Chapter 15).

However, the relevance of papers on storage and viability testing is brought into sharp focus by the growing awareness of the need for the conservation of seeds of the world's vegetation as genetic pools for breeding programmes, for the reconstruction of particular plant communities, and for the reinoculation of natural habitats whose cover has been destroyed. Plant communities may be kept intact by suitable management, but seed banks are the best means of ensuring that replacements are genetically pure. Some laboratory means of storage, as advocated by Frankel (1970) and practised at the Jodrell Laboratory at Kew and the Welsh Plant Breeding Station, is essential for the future, but an immediate answer to the problem of naturally revegetating disturbed ground has been pioneered by Grime and Spray of Sheffield (in preparation). This involves reinoculation of sites using thin layers of seed-bearing topsoil from the original site or an ecologically similar site from which soil is available. Initial viability trials involving different interim storage methods have proved very encouraging.

Seed storage is also important for long-term physiological studies in the laboratory where, year by year, a series of experiments is carried out on the same species using different seed harvests—albeit from the same clones or sites each year. There is, therefore, the need for carefully controlled and reproduceable storage conditions. These must be determined for every ecologically distinct group of species. For example, comparative studies show that seeds of several plants from aquatic habitats cannot survive dry storage (Harrington, Chapter 14). To all this must be added an awareness of the significance of seed polymorphism and the fact that viability is not synonymous with vigour (Perry, Chapter 17; Heydecker, 1972).

CONCLUSIONS

Discussion amongst members of the Easter School has revealed that the term 'seed ecology' extends to different limits in different disciplines.

However, just as Professor Monteith (discussion of Chapter 22, p. 410) mentioned that rate, speed, velocity and resistance are terms with well-known meanings in physical science and should be applied to biological systems accordingly, so the word ecology has meant to biologists, for many years, essentially the study of organisms in the whole environment. It is a holistic philosophy encompassing biotic, climatic and edaphic interactions.

We may choose to isolate single factors, species, even mechanisms within cells, but for it to be an ecological study we must be thinking of these detailed experiments so that ultimately their relevance to the whole pattern of events may be assessed.

There is now much basic information about how environmental factors, both singly and when several are interacting, affect the production and germination of seeds. There is also some basic knowledge of biochemical mechanisms which sense environmental conditions, e.g. phytochrome (Smith, Chapter 12), and regulate plant response. Genetic (Whittington, Chapter 2) and hormonal (Wareing, van Staden and Webb, Chapter 8) control mechanisms are beginning to be understood and so is the synthesis and degradation of specific enzymes, and the role of cellular structures (Roberts and Osborne, Chapter 6 and Hallam, Chapter 7). Encouragingly, some laboratory experiments are proving to have predictive value for field situations (Scott and Longden, Chapter 5 and Hegarty, Chapter 23).

Laboratory-based workers may consider that there is still much to learn about the genetics, biochemistry and physiology of *the seed* regardless of its environment before wider ecological implications are considered, but seed responses can be variable even at the intraspecific, single-plant level and it is vital to know what is typical or atypical. It is wasteful to build an investigation round an environmental factor reported from the field using only plants or tissues readily to hand. A new approach is needed in many cases. It requires a thorough knowledge of the ecology of species so that they are chosen on the basis of their natural distribution. Where possible comparative studies involving species of varying degrees of tolerance to the environmental factor should be attempted—and if possible the environmental factors tested should include levels of the same order of magnitude as those found in the field. Only then can we say with confidence which differences in sensing and regulating mechanisms are concerned with differences in plant distribution.

This fundamental knowledge will ultimately benefit all ecologists who wish to have some control over the growth and development of seeds and seedlings, whether their aim is to achieve the ultimate in uniformity of canned vegetables or the conservation of natural grassland.

What we must now aim for is an increasing awareness of natural processes and a synthesis of ideas.

REFERENCES

BLACK, J. N. (1957). 'Seed size as a factor in the growth of Subterranean Clover (*Trifolium subterraneum* L.) under spaced and sward conditions', *Aust. J. agric. Res.*, **8**, 335–351

CAVERS, P. B. and HARPER, J. L. (1966). 'Germination polymorphism in *Rumex crispus* and *Rumex obtusifolius*', *J. Ecol.*, **54** 367–382

CHRISTENSEN, C. M. (1972). 'Microflora and seed deterioration', in *Viability of Seeds* (ed. Roberts, E. H.), 59–93, Chapman and Hall, London

CUMMING, B. G. (1963). 'The dependence of germination on photoperiod, light quality, and temperature, in *Chenopodium* spp., *Can. J. Bot.*, **41**, 1211–1233

FRANKEL, O. H. (1970). 'Genetic conservation of plants useful to Man', *Biol. Conserv.*, **2**, 162–169

GIGON, A. and RORISON, I. H. (1972). 'The response of some ecologically distinct plant species to nitrate- and to ammonium-nitrogen', *J. Ecol.*, **60**, 93–102

HARLEY, J. L. (1971). 'Fungi in ecosystems', *J. Ecol.*, **59**, 653–669

HARPER, J. L. (1957). 'The ecological significance of dormancy and its importance in weed control', *Proc. 4th Int. Congr. Crop Protection, Hamburg* 1957, **1**, 415–20. Braunschweig, Hamburg

HARPER, J. L. (1967). 'A Darwinian approach to plant ecology', *J. Ecol.*, **55**, 247–270

HARPER, J. L. and BENTON, R. A. (1966). 'The behaviour of seeds in soil, Part 2, The germination of seeds on the surface of a water supplying substrate', *J. Ecol.*, **54**, 151–166

HARPER, J. L., LOVELL, P. H. and MOORE, K. G. (1970). 'The shapes and sizes of seeds', *Ann. Rev. Ecol. Syst.*, **1**, 327–356

S

518   SEED ECOLOGY

HARPER, J. L. and OGDEN, J. (1970). 'The reproductive strategy of higher plants. I. The concept of strategy with special reference to *Senecio vulgaris* L.', *J. Ecol.*, **58**, 681–698

HARPER, J. L. and WHITE, J. (1970). 'The dynamics of plant populations', *Proc. Adv. Study Inst. Dynamics Numbers Popul*, 41–63 Oosterbeek

HARPER, J. L., WILLIAMS, J. T. and SAGER, G. R. (1965). 'The behaviour of seeds in soil. I. The heterogeneity of soil surfaces and its role in determining the establishment of plants from seed', *J. Ecol.*, **53**, 273–286

HEYDECKER, W. (1972). 'Vigour', in *Viability of Seeds* (ed. Roberts, E. H.), 209–286, Chapman and Hall, London

JANZEN, D. H. (1971). 'Seed preparation by animals', *Ann. Rev. Ecol. Syst.*, **2**, 465–492

MACKAY, D. B. (1972). 'The measurement of viability', in *Viability of Seeds* (ed. Roberts, E. H.), 172–208, Chapman and Hall, London

MAUN, M. A. and CAVERS, P. B. (1971a). 'Seed production and dormancy in *Rumex crispus*. I. The effects of removal of cauline leaves at anthesis', *Can. J. Bot.*, **49**, 1123–1130

MAUN, M. A. and CAVERS, P. B. (1971b). 'Seed production and dormancy in *Rumex crispus*. II. The effects of removal of various proportions of flowers at anthesis', *Can. J. Bot.*, **49**, 1841–1848

ROBERTS, E. H. (1972). 'Dormancy: a factor affecting seed survival in the soil', in *Viability of Seeds* (ed. Roberts, E. H.) 321–359, Chapman and Hall, London

RORISON, I. H. (1960). 'Some experimental aspects of the calcicole–calcifuge problem. I. The effects of competition and mineral nutrition upon seedling growth in the field', *J. Ecol.*, **48**, 585–599

RORISON, I. H. (1961). 'The growth of Sainfoin seedlings', *Univ. Nottingham School of Agric. Rept.*, 1960, 44–46

RORISON, I. H. (1967). 'A seedling bioassay of some soils in the Sheffield area', *J. Ecol.*, **55**, 725–741

RORISON, I. H. (1969). 'Ecological inferences from laboratory experiments on mineral nutrition', in *Ecological Aspects of the Mineral Nutrition of Plants* (ed. Rorison, I. H.), 155–175, Blackwell, Oxford

SALISBURY, E. J. (1942). *The Reproductive Capacity of Plants*, Bell, London

SALISBURY, E. J. (1965). 'Germination experiments with seeds of a segregate of *Plantago major* and their bearing on germination studies', *Ann. Bot. N.S.*, **29**, 513–521

SMITH, K. A. and RESTALL, S. W. F. (1971). 'The occurrence of ethylene in anaerobic soil', *J. Soil Sci.*, **22**, 430–443

STEBBINS, G. L. (1971). 'Adaptive radiation of reproductive characteristics of angiosperms. II. Seeds and seedlings', *Ann. Rev. Ecol. Syst.*, **2**, 237–260

STOUTJESDIJK, P. L. (1972). 'Spectral transmission curves of some types of leaf caryopses with a note on seed germination', *Acta Bot. Neerl.*, **21**, 185–191

THOMPSON, P. A. (1970). 'Germination of species of Caryophyllacea in relation to their geographical distribution in Europe', *Ann. Bot.*, **34**, 427–449

VINCENT, E. A. (1970). 'A study of variation in phosphorus availability in calcareous soils', *M.Sc. thesis*, University of Newcastle upon Tyne

DISCUSSIONS

Cavers: In western Ontario results similar to Dr. Rorison's have been obtained with *Amaranthus retroflexus*. Seedlings established early in the season produced a larger total number of seeds, but both early and late established plants first produced large seeds and later small ones. *Plantago* seedlings arising at different times of the year differ in their flowering and seed setting potentials, and only seedlings which have become established within a very restricted period (about one month) have a good chance of successful reproduction in the natural situation.

Rorison: In investigating the ultimate productivity of perennials, we estimate that *Scabiosa* in the field may survive for 30–50 years, although the exact life span is unknown; one plant may produce only 100–200 seeds per year, but the total production may be in excess of 10 000 and only one of the resulting seedlings has to achieve maturity to maintain the numbers of the species in the population. By contrast, in fertile soil single plants have been known to produce as many as 50 000 seeds in five years but have died in the process. Seeds from such plants had twice the weight of those in the field and this indicates that a finite size is much less likely to exist with plants than with animals.

Cavers: Concerning the relation between life span and seed production, there is plenty of evidence that plants which produce vast numbers of seeds every year have a shorter life span than those which produce few seeds.

Brown: Dung incorporated in the seed bed is reported to enhance seedling growth owing to a higher available moisture content and owing to improved potassium nutrition (results of the National Vegetable Research Station).

Rorison: It is difficult to generalise on effects of manuring and fertiliser placement. However, experimental results have shown that one can supply too high a nutrient level too early, especially if the level is increased suddenly.

Whittington: There is evidence that the growth rate of seedlings from small seeds is higher than from large seeds, at least with tomatoes, although it is true that those from small seeds never quite catch up in terms of absolute weight. Our tomato material came from crosses between plants with the same leaf size but vastly different seed sizes.

Rorison: Our evidence comes from J. N. Black on subterranean clover and our own work on other legumes including sainfoin. We have not found any significant differences in relative growth rate between seedlings from fully mature large (25 mg) and small (5 mg) seeds but perhaps our range of sizes was not wide enough.

Mortimer: The effect of the parent on the seed deserves further study; and it might be desirable in a future conference on seed ecology to pay more attention to the way in which seed populations react to the environment and to changes in the environment, with examples taken from the weed/crop situation and from naturally occurring populations.

Appendix 1

PANEL DISCUSSION—PRE-SOWING SEED TREATMENTS

Chairman: R. B. AUSTIN; *Panel:* P. C. LONGDEN, T. W. HEGARTY,
T. H. THOMAS, R. B. MAUDE

Treatment of seeds, before ever they are distributed and incorporated in the seed bed and exposed to the complex and uncontrollable conditions of the field, presents a unique opportunity to the grower interested in giving his crops a good start. The possibility of such treatments have fired the imagination of many, but practical applications are still in their infancy. It was felt that, given the opportunity of contributing their own experiences, conference members might help to construct an up-to-date picture of this exciting and many-sided but little documented subject. The chairman and his invited panel introduced the subject.

Austin: There is room for improvement in at least two respects: in the proportion of viable seeds which establish themselves as seedlings, and in the performance of the seedlings. Regarding this latter aspect, we have found that the nutrition—especially phosphorus nutrition—of the mother plants can affect the nutrient content and the performance of pea, carrot and watercress seeds and increase yields. But the effect has so far been definite and positive only when the basic nutrient supply to the parent plant was decidedly deficient in the respective nutrient element.

Longden: I will confine myself to physical pre-sowing treatments which do not interfere with normal seed storage. Therefore I exclude chitting, stratification and the vernalisation of germinating seeds (as well as of seeds on their mother plants). Here we can distinguish between mechanical treatments and treatments involving water.

Mechanical treatments include grading by density and/or diameter (or thickness) of seeds, in order to eliminate the poorest (lightest) fraction. The proportion rejected depends on how much one can afford to throw away as well as, of course, on how small a seed of high quality can be.

Abrading is beneficial where it is necessary to break a seed coat barrier to water uptake and perhaps occasionally to gas transport. In the special case of *Beta vulgaris*, especially sugar beet, mechanical treatments have been used for both multigerm and monogerm fruits. Removal of part of the fruit tissue ('cortex') results in the removal of a proportion of the endogenous germination inhibitors and on average increases germination by about 5 per cent. The decortication process used for multigerm fruits splits these into single-seeded units, but is a harsh process which can injure seeds, though the resulting fragments are of more or less standardised shape and size for either direct precision drilling or pelleting. In contrast, the 'rubbing' process employed with the modern genetic monogerm fruits has only to smooth the fruits and reduce their volume to a standard size and is much gentler and safer.

Pelleting (or coating) is the process of enclosing a seed into a small quantity of inert material just large enough to produce a globular unit of standard size to facilitate precision drilling. It is essential that pellets should be stable until drilled but should then easily disintegrate to release the seed. Clay minerals (such as bentonite) are probably at present most frequently used but the exact details of the pelleting process are closely guarded trade secrets. In 1972, 90 per cent of the English sugar beet acreage was sown with pelleted seeds. The incorporation of plant protecting and growth stimulating chemicals in the pellets is contemplated.

Regarding treatments involving water, apart from processes which are designed to leach out germination inhibitors by numerous changes of water (such as, again, from *Beta vulgaris* fruits) the process attracting most interest is the one termed 'hardening' by P. A. Henckel in Russia. According to him, repeated imbibition of seeds, short of radicle emergence, followed by redrying, causes the activation of latent physiological mechanisms which enable the resulting plants to withstand certain adverse environmental conditions such as drought and extremes of temperature. We ourselves, using this technique on carrot seeds, have found that in certain seed lots this process increases the number of cells in the embryo before radicle emergence. Total germination was unchanged but the rate of germination of treated seeds was enhanced. As a result, the seedlings were larger at any point in time after sowing and the advantage persisted so as to result in a larger root yield. No evidence was obtained of enhanced performance during drought conditions. We therefore prefer to term the process 'advancing' rather than 'hardening'. In my experience the process occasionally benefits sugar beet seeds, in respect of increased numbers established as well as more rapid establishment; but a fungicidal seed treatment against *Phoma betae* with ethyl mercury phosphate (EMP) 'seed steep' after such treatment has often proved phytotoxic.

Hegarty: I have found that seedling emergence can be improved, at least at low temperatures, by a pretreatment such as outlined by Mr. Longden, but I have experienced no increase in yield. On the other hand, a salt solution (using potassium phosphate) was more effective than water alone for the pretreatment. But the practice cannot be unreservedly recommended for all seed lots of carrots. The effect I have observed is chiefly in the rate of germination (and not on increased embryo growth as reported by Mr. Longden). The effect is especially noticeable with seed lots which germinate poorly at low temperatures (10°C); it remains to be investigated whether such seeds suffer from immaturity (in which case the 'hardening' would result in after-ripening) or from embryo degeneration (in which case such degeneration would be counteracted by the 'hardening' process). Since the appropriate treatment has apparently to be determined for each individual seed lot the practice cannot at present be recommended for commercial use.

Thomas: The commercial treatment of seeds with growth regulators and other chemicals is being investigated. Possibilities are:

1. Application of supplements, e.g. trace elements such as molybdenum to seeds from deficient parent plants or to seeds destined to be sown in soils deficient in these elements.
2. Use of chemicals to break the dormancy of seeds, such as gibberellins and cytokinins (especially to counteract dormancy induced by adverse seed bed conditions).
3. Effects on rate and, more important, uniformity of germination and seedling emergence, with the object of obtaining seedlings of uniform size; this is especially important where once-over harvesting is to be practised.

An example of both the latter points is the benefit conferred upon celery seeds by a combined application of gibberellins 4 plus 7 ($GA_{4/7}$) and of aminozide (B9, Alar) which enables a more reliable germination in the dark, at least under laboratory conditions. GA_3 is much less beneficial than $GA_{4/7}$, and neither is effective by itself. But Palevich, Israel, reports that in the field a continued seed treatment with GA_3 and aminozide has increased emergence from 16 to 60 per cent at high temperature and has produced more uniform of seedlings. Similar results have been experienced on parsnip and parsley seeds which, like celery, are light sensitive at high temperatures. Pretreatment of lettuce seeds with GA_3 or kinetin has enabled these to germinate at 30°C, well above their usual ceiling temperature.

Gibberellic acid (GA_3) and also thiourea have been successfully used as substitutes for chilling, especially on seeds of woody species including conifers. GA_3 also overcomes the inhibitory effects of high osmotic pressure (e.g. on rice and lettuce seeds), of high salt concentration, of water-imposed dormancy (e.g. of barley grains) and of the inhibition caused by mechanical damage (e.g. of *Beta vulgaris* seeds). Auxins have been reported to stimulate the germination of some vegetable seeds, e.g. radish and onion.

One major problem is the selection of the right dosage of the right growth regulator in every instance. Another is the application of the chemical in aqueous solution but this has been largely overcome by using an organic solvent, e.g. dichloromethane, instead of water as a carrier which ensures the incorporation of the chemical into the dry seeds. The solvent is volatile and dries off rapidly after the bulk of the solution has been decanted. The method has been used successfully on lettuce seeds (see Appendix 2, Joshua and Heydecker). Dichloromethane is occasionally slightly beneficial by itself.

Pretreatment of the parent plant (see Chapter 4) may affect the germination of the resulting seeds via their hormone content and furthermore the performance of the resulting plants via the hormone conduct of the seeds. A treatment of carrot plants with CCC (chlormequat) has resulted in an increase of yield from their seeds.

Pelleting should provide a good opportunity for incorporating chemicals such as nutrients, growth regulators and fungicides in larger quantities than was hitherto possible.

Maude (see also Chapter 18): I would stress the importance of eliminating seed-borne fungal diseases at source; otherwise a crop might well suffer right from the outset by either poor emergence or spread of the disease from infected seedlings throughout the crop soon after emergence. Surface treatment of seeds is useful against pathogens carried on the outside of seeds, but where the pathogen penetrates into the seed tissues internal methods are required. The search for these has led to the thiram soak treatment which when combined with a sufficiently high temperature (30°C) for a sufficiently long time (24 h, or in some instances 12 h) has eliminated many seed-borne pathogens and is used commercially on valuable seeds provided they are small enough to be easily dried back. For larger seeds, such as peas and beans, attention is now focused on the systemic fungicides which have begun to appear, e.g. the benzimidazols such as benomyl which, however, are very specific. The combination of conventional, more general fungicides, such as thiram and captan, for external use with more specific systemics for penetration into the seeds holds promise for the future.

Scott: Is it possible at present to give definite recommendations for certain chemical or water treatments for certain purposes on certain seeds, based on fundamental knowledge, or are all recommendations at present merely empirical?

Longden: At present recommendations are empirical. One thing, however, seems clear: pretreatment of seeds by wetting and drying (our 'advancing') uniformly improves the rate of seedling emergence.

Thomas: Light-requiring seeds usually respond to cytokinins and gibberellins, or gibberellins and aminozide, but our knowledge so far is empirical.

Heydecker: It is intriguing to hear that certain pretreatments may have many more than one effect, e.g. 'hardening' of seeds by imbibition and drying is reported to enable them, or their seedlings, to withstand many inimical conditions such as drought, salinity, cold and heat—conditions which only have in common the fact that they are extreme. Is a single fundamental process involved?

Austin: The process enables seeds to utilise favourable conditions while they last and therefore enhances the chances of a good start.

Harrington: An aerated-soak treatment in a hypertonic salt solution, for instance of potassium phosphate—but the kind of salt seems of secondary importance—is now popular in the U.S.A. Work at Purdue University has shown that storage materials are broken down and cell division occurs, as well as many other early steps of germination, but not cell elongation. Treated seeds, when dried after treatment and later sown in the field, germinate distinctly more rapidly. A 7-day treatment of tomato seeds enables them to emerge in the field 9 days (instead of 14–16 days) after sowing. The sowing of pretreated *Capsicum* (Sweet Pepper) seeds is the general rule in Florida now. It should be recognised, however, that the germination percentage of seeds thus treated is sometimes *poorer* and that their storage life is curtailed.

Hegarty: It appears that two to three cycles of 'hardening' with water are equivalent to the 7-day treatment in a hypertonic salt solution.

Storey: A word of warning is needed concerning bacterial pathogens which may be spread by soaking seeds in water. These are few in number but severe. Peas, beans, brassicas and tomatoes are subject to seed-borne bacterial diseases.

Maude: This is important. Thiram soaking does not affect bacteria; on the contrary, if the same fungicidal suspension is used more than once for a seed soak, bacterial populations (even non-pathogenic ones) can increase to germination-inhibiting levels, because of the absence of counter-balancing fungi.

Moore: The water content of seeds is important—seeds store best under dry conditions but certain large seeds, such as *Phaseolus* beans, soya beans and peas may fracture in the soil when exposed to rain within 12 h to 18 h of sowing. If these are conditioned by taking up water slowly before sowing this hazard is avoided. Similarly, acid-delinted cotton seeds respond to the addition of 2 per cent water through an atomiser, followed by 2 weeks' storage, by germinating much faster and more uniformly.

George: Clearly, pretreatment can have a number of purposes: an increase in the number of live plants; a higher rate of emergence; a larger plant size, including leaf size of the plants on any given day after sowing. At Plant Protection Ltd. we are looking for growth regulators to accomplish such improvements. Which of these characters should we chiefly look for?

Austin: The three attributes are not independent. Good or bad seed lots are good or bad in all of these respects.

Longden: The rate of emergence is the characteristic benefiting most clearly, and three to four days' difference in the date of emergence may lead to a 5–10 per cent difference in yield.

Heydecker: Does Mr. Longden imply that the benefit from a pretreatment which causes 4 or 5 days' earlier emergence could be equally obtained by sowing untreated seeds 4–5 days earlier? Or is the benefit more complex?

Longden: A statistical resolution of this question has been attempted, but unsuccessfully. It is unlikely that the benefit can be obtained simply by earlier sowing.

Scott: When beet seeds are graded for size there is no effect on the rate of emergence, but there is an effect on the weight of seedlings and simultaneously on the cotyledon area: the first photosynthetic organs are larger when the seed has been larger.

Harrington: The advantages of rapid emergence are at least twofold: crop plants germinate and stay ahead of weeds and chances are increased that the crop seedlings emerge before a soil crusts. In California, lettuce yields have not differed between rapidly and slowly emerging seeds but the cost of harvesting is greatly reduced by using rapidly emerging seeds, not because of the rate of emergence itself but because such seeds produce crops which can be harvested in one or two operations, as against three or four for the more slowly emerging seeds which produce crops maturing over a much more extended period.

Gordon: Mr. Longden excluded stratification from his pretreatments. With conifer seeds this treatment has the effect of improving the rate of germination, even after the seeds have shed their dormancy so as to become 100 per cent viable. The disadvantage with forestry seeds is that they cannot be subsequently dried without losing the improved rate of germination (see Chapter 22).

Longden: Of course this is an important practice, except that it presents practical difficulties, at least with field crops.

Osborne: Our work on rye and wheat germination (Chapters 6 and 7) suggests the following biochemical basis for the various kinds of procedure which involve preliminary hydration of the seeds: in the early stages of imbibition a series of processes has to be gone through, e.g. protein synthesis which takes place immediately on imbibition and has to take place before other processes can occur. There is good evidence that the kind of proteins made during what we call the 'first phase' are needed for the 'second phase'. One of these proteins is probably DNA polymerase because during the second phase one gets DNA replication. Furthermore, the embryo can be dehydrated at any stage during phase 1, where only proteins are being made, but it is not possible to dehydrate embryos once DNA replication is occurring. So if something has to be synthesised before DNA replication occurs this could be done during some kind of pretreatment phase after which the seeds could still be dried back. One could therefore hydrate seeds for 8 h (at 24°C to 25°C), and perhaps even up to 20 h at lower temperatures, and then safely dehydrate the embryo up to that stage, though not beyond it. This has the advantage of making the first phase independent of the uncertain seed bed environment and prevents it from becoming a long drawn out process by carrying it out under controlled and favourable temperature and moisture conditions. The dried seeds could then be started off later from that stage and would have a 'flying start'.

Austin: Did not Milthorpe dry wheat plants back even after they had produced roots?

Edwards: Yes, but this showed that vacuolated cells died and only meristematic cells survived this dehydration.

Harrington: How does Dr. Osborne's model agree with the experience made with a whole range of species in many genera that seeds can be kept imbibed for 7 days, admittedly without protrusion of the radicles, and can then be dehydrated and stored for several months, to germinate more vigorously later than without treatment?

Osborne: My observations were explicitly confined to rye and wheat. It is very likely that many patterns of events occur in germinating embryos of different species. Is it known whether DNA replication occurs during this early period before radicle protrusion?

Harrington: I have no evidence.

Maguire: Cell division appears to be considerable during this period and one may assume that there has to be DNA replication when there is cell division.

Hegarty: In umbelliferous seeds cell division occurs during the first 24 h of germination at 20°C, at least as early as cell expansion.

Osborne: Dr. Hegarty has himself suggested that certain kinds of seed, e.g. carrots, have after-ripening processes which are very different from those processes in cereal grains which we have studied.

Gray: We have wetted lettuce seeds for periods of 1 h to 24 h and then dried them back. After 6 h imbibition it is no longer possible to dry them back without losing the advantage connected with pre-wetting.

Thurston: Concerning the use of growth substances to obtain 'better seedlings', dormancy of wild oats can be broken by pricking the seeds or by soaking them in gibberellic acid. In the initial stages the gibberellin treated plants are larger and further advanced but the difference soon disappears and at harvest the two types of plant are indistinguishable. It seems that despite early advantages (including earlier germination) the benefit does not persist to the end of a plant's life.

Gordon: With cereals, premature germination on the mother plant is a widely occurring phenomenon. Pre-germinated seeds usually dry again before harvest but are still viable when later sown in the field. This is due to a sequential germination of the seven root initials of cereal grains.

Osborne: It is possible that in this sequential system the root initials follow one another in activity and therefore the later ones remain viable even though the earlier ones die when being dried.

Brown: The available water in the top 1–3 cm of soil varies considerably over the year. Where lettuce seeds are sown in succession, sometimes, but far from always, seedling emergence in the field is close to the laboratory germination percentage. Would it be possible to increase the number of occasions when this happens?

Longden: A system of pelleting is being considered for sugar beet fruits where the pellets incorporate a gel which may be able to conduct moisture to seeds sown in soil which tends to be too dry. At the other extreme, the incorporation of sodium perchlorate which releases hydrogen peroxide may overcome the obstacle to germination presented by an anaerobic situation.

Heydecker: We have been interested in pretreatments which enable seeds to germinate rapidly, so that the presumably favourable conditions under which they have been sown do not deteriorate before the seedlings are established. We have two examples, both related to seedling establishment in tropical countries where the soil water may soon evaporate, even after irrigation. Dr. T. Sivanayagam in Ceylon has sown *Capsicum* seeds pretreated by wetting and drying cycles. He has obtained a marked increase in the rate and in the synchronisation of germination from some of his treatments, provided the treated seeds were not stored for more than 3–4 weeks. After this the effect wore off and after 8 weeks no effect was left. Further storage resulted in a rapid decline in the viability of seeds treated with two wetting and drying cycles of 48 h each, a slight decline in untreated seeds, but no decline in seeds which had received one 12 h cycle of wetting, followed by drying. The other experience, by A. Joshua, concerns an aqueous extract from *Capsicum* (chilli pepper) fruits, a concentrated solution of which the seeds were allowed to imbibe at 30°C for seven days. The seeds were then washed and dried. When put in contact with water all these seeds germinated within 24 h, compared with much less uniform germination, spread over 3–4 days, when seeds had not been pretreated. Treatment at 15°C failed to have this beneficial effect. Treatment with a hypertonic solution of polyethylene glycol was much less effective.

Dean: The Weed Research Organisation is interested in finding material to protect crop seeds before sowing so that pre-emergence herbicides can be applied more safely. Activated charcoal has been used successfully. It may be advantageous to incorporate such a material in pellets surrounding seeds.

Hanson: The Lord Rank Research Centre is interested in techniques involving imbibition and drying back. It seems that the beneficial enzymic changes that occur during the imbibition period can be undone if the drying back treatment is unfavourable. What is the best method of drying back?

Longden: In our experience slow drying has resulted in greater responses than rapid drying, at least in carrot and beet seeds.

Hegarty: Slow drying is likely to continue the 'hardening' process into the drying period.

Woodwark: To what extent does the rubbing of monogerm beet seed exert its beneficial effect through loosening the ovary cap which covers the true seed, and which is reputed to be particularly tightly fixed with monogerm varieties?

Longden: It is likely that any treatment involving water, including wet fungicidal treatment against *Phoma betae*, tends to loosen the caps.

Lexander: We have found that seeds ripening on their mother plants at low temperatures, late in the season, germinate to a very low degree. If however the cap is artificially lifted these seeds germinate very well. Are monogerm seeds in a habit of maturing later than multigerm ones, so that the then prevailing lower temperatures might induce a tighter fit of the caps?

Scott: Our monogerm seeds do not mature noticeably later than multigerm ones but it is worth investigating whether later maturing varieties produce seeds with tighter caps.

Scott: After wetting and drying, the fungicidal ethyl mercury phosphate (EMP) steep appears to be phytotoxic.

Maude: Even with dry, untreated seeds EMP slightly retards the rate of radicle emergence. The seeds are probably made more sensitive by the 'hardening' treatment. The thiram treatment is fundamentally different, not only because it is non-mercurial but because here more water is used for a longer time so that inhibitors are washed out, with the result that radicle emergence is speeded up.

Austin: The many points raised have shown that the greatest and most consistent success to date is being obtained with fungicidal seed dressings. By contrast, our knowledge of the mode of action of physical and biochemical treatments is still scant. Much thought and experimentation are necessary before any such treatment to improve the performance of seeds can be safely recommended.

To sum up, seed treatments may have any or all of the following purposes:

1. Improved percentage germination in the field, leading to a fuller stand.
2. Improved rate and uniformity of seedling establishment.
3. Improved seedling growth.
4. Improved performance of seeds and seedlings under inimical (or at least sub-optimal) environmental conditions.
5. Improved total, or equal earlier, yield.
6. Greater uniformity in the time of, and in the quality at, crop maturity.

At the end of the discussion relatively few practices of immediately practical use had been reported on; but the participants had made each other aware that they might be at the threshold of great events; not only by enabling seeds of doubtful quality to perform reasonably well, but by improving the performance even of high quality seeds. The prospects are enticing.

Appendix 2

EXHIBITS

A

A mathematical model to predict germination and field
emergence of maize (*Zea mays* L.) in an environment of changing
temperatures

W. M. BLACKLOW (*Department of Agronomy, University of Western Australia, Nedlands*)

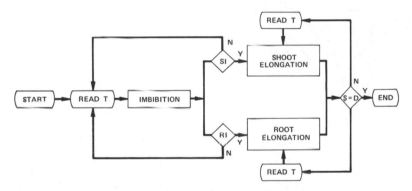

Figure A2.1   The system is visualised as rehydration of the seed by the process of imbibition until sufficient water has been imbibed for germination. Germination is defined as the initiation of linear growth of the radicle (RI) and shoot (SI). Extension growth proceeds until the shoot has reached the depth of planting (S = D) at which time field emergence has occurred. Each hour the temperature (T) in the vicinity of the seed or its elongating axis is sensed and the rates of the processes calculated. At the end of each hour a decision is made on RI, SI or S = D, and if the decision is negative (N) the programme is reiterated. An affirmative answer (Y) transfers control to the next process

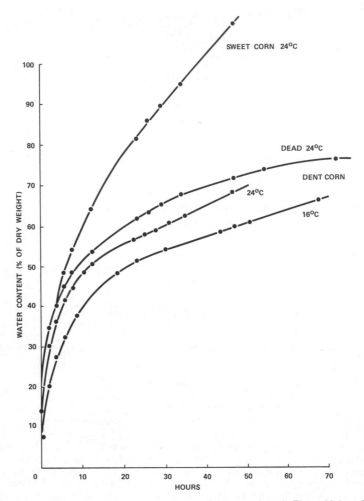

Figure A2.2   The names of the processes and events of the system shown in Figure A2.1 required mathematical descriptions. This figure shows the water content of seed, W, *as a function of time for different varieties and temperatures. It was found that the rate of imbibition, tangent to the imbibition curve, decreased exponentially towards an asymptote that was a linear function of time, i.e.*

$$dW/dt = k(f(t) - W) + b \qquad\qquad (1)$$

where $f(t) = a + bt$ *and is the linear asymptote with intercept* a *and slope* b, *and* k *is a proportionality constant related to the permeability of the seed*

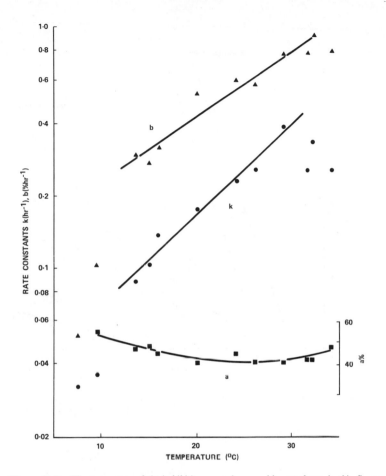

Figure A2.3   The parameters of the imbibition curve, k, a, *and* b, *are shown in this figure as functions of temperature. It can be seen that the rate constants,* k *and* b, *are related exponentially to temperature with* Q_{10}*'s of 2·4 and 1·9, respectively. In contrast, the initial water capacity of the seed,* a, *was little affected by temperature. The figure also shows that values for* k *and* b *below about 10°C departed sharply from the linear relationship that applied between 10°C and 32°C. The increment of water content for 1 h for a particular temperature,* ΔW, *can be calculated by equation (1) after the values of the parameters have been calculated from the relationships shown in this figure. The new water content of the seed is derived by adding* ΔW *to the old value for* W. *In computer language, this is written* $W = W + \Delta W$

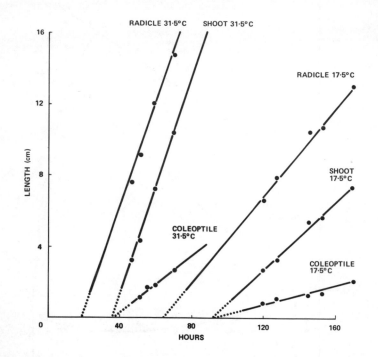

Figure A2.4   The processes of elongation of the radicle and shoot were found to be linear functions of time and were extrapolated to zero length to establish RI and SI for the particular temperature

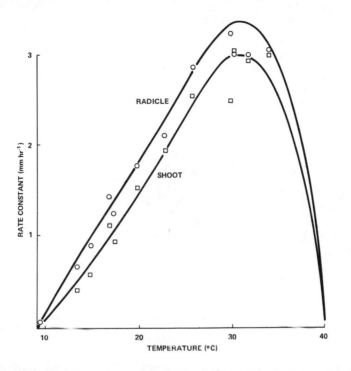

Figure.A2.5 The slopes of the curves shown in Figure A2.4 gave values for the rates of elongation of the radicle, r, and shoot, s. When plotted as a function of temperature the relationship is almost linear between 10°C and 30°C; a constant temperature of 40°C was lethal. The hourly increments of the radicle, r, and shoot, s, for a particular temperature are obtained from the equations for the curves in this Figure. The new values for the lengths of the radicle, R, and shoot, S, are obtained by adding r and s to the old values. In computer language: R = R + r and S = S + s

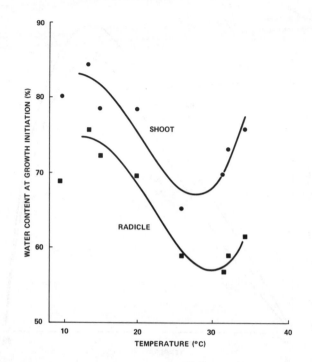

Figure A2.6   The water content of the seed, W, is allowed to increase until RI *and* SI. *The water contents at growth initiation of the radicle and shoot are shown in this figure as functions of temperature. Equations for these curves, using the running average temperature since sowing for the abscissa, are used to test for* RI *and* SI

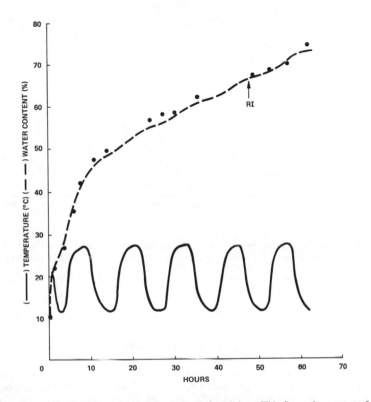

Figure A2.7   The model has been tested for a range of conditions. This figure shows a test of the imbibition phase when the temperature varied from 10°C to 28°C with a period of 12 h. The model predicted the curve; the points are measurements of water content for the same temperature regime

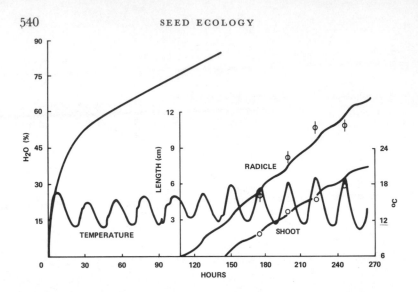

Figure A2.8   This figure shows agreement between prediction by the model, shown by the curves, and observation as shown by the circles for lengths of the radicle and shoot. The temperature varied from 11°C to 20°C on a diurnal cycle

CONCLUSIONS

Successful tests of the model support a number of assumptions on which the model is based. First, the imbibing seed and its elongating axis have no memory of preceding conditions but merely respond to the prevailing environment as though they had experienced a constant environment described by the prevailing conditions. Secondly, the seed and its elongating axis respond to changes in the environment so quickly that when hourly changes are considered the time taken to respond to the new condition is negligible. Thirdly, the seed water content at growth initiation (RI and SI) can be predicted from the average temperature (see *Figure A2.6*). Fourthly, the time of growth initiation, RI and SI, can be predicted from the seed water content.

B

The following brief summaries of other exhibits staged at the Easter School cannot do justice either to the effort that went into them or to their success. Fortunately, the persons who produced them are available for further information. Full addresses are given for those exhibitors who did not attend the conference.

The measurement of mean germination time: applications to the study of changes in wheat grains

J. L. MULTON, E. TRENTESAUX and A. GUILBOT (*Institut National de la Recherche Agronomique, Station de Biochimie et Physicochimie des Céréales, 91305, Massy, France*)

Equipment has been developed which, by means of a camera assembly, records automatically and continually the time course of germination (shoot emergence) of up to 15 samples of 100 seeds each. Mean germination time, thus measured, is a much more sensitive index than germination capacity of physiological changes due to, for example, heat shocks or ageing at different levels of storage temperature and humidity.

REFERENCES

MULTON, J. L. and GUILBOT, A. (1967). 'Méthode de détermination rapide des isothermes de sorption de vapeur d'eau', *Ann. Technol. agric.*, **16**, No. 1, 5–25

MULTON, J. L. and GUILBOT, A. (1968). 'La régulation de l'humidité relative dans les cellules de petit volume', *Industr. aliment. agric.*, **4**, 405–414

MULTON, J. L., GUILBOT, A. and LECOURBE, P. (1967). 'Une nouvelle méthode d'appréciation de la viabilité du blé: la mesure du temps moyen de germination', *C. r. hebd. Séanc Acad. Sci.*, **264**, 1714–1717

MULTON, J. L., LECOURBE, P. and GUILBOT, A. (1968). 'Méthode d'appréciation de la détério ration biologique des grains par mesure du temps moyen de germination', in *Determination of Materials*, Vol. 1, (Walters, A. H. and Elphick, J.) 731–740, Elsevier, London

MULTON, J. L., TRENTESAUX, E. and GUILBOT, A. (1973). 'Technique de détermination du temps moyen de germination: application à la détection des grains de blé altérés au cours de chocs thermiques et de stockages expérimentaux', *Seed Science and Technology*, (in press)

TRENTESAUX, E., MULTON, J. L. and GUILBOT, A. (1972). 'Development of some physiological and biochemical characteristics of wheat grain during storage under controlled conditions', in *Biodeterioration of Materials*, Vol. II, (Walters, A. H., Hueck- van der Plas, E. M.) 499–506, Applied Science Pub. Ltd., London

A general bioassay for germination modifying factors

M. G. WALKER (*Biology Section, Western Australian Institute of Technology, Hayman Road, South Bentley, Western Australia* 6102).

The test is meaningful, reproduceable and capable of statistical analysis. Seeds of the species in question are germinated on RF-unit strips of chromatograms of extracts, in physiological concentrations, from seeds at successive stages of germination. Results from replicated treatment/blank control pairs are compared at the stage when 50 per cent of the (viable) blank control seeds have germinated, and can be expressed in terms of variance ratios. For example, the time course of nett inhibitor or promotor activity within germinating seeds can be traced by this sensitive test.

REFERENCES

WALKER, M. G. (1971). 'A general bioassay for germination modifying factors', *Ann. Bot.*, **35**, 981–989

WALKER, M. G. (1971). 'Changes in germination promotion and inhibition in seed extracts of subterranean clover (*Trifolium subterraneum* L.) related to dormancy and germination', *Aust. J. biol. Sci.*, **24**, 897–903

Automation of seed cleaning and seed counting

I. H. RORISON and F. SUTTON (*Nature Conservancy Grassland Research Unit, Department of Botany, University of Sheffield*)

Inexpensive home-made equipment includes an aspirator, incorporating a variable speed rheostat, which can remove debris from seeds weighing from 0·1 mg upwards. A vibrator supplies a stream of single seeds to be sensed by a photocell and counted electronically. Counts may be either of totals or of predetermined numbers. A pulse from the counter can automatically stop the vibration, divert excessive seeds and place a new receptacle, for a new batch of seeds, into position.

REFERENCES

HERGERT, G., ZILLINSKY, F. J. and KEMP, J. K. (1966). 'An aspirator for cleaning small seed samples', *Can. J. Pl. Sci.*, **46**, 570–572

Ultrastructural cytology of maize (*Zea mays* L.) embryo senescence

PATRICIA BERJAK (*Department of Plant Science, University of Leeds*), and T. A. VILLIERS (*Department of Biological Sciences, University of Natal, Durban*)

The process of senescence of root caps of *Zea mays* radicles following 'accelerated ageing' (storage of grains with 14 per cent water content at 40°C) was followed by cytological, ultrastructural, cytochemical and autoradiographic methods during the 48 h from the onset of imbibition. During the first stage of ageing, cell damage is reversible and germination occurs. However, in a few cases the repaired cells do senesce prematurely. During the second stage, cells can no longer be repaired and lose their regulatory control but still metabolise to some extent. The third, final, stage is one of cell autolysis. It is suggested that the deterioration is due to damage to macromolecules, such as nucleic acids, and membrane components (especially organelles such as mitochondria). In the case of membranes, formation of peroxides of poly-unsaturated lipid components may be implicated.

REFERENCES

BERJAK, P. (1968). 'A lysosome-like organelle in the root cap of *Zea mays*', *J. Ultrastruc. Res.*, **23**, 233

BERJAK, P. and VILLIERS, T. A. (1970). 'Ageing in plant embryos I. The establishment of the sequence of development and senescence in the root cap during germination', *New Phytol.*, **69**, 929

BERJAK, P. and VILLIERS, T. A. (1972). 'Ageing in plant embryos II. Age-induced damage and its repair during early germination', *New Phytol.*, **71**, 135

BERJAK, P. and VILLIERS, T. A. (1972a). 'Ageing in plant embryos III. Acceleration of senescence following artificial ageing treatment', *New Phytol.*, **71**, 511

BERJAK, P. and VILLIERS, T. A. (1972b). 'Ageing in plant embryos IV. Loss of regulatory control in aged embryos', *New Phytol.*, **71**, 1069

BERJAK, P. and VILLIERS, T. A. (1972c). 'Ageing in plant embryos V. Lysis of the cytoplasm in non-viable embryos', *New Phytol.*, **71**, 1075

Leakage from imbibing seeds

E. W. SIMON (*Department of Botany, Queens University, Belfast*)

Dry pea (*Pisum sativum* L.) seeds brought into contact with water lose appreciable amounts of solutes such as sugars and amino acids during the first few minutes of imbibition, especially if the seed coat is damaged or absent. It appears that cell membranes of the maturing embryos lose their integrity during the process of drying and that membrane integrity is re-established during imbibition. Slow imbibition before sowing could help to minimise solute leakage.

REFERENCES

EYSTER, H. C. (1938). 'Conditioning seeds to tolerate submergence in water', *Am. J. Bot.*, **25**, 33–36

LARSON, L. A. (1968). 'The effect soaking peas with or without seed coats has on seedling growth', *Pl. Physiol.* **43**, 255–259

SIMON, E. W. and RAJA HARUN, R. M. (1972). 'Leakage during seed imbibition', *J. exp. Bot.*, **23**, 1076–1085

Field emergence of peas (*Pisum sativum* L.)

S. MATTHEWS (*Department of Biology, University of Stirling*)
M. F. F. CARVER (*Levington Research Station, Ipswich*)
LYNN V. BEDFORD (*Official Seed Testing Station, Cambridge*), and
D. B. MACKAY (*Official Seed Testing Station, Cambridge*)

Low field emergence of seed lots which germinate well in the laboratory is associated with infection by the soil fungus *Pythium ultimum*. When seeds are steeped in water, sugars and electrolytes are leached out. A routine electro-conductivity test of the 'steep water' after 24 h soaking of the seeds can predict their field performance: the higher the loss and hence the conductivity, the greater is the likelihood of fungal attack and failure. Furthermore, the field performance of seed lots is often related to their response to certain levels of temperature and substrate moisture: seed lots which after 24 h in cold, wet sterile conditions show reduced respiration and increased susceptibility to soil *Pythium* spp. tend to emerge poorly in early sowings.

REFERENCES

MATTHEWS, S. and BRADNOCK, W. T. (1968). Relationship between seed exudation and field emergence in peas and French beans', *Hort. Res.*, **8**, 89–93

Hollow heart of peas (*Pisum sativum* L.)

J. G. HARRISON (*Scottish Horticultural Research Institute, Invergowrie*)

This physiological disorder which becomes apparent on the adaxial surfaces of cotyledons of germinating peas is caused during seed maturation by rapid drying of the seeds at high temperatures and becomes most pronounced when imbibition is rapid. Affected seeds are subject to attack by *Pythium ultimum*, germinate more slowly, produce smaller seedlings and yield about 30 per cent less than normal ones. A water-soluble, dialysable and heat-labile extract from hollow heart tissue inhibits pea seedling growth and delays the germination of cress and lettuce seeds. The trouble can be prevented by producing seeds in a cool climate, maturing the seeds on the plants and using a minimum of heat during drying.

REFERENCES

PERRY, D. A. and HARRISON, J. G. (1973). 'Causes and development of hollow heart in pea seeds', *Ann. appl. Biol.*, **73**, (in press)
PERRY, D. A. and HARRISON, J. G. (1973). 'Effects of hollow heart on growth of peas', *Ann. appl. Biol.*, **73**, (in press)

Crop and weed seed survival in soil

J. LEWIS (*Welsh Plant Breeding Station, Aberystwyth*)

Weed seeds are much better adapted to survival in soil than crop seeds and often survive for 20 or more years. Small seeded grass species survive better than large seeded ones and leguminous seeds better than most others (except *Ranunculus repens*, *Chenopodium album* and *Rumex crispus*). Survival is usually better in loam than in acid peat (*Matricaria inodora* is an exception). Germination *in situ* probably accounts for most of the natural losses of the weed seed population in soil, but fewer seeds germinate at deeper soil levels and where the water table is high.

REFERENCES

LEWIS, J. (1958). 'Longevity of crop and weed seeds. 1. First interim report', *Proc. int. Seed Test. Ass.*, **23**, No. 2, 340–354

LEWIS, J. (1961). 'The influence of water level, soil depth and type on the survival of crop and weed seeds', *Proc. int. Seed Test. Ass.*, **26**, No. 1, 68–85

LEWIS, J. (1973). *Weed Res.* (in press)

Comparative germination of species of the genus *Beta*

B. FORD-LLOYD and J. T. WILLIAMS (*Department of Botany, University of Birmingham*)

The 1–7 seeded fruits (seed balls) of all sections of the genus contain ferulic and/or paracoumaric acids in various combinations and probably several other inhibitors. Germination tests on cress seeds in a dilution series (1:1 to 1:16) of aqueous extracts from fruits of *B. trigyna*, *B. lomutogona*, *B. macrorhiza*, *B. patellaris*, *B. procumbens* and cultivated types of *B. vulgaris* suggest that there has so far been little change in the concentration of the inhibitor complex with domestication and selection, despite considerable differences between genetic stocks of individual species in the degree of inhibition exerted by their extracts.

REFERENCES

BATTLE, J. P. and WHITTINGTON, W. J. (1971). 'Genetic variability in time to germination of sugar-beet clusters', *J. agric. Sci., Camb.*, **76**, 27–32

VAN SUMERE, C. F. (1960). 'Germination inhibition in plant material', in *Plants in Health and Disease*, (ed. Pridham, J. G.) 25–33, Pergamon, London

Agronomic effects of fruit size and fruit treatment of sugar beet

D. W. WOOD (*Department of Agriculture and Horticulture, University of Nottingham School of Agriculture*)

Dissections of sugar beet fruits ('seeds') showed that embryo size increased with increasing fruit size. Radiography showed that fewer of the large fruits were seedless and more of their seeds were fully developed. In the field these large fruits gave higher emergence percentages and seedling weights than small fruits or ungraded samples. This improvement persisted, giving quicker leaf development and increases of 10–20 per cent in final root yield. Further improvement in emergence and early growth was obtained by steeping the 'seeds' in water or solutions of the growth regulators gibberellic acid, kinetin or 6-benzylaminopurine.

Radiography of monogerm sugar beet (*Beta vulgaris* L.) fruits

P. C. LONGDEN (*Broom's Barn Experimental Station, Bury St. Edmunds*)

Fruits are arranged in templates, radiographed and then sown. This rapid and non-destructive technique of assessing seed size and quality has been used successfully to predict field emergence, grade by true seed diameter, test the efficiency of seed cleaning machinery (elimination of empty fruits) and find the proportion of fruits with twin embryos.

REFERENCES

LONGDEN, P. C., JOHNSON, M. G. and LOVE, B. (1971). 'Sugar beet seedling emergence prediction from radiographs', *J. int. Inst. Sugar Beet Res.*, **5**, 160–168

Field emergence in barley (*Hordeum vulgare* L.)

M. T. COLLINS (*Department of Biology, University of Stirling*)

Germination of 'seed lots' in water-sensitivity tests was found to be correlated with their field emergence percentage. An improvement in germination in such tests was brought about by certain antibiotics, notably Amphotericin–B and Polymyxin–B–sulphate. Agrosan, an organo-mercurial compound, improved both germination in water-sensitivity tests and field emergence.

REFERENCES

GABER, S. D. and ROBERTS, E. H. (1969). ' "Water-sensitivity" in barley seeds. II. Association with micro-organism activity', *J. Inst. Brew.*, **75**, 303

Inter- and intra-population variation in the germination of *Arabidopsis thaliana* ecotypes

D. RATCLIFFE and JANE A. EVANS (*School of Biological Sciences, University of Leicester*)

Tests were made on seeds from 16 climatically widely different locations over a temperature range from 5°C to 30°C in light and darkness for 21 months. Differences occur in the post-harvest dormancy of seeds from different ecotypes, though also from different individuals of the same ecotype. In general, the germination percentage increases for at least 6 (but in some cases for 21) months after harvest, and sooner at germinating temperatures between 5°C and 15°C than at higher ones. Germination percentages tend to be higher in light, especially at high temperatures. The period of post-harvest dormancy tends to be shorter in ecotypes from regions with damp summers, where seedlings can establish themselves before the end of the growing season, than in dry ones where the soil water is replenished only in winter. Time of germination tends to be inversely related to time of flowering of the parent plants.

REFERENCES

EVANS, J. A. and RATCLIFFE, D. (1972). 'Variability in after-ripening requirements and its ecological significance (in *Arabidopsis thaliana* (L.) Heynh)', *Arabidopsis Information Service*, **9**, 1

Effects of *Capsicum annuum* fruit extract on germination and seedling growth

A. JOSHUA (*Institute of Agricultural Research and Training, University of Ife*)
G. NORTON (*Department of Applied Biochemistry and Nutrition, University of Nottingham School of Agriculture*), and
W. HEYDECKER (*Department of Agriculture and Horticulture, University of Nottingham School of Agriculture*)

Dilute aqueous extract in the germination medium delays germination but speeds up seedling growth of lettuce and tomato. Treatment with *concentrated* extract at 30°C prevents germination of tomato seeds, but seeds removed after 5 days, rinsed, dried and then set to germinate do so in 24 h (instead of 96 h). Treatment at 15°C is not beneficial. Promising effects of modified treatments have been effective on *Capsicum* and carrot but not on onion, seeds. Attempts at isolating and characterising the active principle by ion exchange fractionation, spectrophotometry and bioassays have not so far been successful but indicate that it is not a gibberellin or a cytokinin.

REFERENCES

HEYDECKER, W. and JOSHUA, A. (1972). 'Extract from *Capsicum* hastens seed germination', *Comm. Grower.*, **3999**, 239–240

Germination and anatomy of carrot (*Daucus carota* L.) seeds

T. W. HEGARTY (*Scottish Horticultural Research Institute, Invergowrie*)

In those seed lots which germinate significantly less well at 10°C than at 20°C (but not in others) a linear relationship ($r = -0.88$) exists between the difference in germination percentage and the extent to which the embryo fills the endospermal cavity. The explanation is not clear, but seed maturity or deterioration may be involved.

REFERENCES

BORTHWICK, H. A. (1931). 'Carrot seed germination', *Proc. Am. Soc. hort. Sci.*, **28**, 310–314

FLEMION, F. and UHLMANN, C. (1946). 'Further studies of embryoless seeds in the Umbelliferae', *Contr. Boyce Thomp. Inst. Pl. Res.*, **14**, 283–293

HEGARTY, T. W. (1971). 'A relation between field emergence and laboratory germination in carrots', *J. hort. Sci.*, **46**, 299–305

The effect of seed treatments on the emergence of *Phaseolus vulgaris* L. seedlings

J. ELIZABETH M. BALLANTINE (*Physiology section, National Vegetable Research Station*)

Seeds imbibed in 100 ppm or 500 ppm aqueous solutions of GA_3 at 20°C for 16 h and then dried for 48 h before sowing emerged significantly earlier than water-treated controls. However, these concentrations of GA_3 also stimulated abnormal hypocotyl elongation in the seedling.

REFERENCES

BEREZNEGOVSKAYA, L. N. (1964). 'The effect of gibberellic acid on belladonna seeds', *Fisiologyia Rast.*, **11**, 1081–1082

BOOTHBY, D. and WRIGHT, S. T. C. (1962). 'The effects of Kinetin and other plant growth regulators on starch degradation', *Nature*, **196**, 389–390

HARDWICK, R. C. (1972). 'The emergence and early growth of French and Runner beans (*Phaseolus vulgaris* L. and *Phaseolus coccineus* L.) sown on different dates', *J. hort. Sci.*, **47**, 395–410

Seed dispersal and germination of the desert plant *Blepharis persica* (Burm.) Kuntze

Y. GUTTERMAN (*Department of Botany, Hebrew University of Jerusalem*)

Seed dispersal and germination are regulated by a sequence of three 'water clocks' so that they óccur when water is fully adequate but not excessive; germination then takes as little as 90 min. Mucilage produced by the hairs of the seed coat epidermis helps to create and to perpetuate favourable conditions for establishment of the seedlings.

REFERENCES

GUTTERMAN, Y. (1972). 'Delayed seed dispersal and rapid germination as survival mechanisms of the desert plant *Blepharis persica*', *Oecologia* (Berl.), 10, 145–147

WITZTUM, A., GUTTERMAN, Y. and EVENARI, M. (1969). 'Integumentary mucilage as an oxygen barrier during germination of *Blepharis persica* (Burm.) Kuntze', *Bot. Gaz.*, 130, No. 4, 238–241

Seed ecology of *Rottboellia exaltata* L., a tropical weed in maize and sugar cane

P. E. L. THOMAS (*University of Rhodesia*)

Water relations, light and temperature affect the dormancy of partially or fully imbibed 'in-husk' seeds, buried or near the soil surface. It appears that they all operate via the permeability of the seed coat which thus emerges as the critical factor governing seed dormancy.

REFERENCES

THOMAS, P. E. L. (1972). 'Studies of the biology of *Rottboellia exaltata*, L. f., and of its competition with maize', *Ph.D. Thesis*, University of London. (Copy also available from Weed Research Organisation, Yarnton, Oxford, OX5 1PF.)

Endogenous hormones of *Corylus avellana* in relation to seed dormancy

P. M. WILLIAMS (*Department of Botany, King's College London*)

Abscisic acid (ABA) has been identified and determined quantitatively in hazel seeds by gas–liquid chromatography (GLC–MS) at various physiological stages. The ABA is mainly confined to the seed coats and the highest quantities (4nM/seed) are present in freshly harvested seeds. During chilling of the seeds the ABA content decreases due to leaching. ABA represents approximately 20 per cent of the total inhibitory activity as detected by bioassay techniques. Gibberellin (GA) synthesis occurs when chilled seeds are incubated at 20°C. GA accumulation is mainly confined to the axis. Small increases in the GA content occur in the cotyledons but experiments with chlormequat (CCC) suggest that this may be due to the release of bound gibberellins.

T

The role of soil water in seed germination

J. WILLIAMS (*University of the South Pacific, School of Natural Resources, Suva, Fiji*).

The physical factors which govern the imbibition of water by seeds in the soil are the water potential gradient from soil to seed and the hydraulic conductivities of the soil and the seed. As the gradient decreases and the conductivity of the seed tissues increases, the water potential and the conductivity of the soil in contact with the seed become increasingly critical for successful germination.

REFERENCES

SHAYKEWICH, C. F. and WILLIAMS, J. (1971). 'Resistance to water absorption in germinating rapeseed (*Brassica napus* L.)', *J. exp. Bot.*, **22**, 19–24

SHAYKEWICH, C. F. and WILLIAMS, J. (1971). 'Influence of hydraulic properties of soil on pre-germination water absorption by rapeseed (*Brassica napus*)', *Agron. J.*, **63**, 454–457

WILLIAMS, J. and SHAYKEWICH, C. F. (1971). 'Influence of soil water matric potential and hydraulic conductivity on the germination of rape (*Brassica napus* L.)', *J. exp. Bot.*, **22**, 586–597

Interaction of preparatory imbibition-and-drying cycles with subsequent storage time on the germination of *Capsicum annuum* L. seeds

T. SIVANAYAGAM and W. S. MANOKARAN (*Agricultural Research Station, Maha-Illuppallama, Ceylon*)

Treatments included: storage of seeds before or after extraction from the fruits; imbibition cycles of 12–48 h; 1 or 2 imbibition-and-drying cycles; and storage after treatment for 0–28 weeks. In general, more rapid germination and an improvement in the life-span of the seeds was achieved by pretreatment storage within the fruits; by increasing the imbibition period and the number of cycles (but one long period was better than two half length ones). The time between sowing and germination grew shorter up to 7 weeks of post-treatment storage but then increased again. This was accompanied by a rapid decline in viability of seeds treated for long periods, a much less rapid decline of untreated seeds, but an improvement of the life-span of seeds treated for 12 h or 24 h.

Incorporation of kinetin in dry lettuce (*Lactuca sativa* L.) seeds to permit germination at high temperatures

A. JOSHUA (*Institute of Agricultural Research and Training, University of Ife*), and W. HEYDECKER (*Department of Agriculture and Horticulture, University of Nottingham School of Agriculture*)

High temperatures (around 30°C, but differing with the seed lot) induce thermodormancy in lettuce seeds (at a lower temperature in the dark than in warm white fluorescent light). However, incorporation of kinetin in the seeds raises this ceiling temperature. Lettuce seeds can absorb kinetin from a 7 ppm solution in the organic solvent, dichloromethane, and this affords a practical way of enabling them to germinate rapidly and uniformly at higher temperatures.

REFERENCES

JOSHUA, A. and HEYDECKER, W. (1971). 'Chemicals may aid lettuce germination at high temperatures', *Grower*, **76**, 85?

Effects of conditions during seed production on the following generation of *Cheiranthus cheiri* (wallflower)

D. E. O'CONNOR (*Department of Agriculture and Horticulture, University of Nottingham School of Agriculture*)

Seeds produced in a warm or in a cool environment were set to germinate at 25°C and at 15°C. Seeds germinated better and seedlings grew better at the temperature which resembled the parental environment.

REFERENCES

O'CONNER, D. E. (1972). 'An investigation into some aspects of wallflower seed production', *B.Sc.* (*Honours*) *dissertation*, University of Nottingham Department of Agriculture and Horticulture

Influence of the nutrition of the mother plants on the initial growth and on the yield potential of wheat grains

MARGARETE HOLZMAN (*Institut für kontinentale Agrar-V. Wirtschaftforschung D-63 Giessen, West Germany*)

Transplantation experiments have shown that the better growth from larger grains is due to the size of the embryo rather than of the endosperm. A comparison of grains of equal size shows that increased levels of nutrient supply to the mother plant improve the initial growth of plants from intact grains (though not from excised embryos) and can increase their grain yield.

REFERENCES

HOLZMAN, M. (1972). 'Untersuchungen über den Einfluss der Düngung auf den Saatgutwert und dessen Lokalisation im Getreidekorn', *Z. Acker-u. Pflanzenbau*, **135**, 279–309

Surfaces of seeds

W. HEYDECKER (*Department of Agriculture and Horticulture, University of Nottingham School of Agriculture*), and
L. GREEN (*Wolfson Institute of Interfacial Technology, University of Nottingham*)

The scanning electron microscope reveals patterns in the structure of seed coats which may be important in connection with seed/soil contact and seed/water relations.

REFERENCES

HARPER, J. L. and BENTON, R. A. (1966). 'The Germination of seeds in soil II. The germination of seeds on the surface of a water supplying substrate', *J. Ecol.*, **54**, 155–166
MANOHAR, M. S. and HEYDECKER, W. (1964). 'Effect of water potential on germination of pea seeds', *Nature*, **202**, 22–24

TRADE EXHIBITS

Germain's (U.K.) Ltd. (Mr. J. Simpson): 'Filcoat' pelleted seeds.
Royal Sluis Seed Company (Mr. J. de Boer): The 'Split Pill' pelleted seed.

Appendix 3

GLOSSARY OF TERMS

W. HEYDECKER

Department of Agriculture and Horticulture, University of Nottingham School of Agriculture

Editor's note: Language barriers exist even within a language. Often the same word or phrase is used in a distinctly different sense, or with different connotations and implications, in the jargons of related professional groups. Conversely, a number of apparently unrelated terms may be in use for the same thing, operation or concept, and some terms in common use may perhaps never have been quite thought out.

An attempt is made here to explain, and if possible to standardise, some of the terms concerned with seeds which are used in this volume.

ADVANCING See HARDENING.

AGEING (AGING, U.S.A.) The deterioration of seeds with time. Chronological time is not important to seeds; ageing in the physiological sense can be slowed down by an appropriate environment. (See also VIGOUR.)

CHITTING The first clear manifestation that a seed has germinated, i.e. the emergence of the radicle from the seed. Sometimes used transitively to denote a practice designed to cause seeds to reach this stage before they are sown in the growing medium.

DORMANCY In a broad sense, the state in which ungerminated (or 'resting') seeds survive. This includes not only truly physiological dormancy (see below) but also types of dormancy which only last as long as they are imposed, or enforced, by the environment on a non-dormant embryo, for instance by the absence of sufficient water for germination or by a physical obstacle, such as a 'hard' seed coat which prevents water from reaching the embryo.

In a more narrow sense, a physiological state of the embryo (either innate in the developing seed or induced at a later stage by environmental factors) in which seeds will not germinate under environmental conditions which would favour germination and growth after the dormancy has been 'broken' by appropriate environmental and/or chemical treatments. The gradual cessation of this state of dormancy often manifests itself in a widening of the range of conditions, such as temperature, under which the seeds will germinate, and in an increase in the rate of germination (see Chapter 22). (Distinguish from lack of VIGOUR.)

ECOLOGY   The science of the relationship of organisms with the entirety of their environment, including long-term and high-order interactions.

GERMINATION   The process of activation of a hitherto resting embryo. It begins with the first metabolic processes during imbibition. The decision on the end point depends on the purpose of the definition.

1. Physiologically (biochemically), and strictly speaking, germination ends with the first manifestation of growth, even before the seed coat is ruptured. This is a difficult end point to determine.

2. In germination experiments, the emergence of the *radicle* from the covering structures of the seed is often used as the end point. This point is usually the first clear manifestation that germination (stage 1) has taken place. *Suggestion: call it 'radicle emergence'.* (N.B. On occasion, radicle emergence is not followed by any further growth and is then an erroneous index of success.)

3. Seed analysts consider as the end point of germination the emergence, from the covering structures, of a *seedling* capable of independent growth. This excludes abnormal seedlings (and requires evaluation of what is an abnormal seedling). *Suggestion: call it 'normal germination'.*

4. Farmers and growers regard the emergence of the aerial parts of a seedling above ground as the end point (Scottish synonym: braird). *Suggestion: call it 'seedling emergence' or 'field emergence'.*

5. In practice, it is only the final level of seedling emergence which is valid. *Suggestion: call it 'established stand' or 'field stand'.*

HARDENING OF SEEDS   A treatment preliminary to sowing during which seeds are moistened and dried back (once or a number of times) designed according to P. A. Henckel (=Genkel, Henkel) to activate certain physiological mechanisms which will enable the resulting plants to withstand adverse environmental conditions. ADVANCING: the same process with the restricted aim of enabling the seed to pass through the first stages of germination, short of radicle emergence, before the seeds are actually sown. Hypertonic solutions of salts (e.g. KNO_3, KH_2PO_4) or of osmotically active but chemically inert substances (e.g. polyethylene glycols of high molecular weight) are sometimes employed to prevent radicle emergence. (By contrast, see CHITTING (transitive) where seeds are not dried back.) *Suggestion: to avoid tendentious undertones, call it 'imbibition and drying back'.*

HUMIDITY, RELATIVE (RH)   The *actual* amount of water vapour in a given volume of air is often specified by its vapour pressure (e). The *total* amount of vapour needed to saturate the air determines the saturation vapour pressure (e_s), a quantity which increases at an increasing rate with temperature, e.g. from 6·1 millibars (mb) at 0°C to 17·0 mb at 15°C and 42·2 mb at 30°C. The RH of an air sample is the ratio of the actual to the saturation vapour pressure (e/e_s), expressed as a percentage; the water vapour pressure deficit (v.p.d.) is the difference between the vapour pressure at saturation and the actual vapour pressure $(e_s - e)$.

Seeds in storage gain or lose water by exchange from the surrounding air until they reach an equilibrium water content which depends on the RH of the air. The *rate* at which this equilibrium state is approached is a positive function of the vapour pressure deficit.

At any given water content of a given volume of air any increase in the temperature of the air will increase its v.p.d. and decrease its RH. An increase in v.p.d. will cause the seeds to dry faster; a decrease in RH will cause them to reach a smaller equilibrium water content. For a given change in vapour pressure the change in seed water content will be larger at a low than at a high temperature, because the ultimate seed water content is a function of the ambient RH.

When the air temperature is lowered in a controlled environment, it becomes increasingly difficult (a) to keep the RH low and steady, and (b) to maintain a v.p.d. large enough to expedite the drying of seeds. It is important to realise this, since a constant low water content and low temperature are both conducive to the longevity of most seeds.

LABORATORY GERMINATION TESTS   Generally designed to show what proportion of seeds in a sample is capable of germinating within a reasonable time, generally under near-optimal conditions. This should involve control of temperature and water supply (and sometimes light). A wide range of equipment and methods are used and one should be aware that this in itself may be a source of variation in the results obtained.

LOT, SEED

1. In the seed trade it means a large bulk of seeds (up to 20000 kg), which are tested, sometimes certified, and sold as one unit.
2. In experimental work it has come to mean a distinct group of seeds which, though of the same cultivar as other lots, is likely to have a different history (see PROVENANCE), i.e. details of production, harvest, extraction, processing and/or storage.

MOISTURE   Identical with water. Any distinction is scientifically unjustified though occasionally excusable (e.g. moisture level, cf. water level).

PROVENANCE   The location (and its geographical implications) where a given seed lot was produced.

RATE OF GERMINATION   The reciprocal of *time to germination* (q.v.).

SEEDLING EVALUATION   A somewhat subjective way of estimating the quality of seedlings from comparable seed lots, based on assessing their integrity and measuring their size *under comparable conditions*.

SOAKING   This term should be confined strictly to denote the complete immersion of seeds in water. It is, however, frequently used misleadingly for placing seeds in contact with water in aerobic conditions. Though inducing rapid imbibition, soaking can endanger the germination and further performance of seeds (especially soft-coated ones) unless the period of immersion is very short, when it can indeed be beneficial. In seed testing, a soak is used to induce leakage of cell constituents from seeds such as peas and beans; the 'vigour' of seeds is considered inversely proportional to the quantity of the leachate. On the other hand, a soak in a warm water suspension of a fungicide is sometimes employed to eliminate seed-borne diseases.

SPEED OF GERMINATION   Basically incorrect term often used to denote the reciprocal of TIME TO GERMINATION. *Suggestion: call it 'rate of germination'.*

STAND   See GERMINATION.

STEEP   Soak.

STRESS

1. In terms of soil/plant/water relationship high stress is synonymous with a high environmental water tension (or suction) which the seed has to overcome in order to imbibe water; in the thermodynamic nomenclature now accepted, high water stress is equal to a large negative water potential (for instance, a water potential of − 15 bars (≃ atmospheres) is usually equated with the 'permanent wilting point' of soils). In other words, a high water stress denotes low water availability.

2. In seed testing circles, 'stress' is often understood to be exerted by any environmental condition which imposes a constraint (or difficulty) on the metabolic processes in a seed (e.g. soaking), and 'stress tests' are sometimes employed to assess the performance of seeds near the limit of their endurance. (See also VIGOUR.)

TIME TO GERMINATION

1. Dormant seeds: indeterminable.

2. Non-dormant seeds: the period between sowing or imbibition and the chosen end point of germination (see GERMINATION). To outward appearances, this period is often sub-divided into more or less distinct phases:
   (a) imbibition of water,
   (b) a period when apparently nothing happens,
   (c) a succession of stages of 'germination' (q.v.).

The reciprocal of the time from sowing to germination is the *rate* (inappropriately called 'speed' or 'velocity') *of germination*. It can refer to a single seed or to a seed population.

The mean time taken by a population of seeds to germinate can be expressed in many ways but should never be combined with the percentage of seeds germinating into one meaningless term. Kotowski's (1926) (see references, Chapter 24) so-called 'coefficient of velocity of germination' (CVG) is one useful formula which accurately expresses the mean *rate* of germination (now written CRG) in positive terms. It may be written

$$CRG = \frac{\Sigma n}{\Sigma (D \cdot n)} \cdot 100$$

where n equals the number germinating on day D, and D equals the number of the day, counted from the day of sowing which is 0; the higher the figure, the shorter the time to germination. The CRG is in fact 100 × the reciprocal of the mean time to germination in days. Harrington (1963) (see references, Chapter 24) uses the mean number of days to germination (100/CRG). An alternative measure is based on the time by which an arbitrary proportion of seeds—e.g. 50 per cent of those ultimately germinating—have germinated. (See also UNIFORMITY OF GERMINATION.)

UNIFORMITY OF GERMINATION   A measure of the simultaneity of the germination of members of a population of seeds. (See also TIME TO GERMINATION.) This can be expressed as the variance of the times taken by individual seeds around the mean time to germination. Germination curves are usually positively skewed, i.e. more seeds germinate in the first than in the second half of the germination period. But *assuming* normal distribution of the time to germination, its uniformity can be expressed as the reciprocal of the variance, namely

$$CUG = \frac{\Sigma n}{\Sigma [(\overline{D} - D)^2 \cdot n]}$$

where $\overline{D}$ is (100/CRG), the mean number of days to visible germination, and D is the day on which n seeds germinate.

For field sowings, where stragglers have little chance of becoming useful members of the community, it may be desirable to confine the calculations for time to germination (q.v.) and uniformity of germination to the first 80 or 90 per cent of those seeds which germinate.

VIGOUR A property of seeds which is physiologically imprecise but important in practice—the inherent ability of a seed to survive, germinate and produce a seedling capable of 'doing well' under a wide range of conditions. A gradual loss of vigour of a seed often manifests itself by a growing susceptibility to environmental hazards and/or an increase in its 'lability', i.e. restriction in the range of conditions under which it will germinate and produce a seedling. (Distinguish loss of vigour, which is usually irreversible, from *dormancy*, which can be 'broken'. See also AGEING.)

Appendix 4

LIST OF MEMBERS

ARNOLD, DR. S. M. Biology Department, University of York, Yorks.

ASHBY, J. P. R. Crop Husbandry Department, West of Scotland Agricultural College, Auchincruive, by Ayr.

AUSTIN, R. B. Plant Breeding Institute, Maris Lane, Trumpington, Cambridge.

AYNSLEY, J. S. Department of Agriculture and Horticulture, University of Nottingham School of Agriculture, Sutton Bonington, Loughborough, Leics.

BALINT, A. c/o R. W. Gunson (Seeds) Ltd., 20–21 St. Dunstan's Hill, London, E.C.3.

BALLANTINE, J. E. M. Physiology Section, National Vegetable Research Station, Wellesbourne, Warwicks.

BARBER, A. C. Samuel Dobie & Son Ltd., Upper Dee Mills, Llangollen, Denbighshire.

BEAN, DR. E. W. Welsh Plant Breeding Station, Plas Gogerddan, Aberystwyth, Cards.

BEDFORD, MISS L. V. Official Seed Testing Station, Huntingdon Road, Cambridge.

BEKENDAM, DR. J. Governmental Seed Testing Station, Wageningen, Holland.

BENJAMIN, S. K. Department of Agriculture, University of Reading, Earley Gate, Reading, Berks.

BERJAK, DR. P. Department of Plant Science, University of Leeds, Leeds, Yorks.

BIDDLE, A. J. Pea Growing Research Organisation Ltd., Great North Road, Thornhaugh, Peterborough, Northants.

BOUMAN, IR. P. R. Nunhems Zaden B.V., Haelen.-L, Holland.

BOURNE, DR. D. T. Research Laboratory, A. Guinness Son & Co. (Park Royal) Ltd., Park Royal Brewery, London, N.W.10.

BRADLEY, M. R. Stockbridge House Experimental Horticulture Station, Cawood, Selby, Yorks.

BRIARTY, DR. L. G. Department of Botany, University of Nottingham School of Agriculture, Sutton Bonington, Loughborough, Leics.

BROWN, A. Ministry of Agriculture, Fisheries and Food, Block B., Brooklands Avenue, Cambridge.

BROWN, T. R.   Butterworth & Co. (Publishers) Ltd., 88 Kingsway, London, W.C.2.

BURRELL, N. J.   Pest Infestation Control Laboratory, M.A.F.F., London Road, Slough, Bucks.

BUSZEWICZ, G.   Forestry Commission, Forest Research Station, Alice Holt Lodge, Wrecclesham, Farnham, Surrey.

CARNEGIE, MISS H. M.   Agricultural Botany Division, School of Agriculture, University of Aberdeen, 581 King Street, Aberdeen.

CARVER, M. F. F.   Levington Research Station (Fisons Ltd.), Ipswich, Suffolk.

CAVERS, DR. P. B.   Department of Plant Sciences, University of Western Ontario, London 72, Ontario, Canada.

CHATTERJEE, D. A.   Department of Agriculture and Community Development, Government of West Bengal.

CHERRY, MISS M.   c/o British Broadcasting Corporation, Broadcasting House, Portland Place, London, W.1.

CHURCHMAN, J. L.   Department of Agriculture and Horticulture, University of Nottingham School of Agriculture, Sutton Bonington, Loughborough, Leics.

CLARK, DR. J. A.   Department of Physiology and Environmental Studies, University of Nottingham School of Agriculture, Sutton Bonington, Loughborough, Leics.

CLUCAS, T. M.   Asmer Seeds Ltd., Asmer House, Ash Street, Leicester, Leics.

COCKING, PROF. E. C.   Department of Botany, University of Nottingham School of Agriculture, Sutton Bonington, Loughborough, Leics.

COLE, DR. D. J. A.   Department of Agriculture and Horticulture, University of Nottingham School of Agriculture, Sutton Bonington, Loughborough, Leics.

COLLINS, M. T.   Department of Biology, University of Stirling, Stirling.

CÔME, DR. D.   Laboratoire de Physiologie des Organes Végétaux, CNRS, 4 ter route des Gardes, 92–Meudon, France.

COOPER, S. R.   Department of Agriculture and Fisheries for Scotland, East Craigs, Edinburgh.

COULT, D. A.   Department of Physiology and Environmental Studies, University of Nottingham School of Agriculture, Sutton Bonington, Loughborough, Leics.

COURTNEY, A. D.   Department of Agricultural Botany, Queen's University of Belfast, Elmwood Avenue, Belfast.

COY, R. M. J.   Charles Sharpe & Co. Ltd., Boston Road, Sleaford, Lincs.

CRABB, D.   Brewing Industry Research Foundation, Lyttel Hall, Nutfield, Nr. Redhill, Surrey.

CRAWFORD, DR. D. V.   Department of Physiology and Environmental Studies, University of Nottingham School of Agriculture, Sutton Bonington, Loughborough, Leics.

CROWDER, PROF. A.   Biology Department, Queen's University, Kingston, Ontario, Canada.

CROWHURST, G. J.   Asmer Seeds Ltd., Asmer House, Ash Street, Leicester, Leics.

CURRIE, DR. J. A. Rothamsted Experimental Station, Harpenden, Herts.

DAVIES, D. H. K. Department of Biology, University of Stirling, Stirling.

DEAN, M. L. A.R.C. Weed Research Organisation, Begbroke Hill, Yarnton, Oxford.

DE BEER, J. Royal Sluis, Enkhuizen, P.O. Box 22, Holland.

DE LEE, DR. P. Ricercatore of 'Germ-Plasm' Lab. of C.N.R., Istituto Botanica, Università di Bari, Via Amendola 175, 70126 Bari, Italy.

DENT, DR. K. W. Department of Physiology and Environmental Studies, University of Nottingham School of Agriculture, Sutton Bonington, Loughborough, Leics.

DICK, MISS M. C. D. Botany School, South Parks Road, Oxford.

DULLFORCE, DR. W. M. Department of Agriculture and Horticulture, University of Nottingham School of Agriculture, Sutton Bonington, Loughborough, Leics.

DUNHAM, DR. R. J. Department of Physiology and Environmental Studies, University of Nottingham School of Agriculture, Sutton Bonington, Loughborough, Leics.

DYER, L. Germain's (U.K.) Ltd., P.O. Box 14, Old Meadow Road, Hardwick Industrial Estate, King's Lynn, Norfolk.

EBBELS, MISS S. A. Jodrell Laboratory, Royal Botanic Gardens, Kew, Richmond, Surrey.

EDWARDS, DR. M. Department of Physiology and Environmental Studies, University of Nottingham School of Agriculture, Sutton Bonington, Loughborough, Leics.

ENGLISH, S. D. Department of Agronomy, University of New England, Armidale, N.S.W. 2351, Australia.

ENSOR, H. L. Unilever Research Laboratory, Colworth House, Sharnborook, Beds.

EVANS, MISS J. A. Department of Botany, University of Leicester, Leics.

FEARN, B. Department of Botany, University of Sheffield, Sheffield, Yorks.

GARDINER, MISS K. E. Department of Agriculture and Horticulture, University of Nottingham School of Agriculture, Sutton Bonington, Loughborough, Leics.

GEORGE, DR. E. F. Plant Protection Ltd., Jealotts Hill Research Station, Bracknell, Berks.

GODSALL, J. W. Ministry of Agriculture, Fisheries and Food, Sobraon Barracks, Burton Road, Lincoln.

GORDON, DR. A. G. Forestry Commission Research Station, Alice Holt Lodge, Farnham, Surrey.

GRAHL, DR. A. Institut für Pflanzenbau und Saatgutforschung, Forschungs-anstalt für Landwirtschaft, 3301 Braunschweig, Bundesallee 50, Germany.

GRAY, DR. D. National Vegetable Research Station, Wellesbourne, Warwicks.

GREEN, J. H. Blynkbonnie School, Ringwood, Hants.

GULLIVER, R. L. Department of Agriculture and Horticulture, University of Nottingham School of Agriculture, Sutton Bonington, Loughborough, Leics.

GUNN, J. S.   Agricultural Development and Advisory Service, Ministry of Agriculture, Fisheries and Food, Shardlow Hall, Shardlow, Derby.

GUTTERMAN, DR. Y.   Department of Botany, Hebrew University of Jerusalem, Israel.

HACK, H.   Upperfold Farmhouse, Fernhurst, Nr. Haslemere, Surrey.

HALL, DR. M. A.   Department of Botany, University College of Wales, Aberystwyth, Cards.

HALLAM, DR. N. D.   Botany Department, Monash University, Clayton, Victoria, Australia 3168.

HANSON, DR. A. D.   The Lord Rank Research Centre, Lincoln Road, High Wycombe, Bucks.

HARDING, R. A.   Charles Sharpe & Co. Ltd., Boston Road, Sleaford, Lincs.

HARPER, DR. F.   Essex Institute of Agriculture, Writtle, Chelmsford, Essex.

HARRINGTON, PROF. J. F.   Department of Vegetable Crops, University of California, Davis, California 95616, U.S.A.

HARRISON, DR. J. G.   Scottish Horticultural Research Institute, Mylnefield, Invergowrie, Dundee.

HAWKINS, R. A.   Anglo-Maribo Seed Co. Ltd., Potterhanworth, Lincoln.

HAZZLEDINE, J.   Botany Department, University of Glasgow, University Avenue, Glasgow W.2.

HEBBLETHWAITE, P. D.   Department of Agriculture and Horticulture, University of Nottingham School of Agriculture, Sutton Bonington, Loughborough, Leics.

HEGARTY, DR. T. W.   Scottish Horticultural Research Institute, Mylnefield, Invergowrie, Dundee.

HEYDECKER, DR. W.   Department of Agriculture and Horticulture, University of Nottingham School of Agriculture, Sutton Bonington, Loughborough, Leics.

HIDES, D. H.   Welsh Plant Breeding Station, Plas Gogerddan, Aberystwyth, Cards.

HIGGINS, J.   Department of Agriculture and Horticulture, University of Nottingham School of Agriculture, Sutton Bonington, Loughborough, Leics.

HOLLIWELL, W. A.   Holliwell Seed and Grain Co. Ltd., 5, Oxenturn Road, Wye, Ashford, Kent.

HOLLOWELL, A. J.   Brunel Technical College, Ashley Down, Bristol.

HOLZMAN, MISS M.   Institut für kontinentale Agrar-V. Wirtschaftsforschung, D-63 Giessen, West Germany.

HUDSON, PROF. J. P.   University of Bristol, Department of Agriculture and Horticulture, Research Station, Long Ashton, Bristol.

HUNT, DR. L. A.   University of Guelph, Department of Crop Science, Guelph, Ontario, Canada.

HURD, DR. R. G.   Glasshouse Crops Research Institute, Worthing Road, Rustington, Littlehampton, Sussex.

HUXTER, T. J.   University of St. Andrews, St. Andrews, Fife.

IVINS, PROF. J. D.   Department of Agriculture and Horticulture, University of Nottingham School of Agriculture, Sutton Bonington, Loughborough, Leics.

JACKSON, DR. R. M. Department of Biological Sciences, University of Surrey, Guildford, Surrey.

JAUNCEY, MISS S. Department of Agriculture, University of Reading, Earley Gate, Reading, Berks.

JOHNSON, C. R. Charles Sharpe & Co. Ltd., Boston Road, Sleaford, Lincs.

JONES, L. Grassland Research Institute, Hurley, Maidenhead, Berks.

JOSHUA, A. Institute of Agricultural Research and Training, University of Ife, P.M. Bag. 5029, Ibadan, Nigeria.

JUNTTILA, U. O. Agricultural College of Norway, Box 16, 1432 AS-NLH, Norway.

KADDOU, DR. N. S. University of Baghdad, Iraq.

KARSSEN, DR. C. M. Department of Plant Physiology, Agricultural University, Aboretumlaan 4, Wageningen, Holland.

KEELING, J. G. Elsoms (Spalding) Ltd., Elsom House, Broad Street, Spalding, Lincs.

KERR, M. D. Botany Department, University of Hull, Hull, Yorks.

KING, J. Hill Farming Research Organisation, 29 Lauder Road, Edinburgh.

KIRK, J. Department of Agricultural Botany, Queen's University, Newforge Lane, Belfast.

KLITGÅRD, K. Statsfrøkontrollen, Skovbrynet 20, 2800 Lyngby, Denmark.

KOGBE, J. O. S. Institute of Agriculture, Research and Training, Moor Plantation, Ibadan, Nigeria.

LEES, P. T. Hurst Gunson Cooper Taber Ltd., Avenue Road, Witham, Essex.

LEWIS, J. Welsh Plant Breeding Station, Plas Gogerddan, Aberystwyth, Cards.

LEXANDER, DR. K. A. Institute of Plant Physiology, University of Lund, Sölvegatan 35, S-22362 Lund, Sweden.

LINDSAY, J. Anglo-Maribo Seed Co. Ltd., Potterhanworth, Lincoln.

LONGDEN, P. C. Broom's Barn Experimental Station, Higham, Bury St. Edmunds, Suffolk.

MACCONNELL, DR. J. T. Department of Biological Sciences, Glasgow College of Technology, North Hanover Place, Glasgow C.4.

MCDERMETT, W. G. Hill Farming Research Organisation, 29 Lauder Road, Edinburgh.

MACKAY, D. B. Official Seed Testing Station, Huntingdon Road, Cambridge.

MCWHA, J. A. Botany Department, University of Glasgow, University Avenue, Glasgow, W.2.

MADSEN, E. Statsfrøkontrollen, Skovbrynet 20, 2800 Lyngby, Denmark.

MAGUIRE, PROF. J. D. Department of Agronomy and Soils, Washington State University, Pullman, Washington 99168, U.S.A.

MARSTON, DR. M. E. Department of Agriculture and Horticulture, University of Nottingham School of Agriculture, Sutton Bonington, Loughborough, Leics.

MARTIN, A. R. Ministry of Agriculture, Fisheries and Food, Regional Office, Block 2, Chalfont Drive, Nottingham.

MASSET, A. R. Hurst Gunson Cooper Taber Ltd., Avenue Road, Witham, Essex.

MATTHEWS, DR. S. Department of Biology, University of Stirling, Stirling.

564 SEED ECOLOGY

MAUDE, R. B. National Vegetable Research Station, Wellesbourne, Warwicks.

MAY, MISS M. R. Ministry of Agriculture, Fisheries and Food, Fairfax House, North Station Road, Colchester, Essex.

MELICAN, N. J. T. Unilever Research Laboratory, Colworth House, Sharnbrook, Beds.

MELLARD, D. Ministry of Agriculture, Block 2, Government Buildings, Leeds, Yorks.

MILES, DR. J. The Nature Conservancy, Blackhall, Banchory, Kincardineshire.

MILOHNIC, PROF. J. Agricultural Faculty, Department of Plant Breeding, Zagreb, Maksimir, Yugoslavia.

MOISEY, F. R. Department of Agriculture and Horticulture, University of Nottingham School of Agriculture, Sutton Bonington, Loughborough, Leics.

MONTEITH, PROF. J. L. Department of Physiology and Environmental Studies, University of Nottingham School of Agriculture, Sutton Bonington, Loughborough, Leics.

MOORE, PROF. R. P. Crop Science Department, North Carolina State University, Raleigh, N.C. 27607, U.S.A.

MORTENSEN, G. Royal Veterinary and Agricultural University, Department of Crop Husbandry and Plant Breeding, Højbakkegaard, 2630, Taastrup, Denmark.

MORTIMER, A. M. School of Plant Biology, University College of North Wales, Bangor, Caerns.

MOUNTER, B. E. Charles Sharpe & Co. Ltd., Boston Road, Sleaford, Lincs.

NAING ZAW Department of Botany, Arts and Science University, Rangoon, Burma.

NAYLOR, R. E. L. Agricultural Botany Division, School of Agriculture, University of Aberdeen, 581 King Street, Aberdeen.

NEWTON, DR. P. Botany Department, University of Manchester, Manchester.

NORTON, DR. G. Department of Applied Biochemistry and Nutrition, University of Nottingham School of Agriculture, Sutton Bonington, Loughborough, Leics.

O'CONNOR, D. E. Department of Agriculture and Horticulture, University of Nottingham School of Agriculture, Sutton Bonington, Loughborough, Leics.

OLUSUYI, S. A. Department of Agriculture, University of Reading, Earley Gate, Reading, Berks.

OSBORNE, DR. D. J. A.R.C. Unit of Development Botany, University of Cambridge, 181a Huntingdon Road, Cambridge.

OVERGAARD, S. The Breeding Station 'Maribo', Holeby, Denmark.

OWENS, S. J. Jodrell Laboratory, Royal Botanic Gardens, Kew, Richmond, Surrey.

ÖZKAN, DR. H. Institutionen för Växtodling, 750 07 Uppsala, 7-Sweden.

PAGE, J. B. Agricultural Development and Advisory Services, 'Woodthorne', Wolverhampton, Staffs.

PALMER, DR. A.   Sir John Cass School of Science and Technology, City of London Polytechnic, London, E.C.3.

PAMMENTER, N. W.   University of Leeds, Leeds, Yorks.

PEARSON, DR. M. C.   Department of Botany, University of Nottingham School of Agriculture, Sutton Bonington, Loughborough, Leics.

PERRY, DR. D. A.   Scottish Horticultural Research Institute, Mylnefield, Invergowrie, Dundee.

PERSSON, P. I.   Hilleshögs Fro AB, Box 86, S-261 22 Landskrona, Sweden.

PUGH, DR. G. J. F.   Department of Botany, University of Nottingham School of Agriculture, Sutton Bonington, Loughborough, Leics.

RATCLIFFE, D.   Department of Botany, University of Leicester, Leics.

RATCLIFFE, T. H.   Suttons Seeds Ltd., London Road, Reading, Berks.

RAWLINSON, MISS H. A.   Department of Genetics, University of Aberdeen, 2 Tillydrone Avenue, Aberdeen.

RENARD, H. A.   Station Nationale d'Essais de Semences, La Minière, 78 Versailles, France.

RENNIE, W. J.   Department of Agriculture and Fisheries for Scotland, East Craigs, Edinburgh.

REYNOLDS, T.   Jodrell Laboratory, Royal Botanic Gardens, Kew, Richmond, Surrey.

RICHARDSON, P.   National Institute of Agricultural Engineering, Wrest Park, Silsoe, Beds.

RIEDER, DR. G.   Ciba–Geigy AG, Basel, Switzerland.

ROBERTS, B.   A.R.C. Unit of Developmental Botany, University of Cambridge, 181a Huntingdon Road, Cambridge.

ROBERTS, PROF. E. H.   Department of Agriculture, University of Reading, Earley Gate, Reading, Berks.

ROBERTSON, J.   Agricultural Development and Advisory Service, Government Buildings, Brooklands Avenue, Cambridge.

ROEBUCK, J.   Agricultural Development and Advisory Service, Government Buildings, Coley Park, Reading, Berks.

ROGERS, DR. J. A.   Hill Farming Research Organisation, 29 Lauder Road, Edinburgh.

RORISON, DR. I. H.   Nature Conservancy Grassland Research Unit, Department of Botany, University of Sheffield, Sheffield, Yorks.

SCOTT, R.   Coastal Ecology Research Station, Colney Lane, Norwich, Norfolk.

SCOTT, DR. R. K.   Department of Agriculture and Horticulture, University of Nottingham School of Agriculture, Sutton Bonington, Loughborough, Leics.

SHELDON, DR. J. C.   Department of Botany, University of Liverpool, P.O. Box 147, Liverpool.

SIMON, PROF. E. W.   Department of Agricultural Botany, Queen's University, Newforge Lane, Belfast.

SIMPSON, J.   Germain's (U.K.) Ltd., P.O. Box 14, Old Meadow Road, Hardwick Industrial Estate, King's Lynn, Norfolk.

SKINNER, R. J.   Agricultural Development and Advisory Service, Shardlow Hall, Shardlow, Derbys.

SMED, E.   The Breeding Station 'Maribo', Holeby, Denmark.

SMITH, DR. C. J.   Department of Rural Studies, Newland Park College of Education, Chalfont St. Giles, Bucks.

SMITH, PROF. H.   Department of Physiology and Environmental Studies, University of Nottingham School of Agriculture, Sutton Bonington, Loughborough, Leics.

SMITH, J. M.   Agrochemical Division, Ciba–Geigy (U.K.) Ltd., Whittlesford, Cambridge.

SMITH, J. R.   Miln-Marsters Laboratories, Station Road, Docking, Norfolk.

SMITH, R. D.   Department of Agriculture, University of Reading, Earley Gate, Reading, Berks.

SPURNÝ, DR. M.   Institute of Experimental Botany, Brno, Czechoslovakia.

STEIN, DR. M.   Department of Applied Biochemistry and Nutrition, University of Nottingham School of Agriculture, Sutton Bonington, Loughborough, Leics.

STEINKE, DR. T. D.   Department of Botany, University of Durban–Westville, P/Bag 4001, Durban, South Africa.

STONE, D. A.   National Vegetable Research Station, Wellesbourne, Warwicks.

STOREY, DR. I. F.   Agricultural Development and Advisory Service, Shardlow, Derbys.

STREET, PROF. H. E.   Botanical Laboratories, University of Leicester, Adrian Building, University Road, Leicester.

SUTTON, F.   Nature Conservancy Unit, Botany Department, University of Sheffield, Sheffield, Yorks.

SYLVESTER-BRADLEY, R.   Agricultural Botany Department, School of Agriculture, 581 King Street, Aberdeen.

TADRIA, C.   Cotton Research Station, Namulonge, P.O. Box 7084, Kampala, Uganda.

TATHAM, P. B.   Agricultural Development and Advisory Service, Efford Experimental Horticulture Station, Lymington, Hants.

TAYLOR, P. A. S.   Charles Sharpe & Co. Ltd., Boston Road, Sleaford, Lincs.

THEVENET, MRS. C.   Laboratoire de Physiologie des Organes Végétaux, CNRS, 4 ter route des Gardes, 92–Meudon, France.

THOMAS, DR. P. E. L.   University of Rhodesia, Box MP 167, Mount Pleasant, Salisbury, Rhodesia.

THOMAS, DR. T. H.   National Vegetable Research Station, Wellesbourne, Warwicks.

THOMPSON, DR. P.   Jodrell Laboratory, Royal Botanic Gardens, Kew, Richmond, Surrey.

THOMSON, D. C. G.   British Sugar Corporation, P.O. Box 26, Central Offices, Peterborough, Northants.

THURSTON, MISS J. M.   Rothamsted Experimental Station, Harpenden, Herts.

TISSAOUI, T.   Laboratoire de Physiologie des Organes Végétaux, CNRS, 4 ter route des Gardes, 92–Meudon, France.

TJEERTES, P.   Sluis en Groot, P.B. 13 Enkhuizen, Holland.

TOMASSON, T.   Agricultural Research Institute, Keldnaholti, Reykjavik, Iceland.

LIST OF MEMBERS567

TOWN, P. A.   Food and Agricultural Organisation of the United Nations, P.K.8, Yalova, Turkey.

TRETHEWAY, D.   Charles Sharpe & Co. Ltd., Boston Road, Sleaford, Lincs.

TUPPEN, R. J.   Plant Pathology Laboratory (M.A.F.F.), Hatching Green, Harpenden, Herts.

TURNER, MISS Y. J.   Department of Agriculture and Horticulture, University of Nottingham School of Agriculture, Sutton Bonington, Loughborough, Leics.

VALETTE, R.   Comité Culture maraîchère, Chaussée de Charleroi 237, 5800 Gembloux, Belgium.

VAN VLIET, IR. G.   Sluis en Groot, P.B. 13 Enkhuizen, Holland.

VILLIERS, PROF. T. A.   Biological Sciences, University of Natal, Durban, South Africa.

VINCENT, MRS. E. M.   Department of Agriculture, University of Reading, Earley Gate, Reading, Berks.

WADDOUPS, D. R.   Department of Plant Biology and Microbiology, Queen Mary College, London, E.1.

WALKER, DR. J. T.   Rothwell Plant Breeders Ltd., Rothwell, Near Caister, Lincs.

WAREING, PROF. P. F.   Department of Botany, University College of Wales, Aberystwyth, Cards.

WATERMAN, J.   Germain's (U.K.) Ltd., P.O. Box 14, Old Meadow Road, Hardwick Industrial Estate, King's Lynn, Norfolk.

WEATHERITT, N. T.   Ministry of Agriculture, Fisheries and Food, 6 Guys Cliffe Avenue, Leamington Spa, Warwicks.

WEBB, D. P.   Department of Botany, University College of Wales, Aberystwyth, Cards.

WEBSTER, T. I.   D. R. Colegrave Seeds Ltd., West Adderbury, Banbury, Oxon.

WELLS, D. A.   Monks Wood Experimental Station, Abbots Ripton, Huntingdon.

WELLS, G. J.   Weed Research Organisation, Yarnton, Oxford.

WELLS, T. C. E.   Monks Wood Experimental Station, Abbots Ripton, Huntingdon.

WHEATON, MISS O. E.   Hurst Gunson Cooper Tabor Ltd., Avenue Road, Witham, Essex.

WHITTINGTON, DR. W. J.   Department of Physiology and Environmental Studies, University of Nottingham School of Agriculture, Sutton Bonington, Loughborough, Leics.

WILLIAMS, E. D.   Botany Department, Rothamsted Experimental Station, Harpenden, Herts.

WILLIAMS, DR. J. T.   Department of Botany, University of Birmingham, Birmingham 15.

WILLIAMS, DR. P. M.   King's College London, Botany Department, 68 Half Moon Lane, London, S.E.24.

WILLIAMS, S.   Welsh Plant Breeding Station, Plas Gogerddan, Aberystwyth, Cards.

WOOD, D. W.   Department of Agriculture and Horticulture, University of Nottingham School of Agriculture, Sutton Bonington, Loughborough, Leics.

WOODWARK, W.   British Sugar Corporation Ltd., Central Laboratory, P.O. Box 35, Wharf Road, Peterborough, Northants.

Index